6th International Symposium on Resource Exploration and Environmental Science (REES 2023)

E3S Web of Conferences Volume 394 (2023)

Ordos, China
28 - 30 April 2023

Editors:

Zhibin You
Jianjun Xu

ISBN: 978-1-7138-8357-9

Printed from e-media with permission by:

Curran Associates, Inc.
57 Morehouse Lane
Red Hook, NY 12571

Some format issues inherent in the e-media version may also appear in this print version.

For additional information, please contact EDP Sciences – Web of Conferences
at the address below.

EDP Sciences – Web of Conferences
17, Avenue du Hoggar
Parc d'Activité de Courtabœuf
BP 112
F-91944 Les Ulis Cedex A
France

Phone: +33 (0) 1 69 18 75 75
Fax: +33 (0) 1 69 28 84 91

contact-edps@webofconferences.org

Additional copies of this publication are available from:

Curran Associates, Inc.
57 Morehouse Lane
Red Hook, NY 12571 USA
Phone: 845-758-0400
Fax: 845-758-2633
Email: curran@proceedings.com
Web: www.proceedings.com

TABLE OF CONTENTS

Preface

This proceeding includes original and peer-reviewed research papers from 6th International Symposium on Resource Exploration and Environmental Science (REES 2023), which held on April 28-30, 2023 in Ordos, China.

The REES 2023 conference had four main topics: 1). Resource Exploration and Utilization, Geography and Geological Engineering; 2). Environmental Engineering and Sustainable Development; 3). Resources and Energy, Power Engineering; 4). Environmental Chemistry, Environmental Bioengineering and Related Technologies. REES conference is a platform for presenting excellent results and new challenges facing the field of earth and environmental sciences, resources and environmental engineering, energy and electricity engineering and related topics. It brought together experts from industry, governments and academia, experienced in engineering, design and research.

REES 2023 received 81 manuscripts, and 32 submissions had been accepted by our reviewers. By submitting a paper to REES 2023, the authors agreed to the review process and understood that papers undergo a peer-review process. Manuscripts were reviewed by appropriately qualified experts in the field selected by the conference committees, who gave detailed comments and-if the submission was accepted the authors would submit a revised version that took into account this feedback. All papers were reviewed using a double-blind review process: authors declared their names and affiliations in the manuscript for the reviewers to see, but reviewers did not know each other's identities, nor did the authors receive information about who had reviewed their manuscript. The Committees of REES 2020 invested great efforts in reviewing the papers submitted to the conference and organizing the sessions to enable the participants to gain maximum benefit.

Hopefully, all participants and other interested readers will benefit scientifically from the proceedings and also find it stimulating in the process.

With our warmest regards,

Conference Organizing Committees
Ordos, China

Statement of Peer review

In submitting conference proceedings to *Web of Conferences*, the editors of the proceedings certify to the Publisher that

1. They adhere to its **Policy on Publishing Integrity** in order to safeguard good scientific practice in publishing.
2. All articles have been subjected to peer review administered by the proceedings editors.
3. Reviews have been conducted by expert referees, who have been requested to provide unbiased and constructive comments aimed, whenever possible, at improving the work.
4. Proceedings editors have taken all reasonable steps to ensure the quality of the materials they publish and their decision to accept or reject a paper for publication has been based only on the merits of the work and the relevance to the journal.

Title, date and place of the conference

6th International Symposium on Resource Exploration and Environmental Science (REES 2023)
April 28-30, 2023 in Ordos, China

Proceedings editor(s):

Zhibin You, Jianjun Xu

Date and editor's signature May 25, 2023

Characteristics of Gas Seepage and Pore Structure Response in CO_2-ECBM Process of Low Permeability Coal Seam

Junlin Liu*, Kun Zhang and Huihu Liu

School of Earth and Environment, Anhui University of Technology, Huainan, China

Abstract. CO_2-ECBM is a method of enhanced coalbed methane extraction followed by cutting greenhouse gas emissions and new energy development. In order to reveal the characteristics of gas flow in porous media and the pore structure response characteristics of coal rocks, the experiments were carried out to simulate the process of CO_2 displacement of N_2 at a buried depth of 900 m, including monitoring the changes in gas permeability and strain of coal samples along with a comparison of the pore structure of low-temperature liquid nitrogen adsorption on coal samples both before and after displacement were both done. The findings of the experiment are listed below. The N_2 permeability of the LiuZhuang sample ranges from 0.0008mD to 0.0014mD, whereas the permeability of QiDong is around 0.0003mD. With an increase in gas injection duration and an expansion of the coal matrix for N_2 adsorption, the permeability steadily decreases. The efficient stress compression of the coal pore fracture structure during sample preparation and testing avoids the visible fracture region, which results in poor permeability. The displacement stages of CO_2 can be divided into three phases. Free nitrogen flows from the end of the position and the permeability diminishes during the phase of free nitrogen. When CO_2 is introduced into the penetration stage, the permeability tends to rise, however when there is no penetration, the permeability test values are frequently low. During the CO_2 steady displacement phase, gas permeability gradually declines. Axial and radial strains are progressively raised during the initial stage of the CO_2 injection whereas they are gradually reduced during the initial stage of the N_2 injection. While CO_2 is continuously supplied through the coal body stage, there are modest axial and radial strain changes. The axial and radial stresses are stabilized by the CO_2 displacement. The overall pore volume of the coal significantly rises following the displacement. The increase part of the pore volume is primarily focused on the pore of absorption and filling (aperture < 10nm), whereas the decreased part is mainly concentrated in the diffusion pore of the Fick type and the permeability part (aperture > 50nm). The increased in pore volume ratio surface area is centered mostly in the fill pore region (aperture 10 nm) and is four times greater than it was before the displacement. The CO_2 injection exerts an expansion impact on the adsorption-filled and diffusion pores during the CO_2-ECBM process, whereas the compression effect on the percolation pores results in a reduction in permeability.

Keywords: CO_2-ECBM, characteristics of gas flow in porous media, characteristics of pore structure response.

1. Introduction

The low-carbon economy has taken on the role of the trend of the sustainable and healthy growth of human civilization in light of the growing global greenhouse effect and the harm to the natural environment. Coal has played a significant role in the energy structure of China, which accounts for up to 70% of total energy consumption. Therefore, coal forms the bulk of the energy structure of China. However, compared to clean energy like natural gas, coal is less efficient to burn and more polluting. As a method of enhanced coalbed methane extraction followed by greenhouse gas emissions and new energy development, CO_2-ECBM is in line with the new strategy for energy security in China in light of the new alterations in energy supply and demand patterns as well as the new trends in international energy development[1].

CO_2[2], N_2[3], or a mixture of the two gases[4] are the gases that are typically employed in the ECBM process. Instead of CH_4, CO_2 is initially adsorbed when the coal seam is absorbed[5]. Hence, the injected CO2 is more simply adsorbed by the coal seam. N_2 is less effective than CO_2 in coal seam displacement, according to research by Jiang et al. Although N_2 can promote coal seam permeability, CO_2 injection causes a reduction in coal seam permeability[6]. With the application of modified triaxial loading rigs, Liu et al. discovered that the injection of CO_2 into the coal seam dramatically enhanced the

* Corresponding author: 603093502@qq.com

expansion of the coal substrate, resulting in a decrease in permeability[7].

Most researchers have used physical simulation studies to analyze the features of the fluctuation of gas permeability in coal samples that have been injected with gas. As CO_2 displaces CH_4, Li et al. analyzed the pattern of variation in coal seam permeability and discovered that CO_2 adsorption rises with falling pore pressure and falls with rising sealing pressure[8]. The change in coal permeability varies with the change in pore pressure and surrounding pressure throughout the process of displacing methane with CO_2[9]. Non-isothermal CO_2 gas flow in porous media experiments were conducted by Wu et al. under the influence of thermodynamics. According to the study, pore pressure and CO_2 and CH_4 permeability are positively associated and the temperature has a negative exponential association with volume stress change[10]. In the gas permeability-stress coupling experiment of coal under the heat flow field, it was discovered that under the same pressure conditions, CH_4 displayed a strong permeability at high temperatures. As effective stress increases, permeability decreases[11–12]. Yin used a three-axis experimental setup to investigate the impact of CO_2 injection pressure on the desorption of CH4 and to examine variations in gas concentration and CH_4/CO_2 exit flow. According to the experimental findings, CH_4 generation efficiency decreases with increasing CO_2 injection pressure[13]. By using a gas-containing coal thermofluid-solid coupled triaxial servo seepage device to conduct experiments on the CO_2 gas flow in porous media characteristics in coal seams under various pressure conditions, Xu et al. discovered that the higher the ambient pressure, the slower the CO_2 permeation rate under the same axial pressure condition. The higher the axial pressure, the lower the rate of CO_2 infiltration under the same circumferential pressure. The stress sensitivity coefficient steadily declines with increasing stress under various axial and circumferential pressure settings, i.e., the greater the stress, the less sensitive the coal sample permeability rate is to stress[14]. In order to track changes in gas concentration and pore pressure, Wang et al. employed rectangular coal samples with dimensions of 300*50*50 mm and discovered that the injection of CO_2 might considerably increase the recovery of CH_4. The breakthrough time of the injected gas is nonetheless shorter the greater the injection pressure[15]. Previous studies have demonstrated that factors including temperature, pressure, adsorbed gas, varying coal ranks, and water saturation all have an impact on how much CO_2 is displaced and absorbed[16–17]. However, the relevant alterations in stress-strain conditions, adsorption/desorption and permeability need to be further investigated[18-20].

We have utilized an experimental device to model the displacement of CH_4 by CO_2 under conditions of a coal reservoir in light of the aforementioned research state. The innovation of this paper lies in the following three aspects. Firstly, the experimental temperature and pressure settings are set up in accordance with the ground temperature and recurrence pressure gradient of the coal seam, which is the closest representation of the genuine coal reservoir. Secondly, the entire experiment—from the CO_2 repulsion process to the CH_4 adsorption process—is fully and dynamically monitored, and the gas flow rate is tracked with the strain on the coal. Thirdly, analysis and research were done on the pore structure of coal samples both before and after the displacement experiment.

The purpose of the paper is to investigate typical mines in low permeability coal reservoirs in mining regions in the Huainan and Huaibei areas of China. Using self-made CO_2 displacement experiment equipment, an experimental examination of CO_2 displacing CH_4 was conducted. By using an experimental setup, the alternations in gas flow during CO_2 displacement were observed, and the coal-like pore structure before and after CO_2 displacement was examined.

2. Test Method

2.1 Sample Acquisition and Preparation

The No.13 coal seam of LiuZhuang and the No.7 coal seam of QiDong in Huainan and Huaibei areas of China were used as the main structural coal sources for this study. The results of the basic physical properties analysis of the collected coal samples are demonstrated in Table 1. Liu Zhuang No.13 and Qi Dong No.7 coal seams are located at depths of 763 and 1044 meters, respectively, with moisture contents of 1.25% and 0.8% and mirror grade group reflectance of 0.7252% and 0.7558%.

Table 1. Basic physical properties of coal samples

Industry analysis of coal samples		
Name of sample	13 coal seam of LiuZhuang	7 coal seam of QiDong
Type of coal	Gas coal	Fat coal
Depth of sampling	763m	1 044m
Moisture content(M_{ad})	1.25%	0.88%
Volatile matter(V_{daf})	39.34%	38.59%
Fixed carbon(FC_{ad})	39.81%	49.51%
Ash content(A_{ad})	19.60%	11.02%
Average True relative density (g/cm^3)	1.36	1.45
Vitrinite reflectance(R_O)	0.7252%	0.7558%

2.2 Experimental Apparatus

The schematic diagram of the experimental protocol of this study is illustrated in figure 1. To ensure that the pressure at the inlet meets the experimental requirements, gas booster pumps, and air compressors are used to pressurize the injected gas. A pressure-reducing valve regulates the pressure at the inlet, and a pressure-return valve controls the pressure at the outlet. To maintain temperature stability, a thermostat heats the sample holder. Pressure sensors are positioned at the inlet, outlet and both sides of the holder to monitor the inlet, outlet, perimeter and axial pressure as well as to transfer the collected pressure data to a computer. The rubber tube for gas conduction is linked to the bottle with clarified lime water, which is used to check the gas. The position of it is at the end of the pipe, where the gas flow meter continuously checks the gas flow at the outlet end.

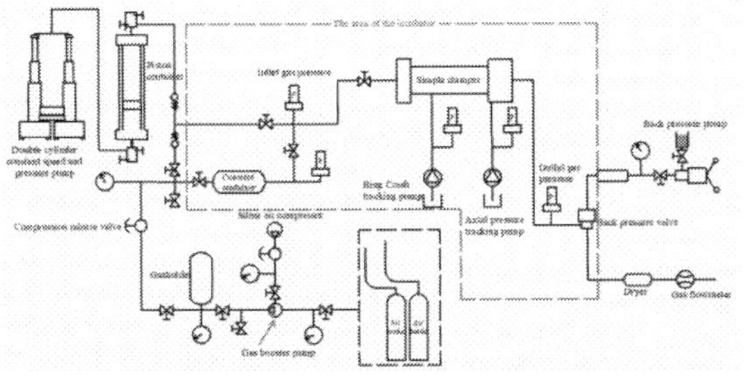

Figure 1. Experimental flow diagram

2.3 Experimental Method

The studies tested the change in permeability and the degree of expansion and contraction of coal samples before and after CO_2 displacement under conditions of constant temperature, varied gas injection pressure, and varying surrounding pressure.

2.3.1 Determination of Gas Permeability in Single Component during Pressure Change.

The temperature for the experiment is 36℃. Pressure conditions at the inlet are 2MPa, 3MPa, 4MPa and 5MPa. The experimental samples were 50mm x 100mm cylindrical samples from the coal seam of LiuZhuang No.13. N_2 and CO_2 are used as the experimental gas (Considering the safety of the experiment. N_2, which has similar properties, was used instead of CH_4 for the experiments). The specific steps of the experiment are as follows:

(1) The coal pillar sample is dried and fixed to one end of the holder with adhesive tape. The coal pillar is then covered to enhance axial pressure and placed into the holder with the rubber cylinder. Ultimately, vacuum extraction and closure detection were performed.

(2) To obtain data continuously until the current gas pressure at the inlet is complete, preload the axial and surrounding pressure to 4 MPa, control the loading rate at roughly 0.2 MPa/s, adjust the N_2 gas pressure at the inlet to 2 MPa, and keep the axial pressure tracking pump and surrounding pressure tracking pump always higher than the gas pressure at the inlet by 2 MPa. The pressure was then changed to 3 MPa, 4 MPa, and 5 MPa. When the experiment is complete, the vacuum is removed from the holder and the pipework and substituted with CO_2 gas to repeat the experiment using the same procedures as before.

2.3.2 Simulating CO_2 Displacement under Deep Buried Conditions

In order to simulate the stratum conditions at an embedded depth of 900m, the simulated experimental temperature was set to 36℃, the gas pressure at the inlet end was set at 9 MPa, the effective stress of the coal pillar was set at 2 MPa, and the samples were taken from the coal pillar of the LiuZhuang No.13 coal seam and the QiDong No.7 coal seam in accordance with the progressive gradients of temperature and pressure of the coal strata. Below is the particular experimental protocol:

(1) The coal pillar with strain gauges attached were loaded into the holder with heat-shrinkable tubes, and the seal was examined by He. The thermostat was set to a temperature of 36℃. The next experiment was launched following the stabilization of the temperature.

(2) Vacuum each experimental module for at least an hour, as well as the holder and the pipes.

(3) The N_2 gas pressure at the inlet was set and regulated at 9 MPa. Once the flow rate had stabilized, the gas pressure at the outlet was changed to detect and monitor the gas flow rate at the tail end. Based on the recorded data for the gas input and outlet pressure and flow rate, the N_2 gas permeability of the coal pillar was then estimated.

(4) To guarantee that the coal sample adsorbs N_2 for more than 24 hours, close the gas valve at the outlet end and regulate the N_2 gas pressure at the intake. According to the adsorption equilibrium standard, stable adsorption of a coal pillar is defined as a pressure change in the gas buffer tank within two hours of less than 0.01 MPa.

(5) CO_2 is loaded and injected at a particular pressure, typically 0.5 MPa above the N_2 injection pressure, following vacuum treatment of the pipeline and storage tank. Record the permeability change and stress strain during this process, and make sure that CO_2 enters the coal sample to complete adsorption and desorption.

3. Results and Discussion

3.1 Changes in Permeability of Single Component Gases

The calculation formula of penetration is as follows [21-24]:

$$k_{ge} = \frac{2p_d q_{ge} \mu_g h}{A(p_u^2 - p_d^2)} \times 100 \tag{1}$$

where k_{ge} is permeability (mD), p_u and p_d are the pressures upstream and downstream of the sample (MPa), q_{ge} is velocity (cm^3 /s), μ_g is viscosity (mPa · s), h is the sample length (cm), and A is the sample cross-sectional area (cm^2). It can be observed from figure 3 that the gas permeability usually grows as the gas pressure at the inlet rises from 2MPa to 5MPa. The enlargement of pore fissures in the coal pillar brought on by the rise in gas pressure at the entry may be the cause of the growing permeability. Simultaneously, the increase in porosity between 2MPa and 3MPa is more noticeable. N_2 and CO_2 have a permeability of between 0.0006 and 0.0007mD, with CO_2 having a considerably lesser permeability than N_2. The explanation could be that the coal matrix absorbs CO_2 more readily and that the impact of matrix expansion on pore fracture compression is more pronounced, leading to reduced permeability.

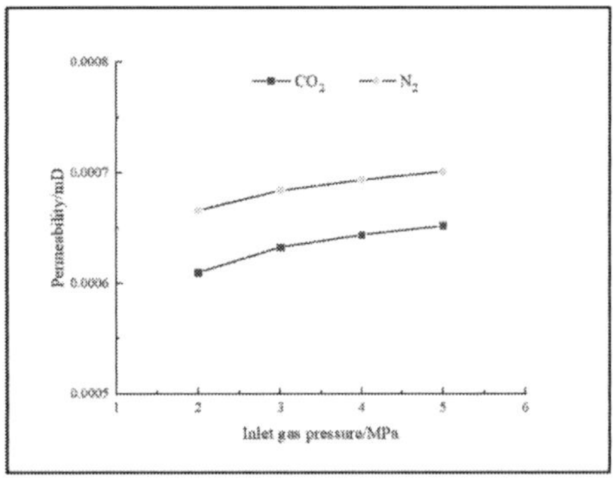

Figure 2. Changes in CO_2 and N_2 permeability with pressure

3.2 Characteristics of Gas Permeability under Simulated Reservoir Condition

3.2.1 Original Penetration

As shown in figure 3, the flow rate of N_2 gas at 9MPa injection pressure is calculated and the point line diagram of coal permeability which changes with time is obtained by employing Darcy's Law. The computed samples of Liu Zhuang had total permeability ranging from 0.0008mD to 0.0012mD (1mD=9.869*10^{-4}μm^2), whereas Qidong's was around 0.0003mD. At the temperature and pressure conditions of 36°C and 9 MPa, both samples showed low permeability. Additionally, when the N_2 transit duration increased, the permeability continued to decrease. Generally speaking, there are several variables that impact how permeable coal seams are. While geological structure, stress condition, and other factors all have an impact on permeability, coal seam pores and fissures have the greatest influence. In the research area, low-permeability coal from China's Huainan and Huaibei mining regions was chosen, and the sample and permeability test was chosen to avoid the evident fracture area of the sample. It results in a significantly low permeability of the sample. The pore structure of the coal will also be compressed by the effective stress during the permeability test, which will lower the permeability. The coal matrix, on the other hand, has the ability to expand during gas adsorption and contract during gas desorption. The volume expansion of the experiment, which was brought on by the coal matrix's adsorption of N_2 gas, constricted the coal's linked pore fissures and led to a steady drop in coal permeability with N_2 injection.

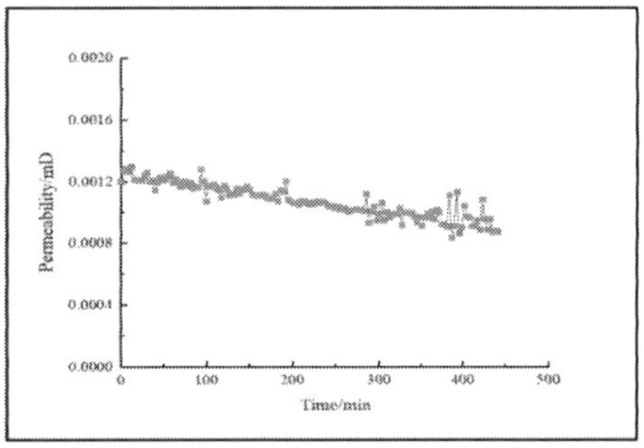

(a)

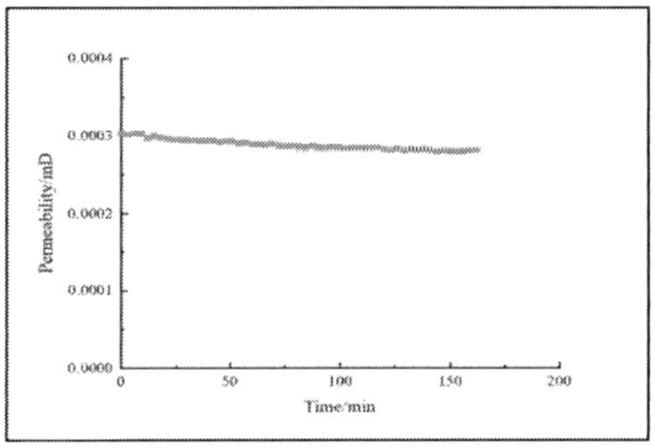

(b)

Figure 3. Variation of N_2 permeability of samples in the study area with time (a: LiuZhuang No.13 Coal; b: QiDong No.7 Coal)

3.2.2 Changes in Permeability of CO_2 Displacement.

When adsorbed N_2 had saturated the coal pillar, CO_2 was utilized to displace N_2 from the coal using a repelling pressure that was 0.5MPa greater than the fluid pressure of the coal pillar. Figure 4 depicts the estimated variation in coal pillar permeability over time throughout the displacement process. It is evident that both samples exhibit a comparable fluctuation in permeability. Based on the features of the sample permeability changes, the CO_2 displacement process may be separated into three stages:

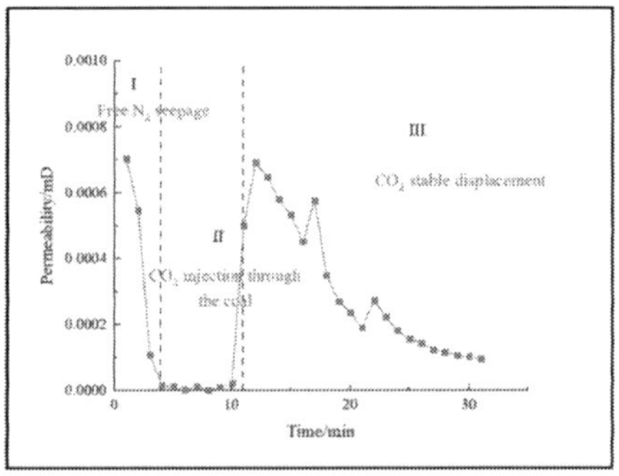

(a)

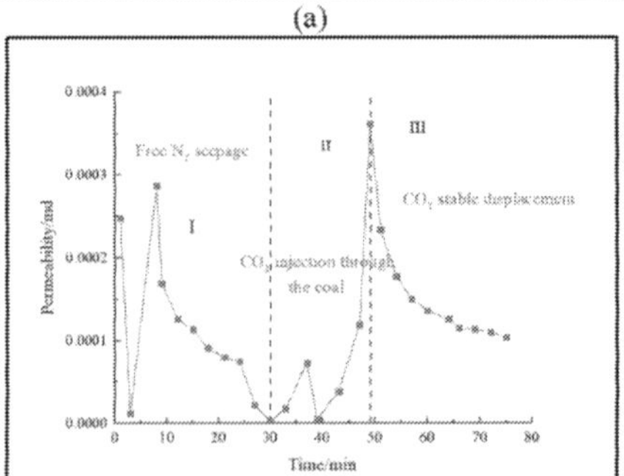

(b)

Figure 4. Change of CO_2 permeability over time in study area (a: Liu Zhuang No.13 coal; b: Qidong No.7 coal)

https://doi.org/10.1051/e3sconf/202339401001

①Stage I: Free N_2 seepage stage. This stage occurs when the free nitrogen gas in the coal pillar connecting the fissure is first detected to flow out by the tail end and the CO_2 gas at the intake end just begins to be injected. The gas flow rate at the outlet end and the permeability of the recorded data point both steadily decrease as a result of the slow outflow of free nitrogen in the coal pillar's connecting pore fissure. The measured permeability of the coal pillar is now inaccurate and only accurately measures the size of the gas flow rate at the output end. According to the comparison between LiuZhuang No.13 and QiDong No.7 samples in Figure 4, the permeability of LiuZhuang decreases at a faster rate. It is obvious that the LiuZhuang mine sample has higher a pore connection and faster free gas discharge.
②Stage II: Stage of CO_2 injection through the coal. During this phase, CO_2 slowly permeates the coal pillar and enters the coal matrix, and the permeability will exhibit an upward trend. CO_2 initially penetrates the coal pillar's continuous pore-fissure before slowly making its way into the coal matrix, where it competes with the adsorbed N_2 there. The gas flow rate at the end is very low and the permeability measurement value of the coal pillar is comparatively low because it does not enter the coal pillar.
③Stage III: Stage of CO_2 stable displacement. At this point, CO_2 penetrates the coal pillar, and the gas flow rate at the exit end causes the test permeability to gradually drop. CO_2 enters the coal matrix as it is injected to permeate the coal, competing with N_2 for adsorption and promoting N_2 gas desorption. The gas flow rate at the intake end is greater than that at the output end because the coal matrix can adsorb more CO_2 than N_2. On the other hand, during the displacement process, the coal matrix expands in volume as a result of CO_2 adsorption, which causes the gas permeability to steadily decrease over time.

3.2.3 *Characteristics of Pore Structure Response after CO_2-ECBM Process.*

3.2.3.1. *Strain Characteristics of Coal.*

At room temperature, coal is a type of vitreous material with a linked molecular structure system, and it has the properties like brittleness, hardness, and tensile. The initial stress condition of the original coal seam will alter as a result of the CO_2 injection displacement. The desorption of CH_4 and re-adsorption of CO_2 in the coal reservoir will take place concurrently with the injection of CO_2, causing contraction and expansion of the coal matrix. Accordingly, using the LiuZhuang mine sample as an example, strain monitoring experiments of the coal matrix modeling the CO_2 displacement process are carried out to investigate the strain variations of the coal matrix.
(1) Adsorption Strain
As seen in Figure 5, a sample adsorption experiment using N_2 in the coal pillar is carried out while the axial and radial strain of the coal pillar is being measured. After loading the surrounding and axial pressure with the sample and bench-marking the coal substrate, the compression and stretch should be negative and positive, respectively. As can be seen, axial and radial strain is somewhat decreased and then steadily raised throughout the processes of gas injection and coal substrate adsorption. This is because during the first stage of the N_2 injection to link the coal, axial tension was applied to the coal pillar to compress the coal. The coal matrix then starts to absorb the gas that is expanding as N_2 slowly permeates the coal, gradually increasing the axial and radial strain.

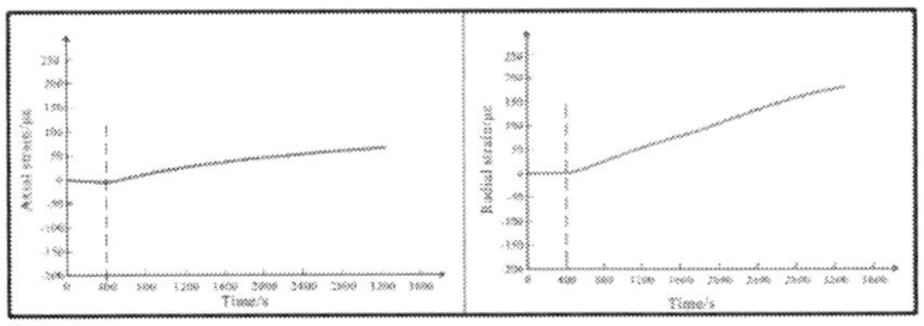

(a) (b)

Figure 5. Strain variation with time during adsorption of Liu Zhuang coal sample (a: Axial Strain; b: Radial Strain)

(2) Displacement Strain
Once the N_2 adsorption test was finished, CO_2 was introduced into the coal pillar to begin the displacement while the volumetric strain of the coal pillar was monitored. While the effective tension on the coal pillar in the holder remained constant, the adsorption/desorption of the coal matrix was primarily responsible for controlling the strain of the coal matrix. According to the operation flow, the entire experimental procedure may be separated into three stages. The corresponding strains are presented in Fig. 6, and the strain at the start of the CO_2 injection is set as the baseline strain.
①Stage I: Phase of CO_2 injection and adsorption. At this point, CO_2 was fed into the coal pillar to ensure that it was fully in touch with the coal matrix, but the gas valve at the outlet end was left closed to prevent leaks. As more CO_2 is absorbed by the coal matrix than N_2, this causes a comparatively substantial amount of adsorption expansion, which causes both axial and radial stresses to continuously grow at this time.
②Stage II: Phase of CO_2 displacement and flow. The gas valve at the outlet end is opened at the start of this stage, and the quick flow of gas from the fissure of the coal pillar causes a rapid drop in fluid pressure, which causes significant desorption of gas from the coal matrix and a rapid drop in coal matrix strain. After that, the coal sample's CO_2 gas progressively enters the coal, raising the fluid pressure and resulting in the slow adsorption of the coal matrix. The axial and radial strains of the coal pillar fluctuate unpredictably. The adsorption and expansion of the coal matrix cause a minor rise in the strain of the coal matrix when CO_2 gas is gradually fed into the coal pillar.
③Stage III: Phase of CO_2 stable displacement. The CO_2 adsorption of the coal matrix at the same time displaces the CH_4 output, and the strain of the coal matrix gradually stabilizes.

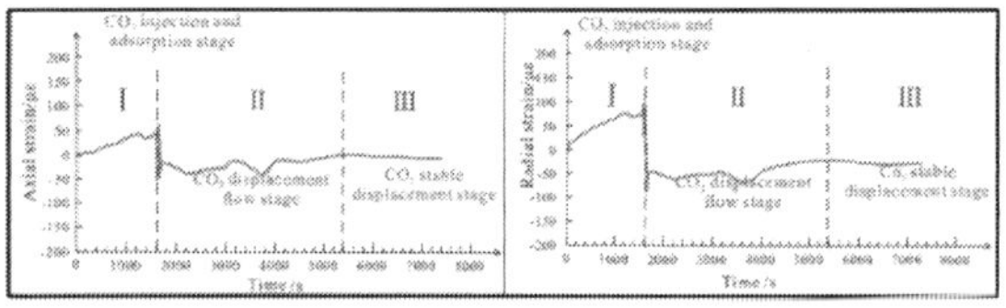

(a) (b)

Figure 6. Strain variation with time during Liu Zhuang coal displacement (a: Axial Strain; b: Radial Strain)

3.2.3.2. Strain Characteristics of Coal.

Coal is an example of a typical porous medium with a complicated structure, mostly composed of matrix pores and fracture systems. The high-pressure fluid will enter the coal pore fissure, enlarge it, and improve the coal permeability while simultaneously compressing the coal sample under high-pressure CO_2 injection. On the other hand, during the CO_2 injection to displace N_2, the coal matrix may potentially alter the pore-fissure structure of the coal. Based on various study objectives and levels of testing accuracy, domestic and international academics have developed numerous criteria for classifying the pore structure of coal (Table 2).

Table 2. Comparison of coal pore size structure classification schemes (Diameter unit: nm)

B. B. Ходот (1961)	Dubinin (1966)	IUPAC (1966)	H. Gan (1972)	Coal Research Institute of FuShun (1985)	Sijing Yong et al. (1991)
Micropore <10	Micropore <2	Micropore <2	Micropore <1.2	Micropore <8	Micropore <10
Transitional pore 10-100	Transitional pore 2-20	Transitional pore 2-50	Transitional pore 1.2-30	Transitional pore 8-100	Transitional pore 10-50
Mesopore 100-1000					Mesopore 50-750
Macropore >1000	Macropore >20	Macropore >50	Macropore >30	Macropore >100	Macropore >750

The process of gas movement during CO_2 displacement in coal seams is the focus of the current investigation. In light of the results of the analysis of the diffusion-percolation characteristics of CH_4 and CO_2 gases in coal and taking into account the opposite diffusion and competitive adsorption of CH_4 and CO_2, the pores in this study were classified into the following categories: adsorption-filled pores (<10nm), transition diffusion pores (10-20nm), Fick diffusion pores (20-100nm) and percolation pores (>100nm). The most widely used techniques for studying the pore structure of coal are mercury porosimetry, low-temperature nitrogen adsorption, and CO_2 adsorption. The BET technique is one of them and can measure pores up to 0.35 and 200nm in size, which can measure practically all pore types. Consequently, using a Liuzhuang sample as an example, the structural alterations of the pores and fractures of the coal body before and after the CO_2 displacement treatment were examined in this work using the low-temperature nitrogen adsorption technique.

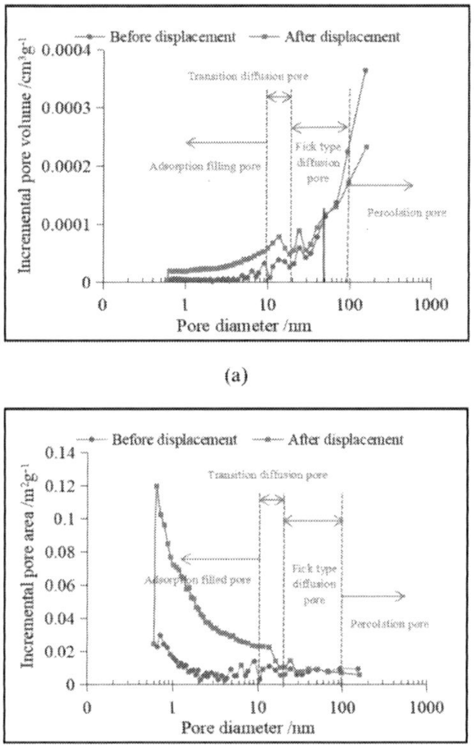

(a)

(b)

Figure 7. Pore volume changes of LiuZhuang coal sample before and after the experiment of using CO2 to displace N2 (a: Pore volume increment; b: Total pore volume)

Pore volume is the total volume of tiny pores within the mass of a porous solid. Each of these crucial factors helps determine the pore fracture structure of coal. Specific surface area is the total area per unit mass of material. Figure 8 compares the findings of the LiuZhuang coal sample's pore volume before and after the experiment in which N_2 was replaced by CO_2. As can be observed, the displacement resulted in a large increase in the coal's overall pore volume. While the pore volume bigger than 50nm declines, concentrating in some Fick-type diffusion pores and percolation pore sections, the rise in pore volume is primarily seen in the pore size less than 50nm, of which the volume and proportion of adsorption-filled pores (10nm) increase the most.

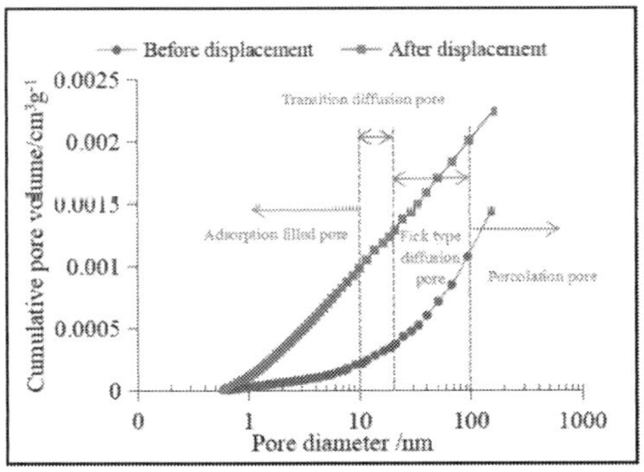

(a)

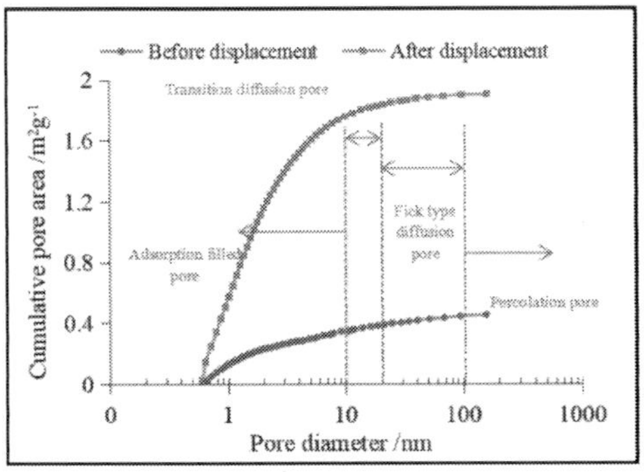

(b)

Figure 8. Change of pore specific surface area before and after the experiment of N_2 displacement by CO_2 in low permeability coal seam (a: Increment of pores pecific surface area; b: Total specific surface area)

It can be shown in Fig. 8 that the displacement process greatly increased the specific surface area of the micro-pore section of coal samples (approximately 4 times greater after displacement than before displacement) and that the increase is mostly concentrated in the adsorption-filled pore section (< 10nm). This suggests that the CO_2 displacement helps to significantly boost the coal sample's ability to adsorb gas.

Hence, the CO_2 injection procedure will significantly contribute to changing the pores of the coal during CO_2-ECBM in low-permeability coal seams. Percolation pores primarily regulate changes in coal body permeability, while adsorption-filled and diffusion pores control gas exchange and sequestration in the coal matrix. The expansion of sorption-fill and diffusion pores after CO_2 displacement aids in the complete exchange of CO_2 and CH_4, increasing the ability of the coal matrix to contain the gas. However, the compression of percolation pores may decrease the permeability of the coal matrix, which would then make it more difficult to inject CO_2.

4. Conclusion

By conducting the CO_2-N_2 displacement tests and low-temperature liquid nitrogen adsorption experiment, research the characteristics of gas flow in porous media and pore structure response characteristics during the CO_2-ECBM process.

The total N_2 permeability of the Liu Zhuang sample varies between 0.0008mD and 0.0014mD, while the permeability of the Qi Dong sample is around 0.0003mD. When the expansion caused by N_2 adsorption by the coal matrix increases gas injection time, the permeability steadily decreases. Low permeability is a result of the samples' preparation, which avoided apparent fracture zones, and the effective stress of the test, which compressed the pore-fracture structure of the coal.

There are three stages to the CO_2 displacement phase. Free nitrogen exits from the tail end during the free nitrogen stage, and the permeability constantly drops. When there is no penetration, the permeability tends to be lower, but it tends to rise throughout the CO_2 injection penetration stage. In the stage of steady displacement of CO_2, the gas permeability steadily decreases over time.

At the initial stage of N_2 injection, the axial and radial strain decreased slightly; at the initial stage of CO_2 injection, the axial and radial strain increased gradually; at the stage of continuous CO_2 injection and penetration into the coal, the axial and radial strain fluctuated and increased slightly overall; at the stable stage of CO_2 displacement, the axial and radial strain became stable.

Following the displacement, the total pore volume of the coal dramatically rises, with the majority of the growth occurring in the adsorption-filled pore (pore diameter <10nm). The Fick-type diffusion and permeation pore sections (pore diameter >50nm) are where the majority of the reduction in pore volume is focused. After the displacement, the specific surface area of the pore volume increased four times more than it had previously, with the majority of the increase occurring in the filled pore section (pore diameter <10nm). The CO_2 injection in the CO_2-ECBM process causes the diffusion and adsorption-filled pores to expand, while the compression of the permeation pores causes a reduction in permeability.

References

1. Junlin Liu, Huihu Liu, Kun Zhang, et al. Diffusion characteristics of CO2 and CH4 in CO2-ECBM process of low permeability coal seam[J/OL]. Coal Science and Technology :1-11[2023-03-30].
2. Busch A, Gensterblum Y. CBM and CO2-ECBM related sorption processes in coal: a review[J]. Coal Geol 2011; 87:49e71.
3. Lin J, Ren T, Wang GD, Nemcik J. Simulation investigation of N2-injection enhanced gas drainage: model development and identification of critical parameters[J]. Nat Gas Sci Eng 2018; 55:30-41.
4. Fan N, Wang JR, Deng CB, et al. Numerical study on enhancing coalbed methane recovery by injecting N2/CO2 mixtures and its geological signifi- cance[J]. Energy Sci Eng 2019.
5. Bai G, Zeng XK, Li XM, et al. Influence of carbon dioxide on the adsorption of methane by coal using low-field nuclear magnetic resonance[J]. Energy Fuel 2020;34(5):6113-23.

6. Wenping Jiang, Yongjun Cui. A Discussion on Main Geologic Controlling Factors of CO2 Sequestration in Deep Coal Seams[J]. Coal Geology of China, 2010, 22 (11): 1-6.
7. Lin J, Ren T, Wang GD, et al. Experimental study of the adsorption-induced coal matrix swelling and its impact on ECBM[J]. Earth Sci 2017;28(5):917-25.
8. Li Y, Wang Y, Wang J, Pan Z. Variation in permeability during CO2-CH4 displacement in coal seams: Part 1-Experimental insights[J]. Fuel 2019:116666.
9. Gang Bai, Jun Su, Zunguo Zhang, Anchang Lan, Xihua Zhou, Fei Gao, Jianbin Zhou, Effect of CO2 injection on CH4 desorption rate in poor permeability coal seams: An experimental study[J]. Energy, Volume 238, Part A,2022,121674.
10. Di Wu. The reserch on rule of CH4/CO2 seepage-adsorption-desorption and CO2 displacement CH4 by injiecting CO2 in residual coal seams under the thermo-mechanical action[D]. LiaoNing Technical University, 2013.
11. Li Y., Tang D., Xu H., et al. Experimental research on coal permeability: The roles of effective stress and gas slippage[J]. Journal of Natural Gas Science & Engineering, 2014, 21: 481-488.
12. Zou J., Chen W., Yang D., et al. The impact of effective stress and gas slippage on coal permeability under cyclic loading[J]. Journal of Natural Gas Science & Engineering, 2016, 31: 236-248.
13. Yin GZ, Deng BZ, Li MH, et al. Impact of injection pressure on CO2-enhanced coalbed methane recovery considering mass transfer between coal fracture and matrix[J]. Fuel, 2017; 196: 288-97.
14. Jiajun Xu, Yulong Chen, Chengrong Jiang, et al. Test on carbon dioxide permeability in coal seam[J]. Mining Safety & Environmental Protection , 2017, 44(04): 10-13.
15. Wang LG, Wang ZF, Li KZ, Chen HD. Comparison of enhanced coalbed methane recovery by pure N2 and CO2 injection: experimental observations and numerical simulation[J]. Nat Gas Sci Eng 2015; 23:363-72.
16. Metz B. IPCC Special Report on Carbon Dioxide Capture and Storage[J]. New York: Cambridge University Press; 2005.
17. Li Y, Wang Z, Pan Z, Niu X, Yu Y, Meng S. Pore structure and its fractal dimensions of transitional shale: A cross section from east margin of the Ordos Basin, China[J]. Fuel 2019; 241:417-31.
18. Kim HJ, Shi Y, He J, Lee HH, Lee CH. Adsorption characteristics of CO2 and CH4 on dry and wet coal from subcritical to supercritical conditions[J]. Chemical Engineering Journal, 2011;171(1):45-53.
19. Hemant K, Derek E, Liu J, Pone D, Mathews JP. Permeability evolution of propped artificial fractures in coal on injection of CO2[J]. Petrol Sci Eng 2015; 133:695-704
20. Li Y, Zhang C, Tang D, Gan Q, Niu X, Shen R. Coal pore size distributions controlled by the coalification process: an experimental study of coals from the Junggar, Ordos, and Qinshui basins in China[J]. Fuel 2017; 206:352-63.
21. Izadi G, Wang S, Elsworth D. Permeability evolution of fluid-infiltrated coal containing discrete fractures[J]. Coal Geol 2011;85(2):202-11.
22. Skurtveit E, Aker E, Soldal M, Angeli M, Wang Z. Experimental investigation of CO2 breakthrough and flow mechanism in shale[J]. Pet Geo Sci 2012; 18:3–15.
23. Wu Y, Pruess K, Persoff P. Gas flow in porous media with Klinkenberg effects[J]. Transp Porous Media 1998; 32:117-37.
24. Kollek JJ. The determination of the permeability of concrete to oxygen by the cembureau method-a recommendation[J]. Mater Struct 1989;22(3):225-30.
25. Shuxun Sang, Yanmin Zhu, Jing Zhang, et al. Solid-gas interaction mechanism of coal-adsorbed gas(III)& Physical processes and theoretical models of coal adsorption of gas[J]. natural gas industry. 2005, 25(1):4.
26. Yuanping Cheng, Biao Hu. A new pore classfication method based on the mechane occurrence and migration characteristics in coal[J]. Journal of China Coal Society, 2023, 48(01): 212-225.

Comprehensive Review of Machine Learning-Based Methods for Electricity Load Forecasting

Zixu Zhao*

Jiangnan University, Wuxi 214122, China.

Abstract: With the improvement of data processing power and the continuous development of modern power grids, there is an increasing demand for accuracy in predicting power load. To study the field of power load forecasting, this article summarizes and categorizes different models into three types: traditional models, single machine learning models, and hybrid models, based on previous literature. Firstly, a general overview is provided of the application of different models in power load forecasting. Secondly, typical models from three categories are selected for a detailed introduction. In traditional models, the ARIMA model is chosen, while in single machine learning models, CNN, and LSTM are chosen. For the hybrid model, the ResNet-LSTM mixed neural network is selected for the introduction. Finally, four different datasets were used to test different models. The differences and patterns of the models were summarized, and suggestions were proposed for future research directions in the field of power load forecasting.

Keywords: Power load forecasting, ARIMA, CNN, LSTM, ResNet-LSTM.

1. Introduction

With the development of the economy, people's reliance on electricity is increasing. Due to the difficulty of storing electric energy, the output of power plants should be kept as consistent as possible with the power consumption. This means that accurate predictions of power loads can help relevant departments develop reasonable power dispatching plans. This ensures a reasonable arrangement of unit maintenance plans, guarantees normal production and life of society, effectively reduces power generation costs, and improves economic and social benefits. According to the predicted time, load forecasting can be divided into three types: short-term forecasting, medium-term forecasting, and long-term forecasting. Short-term forecasting can be provided one day to one week in advance, medium-term forecasting can be provided one week to several months or even a year in advance, and long-term forecasting can be provided more than one year in advance. The emphasis varies at different times. Short-term forecasts are mainly used to arrange power generation and transmission, medium-term forecasts focus on arranging fuel purchases. And long-term forecasts focus on overall improvements to power generation units, transmission systems, and distribution systems [1]. The electric power system requires strict control functions to produce high-quality electrical energy, which places high demands on the speed of correcting related parameters [2], therefore load forecasting is of great significance. This article will focus on discussing the relevant model methods for electric load forecasting, rather than excessively classifying them according to the length of the forecasting period.

There are roughly three types of models for electrical load forecasting: traditional models, single machine learning models, and hybrid models. Traditional models apply traditional mathematical, statistical, and operations research methods to electrical load forecasting, while making certain improvements on this basis. For example, Jin [3] et al. proposed a new grey correlation competition model based on the general GM (1,1) model, which improves the problem of decreased accuracy in traditional grey prediction models when dealing with significant load mutations. Nepal B [4] et al., on the other hand, combined clustering and ARIMA models for prediction. Furthermore, regression, multiple regression, exponential smoothing, and iterative re-weighted least squares techniques [5] are all traditional load forecasting methods. These methods are relatively simple, but at the same time, their drawbacks are also very obvious. They cannot accurately predict power load that exhibits non-linear changes and is influenced by multiple factors; thus the upper limit of predictive accuracy cannot be exceeded. With the improvement of data computing power and the continuous development of modern power grids, machine learning is gradually entering people's vision. Compared to traditional models, machine learning can better handle non-linear data. Various machine learning methods are also being applied to power load prediction.

Specifically, a single machine learning model includes models such as least squares support vector machine (LSSVM), artificial neural networks (ANN),

* Corresponding author: zzx_5656@163.com

convolutional neural networks (CNN), and so on. The following are examples. Yang et al. [6] combined the autocorrelation function (ACF) with least square support vector machines (LSSVM) to establish a model called AS-GCLSSVM for power load forecasting. The study aimed to select optimal input features (Feature Selection, FS) and uncover potential feature variables. Bakirtzis A G et al. [7] proposed a short-term load forecasting model based on an artificial neural network (ANN), which was applied to the energy control center of the Greek Public Power Corporation (PPC) using 63 input neurons, 24 hidden neurons, and 24 outputs for prediction. Imani M [8] mainly uses Convolutional Neural Networks (CNN) for extracting non-linear features of the load. Muzaffar S [9] et al. collected electric load data with external variables such as temperature, humidity, and wind speed, and used this data to train LSTM networks. Due to the high number of parameters and training costs of LSTM, a variant of LSTM called GRU was designed. For instance, Zhu [10] et al. applied the Gated Recurrent Unit (GRU) model to short-term electric vehicle load forecasting scenarios and achieved good performance. Whether using supervised learning support vector machines or neural networks such as GRU, CNN, and LSTM, each of them has its advantages. Compared to traditional models, they exhibit superior performance. However, each method also has its limitations to varying degrees. Therefore, to avoid their respective limitations, hybrid models that combine multiple methods have also begun to emerge continuously.
A hybrid model combines multiple neural networks or other algorithms to optimize the model. For example, as mentioned in the text, combinations of GRU, LSTM, and CNN have shown better results than using a single network. For instance, Tang et al. [11] proposed a multi-layer bidirectional recursive neural network model based on LSTM and GRU to predict short-term power load. The model was validated on two datasets, the European Intelligent Technology Network Competition data and Chongqing Power Company data, and achieved good results. Guo et al. [12] fused CNN and LSTM to create a multiscale CNN-LSTM hybrid neural network model for short-term load forecasting. Wu et al. [13] combined CNN and GRU for a similar purpose.
We can see that the combination of various models is flexible and diverse. Not only can networks mentioned in a single machine learning model be mixed, but other neural networks and algorithms can also be integrated. For example, Xie et al. [14] proposed a short-term power load forecasting method that combines the Elman neural network (ENN) and particle swarm optimization (PSO). Nie [15] proposed an RBF-GRNN-ELM composite prediction model. Dai et al. [16] transformed traditional support vector machines and proposed a hybrid model that combines intelligent feature selection and parameter optimization methods. Lv et al. [17] proposed a hybrid model based on the variational mode decomposition (VMD) and long short-term memory (LSTM), and seasonality elimination and error correction were also performed. Wang et al. [18] proposed an industrial user short-term load forecasting model based on Temporal Convolutional Network (TCN) and Light Gradient Boosting Machine (LightGBM), which was applied to data from multiple countries and validated for its accuracy compared to other models. Chen [19] et al. developed a hybrid model that combines Complete Ensemble Empirical Mode Decomposition with Adaptive Noise (CEEMDAN), Gated Recurrent Unit (GRU), and an improved Grey Wolf Optimizer (IGWO), named CEEMDAN-IGWO-GRU (CIG) hybrid algorithm. Huang [20] et al. proposed a comprehensive short-term power load prediction method based on Complete Ensemble Empirical Mode Decomposition with Adaptive Noise (CEEMDAN), Sample Entropy (SE), Back-Propagation Neural Network (BPNN), and Transformer Model.
Neural networks are rapidly evolving, and various new models combining different types of neural networks are constantly emerging. Chen et al. [21] proposed a hybrid model based on Residual Neural Network (ResNet) and Long Short-Term Memory (LSTM) to fully utilize the time characteristics. Liu et al. [22] proposed a short-term load forecasting method based on an improved Temporal Convolutional Network (TCN) and Dense Connection Convolutional Network (DenseNet), where parallel pooling was introduced based on traditional TCN to extract features from time series. The effectiveness of the model was demonstrated after applying it to four datasets, including Singapore and the United States. Some researchers have also attempted to incorporate other algorithms into models that have already combined multiple neural networks. Su et al. [23] proposed a variational mode decomposition optimization algorithm (PSVMD-CGA) that combines CNN, GRU, attention mechanism, and mahjong algorithm. Cai et al. [24] introduced a short-term load forecasting approach that integrates variational mode decomposition (VMD), gated recurrent unit (GRU), and temporal convolutional network (TCN) into a hybrid network. Yi et al. [25] optimized the CNN-LSTM model by introducing a self-attention mechanism (SAM). The hybrid models can be further divided into three types: combinations of multiple neural networks, combinations of a single neural network and other algorithms, and combinations of multiple neural networks and other algorithms.
However, after summarizing the literature, it was found that these models mainly have the following shortcomings:
1. The application scenarios are limited and only tested in a single scenario [8,10], not tested in multiple scenarios.
2. The computation time is comparatively longer and consumes more time than other models [15].
3. The testing and training data are insufficient [10], which can be further optimized.
4. To predict power load, it is necessary to add more irrelevant relevant variables, such as weather factors, temperature, holidays, and other events [6].
5. Optimizing the internal structure of the neural network, such as selecting the scale of the CNN convolution layer, affects the feature dimension and extraction accuracy, and further affects the accuracy of load forecasting [12].
In the next section of this article, we will explain typical models selected for traditional models, single machine learning models, and hybrid models separately. In section 3, we will compare and evaluate different models. Finally,

we will summarize and offer suggestions based on multiple experiments.

2. Traditional Models

2.1 ARIMA Model based on K-means Clustering

The ARIMA model based on K-means clustering belongs to traditional models. This article will introduce the ARIMA model and K-means clustering method separately, and then introduce the ARIMA model incorporating K-means.

2.1.1 ARIMA

The ARIMA model is a famous time series prediction method proposed by Box and Jenkins in the early 1970s [26]. The ARIMA model is represented as ARIMA (p, d, q). ARIMA is essentially divided into three parts: AR, I, and MA. AR stands for autoregression, MA stands for moving average, and I stands for differencing. p is the order of autoregression, q is the order of moving average, and d is the order of differencing.

The ARIMA model requires data to be stationary. Before modeling, a stationarity test should be performed. If data is non-stationary, differencing can be performed to make it stationary (as mentioned earlier, d is used for differencing). If data is already stationary, then d is 0 and differencing is not needed.

The AR model is an autoregressive model that focuses on the relationship between current and past values, which is shown in formula (1):

$$y_t = \mu + \sum_{i=1}^{p} r_i y_{t-i} + \varepsilon_t \tag{1}$$

Where y_t is the current value, μ is the constant term, r_i is the autocorrelation coefficient, and ε_t is the error term. As for the autocorrelation function, it is shown in formula (2):

$$r_K = \frac{\sum_{t=1}^{n-k}[(X_t-\bar{X})(X_{t+k}-\bar{X})]}{\sum_{t=1}^{n}(X_t-\bar{X})^2}, k = 1,2,\cdots K \tag{2}$$

Where n is the sample size, and $\bar{X}$ is the sample mean. The MA model stands for moving average, which focuses on the accumulation of the error term, as shown in formula (3):

$$y_t = \mu + \sum_{i=1}^{q} \theta_i \epsilon_{t-i} + \varepsilon_t \tag{3}$$

The ARMA model is a combination of the AR model and MA model, as shown in formula (4):

$$y_t = \mu + \sum_{i=1}^{p} r_i y_{t-i} + \varepsilon_t + \sum_{i=1}^{q} \theta_i \epsilon_{t-i} \tag{4}$$

The ARIMA model is an ARMA model that has been differenced d times. The values of p and q in the ARIMA model are determined based on the autocorrelation function (ACF) and partial autocorrelation function (PACF). Then, the model fitness is evaluated by checking the Akaike information criterion (AIC) and other criteria. Finally, the residuals ϵ is examined. If the residual ϵ is white noise, it indicates that all useful information in the time series has been extracted and the modeling process can be terminated.

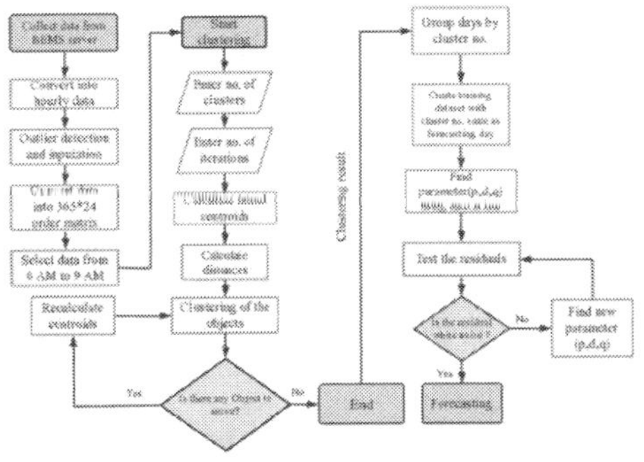

Fig.1 Flowchart of the model combining K-means and ARIMA. BEMS stands for the Building Energy Management System

2.1.2 Combination of K-means clustering and ARIMA

In reference [4], the ARIMA model is combined with K-means clustering[27] to predict the power consumption of buildings. The process is shown in Figure 1.

First, extract the existing data from the Building Energy Management System (BEMS) and process the data. Then, use K-means clustering to cluster the data into 6 classes as in [4]. After the clustering is completed, create a dataset that belongs to the same cluster as the predicted values, and then build an ARIMA model and select suitable p, d, and q values. Check whether the residuals ϵ are white noise during the process. And after the test, use the set ARIMA model for prediction.

2.2 CNN-Based Nonlinear Feature Extraction Model for Power Load

This model belongs to a single machine learning model, which accomplishes nonlinear feature extraction for power load prediction by constructing a convolutional neural network(CNN).

The first implementation of Convolutional Neural Networks (CNN) was carried out by LeCun Y [28] and his colleagues in 1989. CNN has shown excellent capability in capturing load data trends for load forecasting [29] and performs well in extracting and analyzing multidimensional data.

In literature [8], the prediction of electricity consumption for the next hour is finally made. To achieve this goal, two CNN networks are set up to extract load characteristics and load-temperature characteristics respectively. The sampling points for load and temperature are set to be collected once every hour, and the data collected for the previous two weeks for load is 24×7×2, while for temperature it is 24×7×1 for the previous week. The two CNN networks are named Load-CNN and Load-Temperature CNN respectively. The input for Load-CNN and Load-Temperature CNN is load data from the previous two weeks, as shown in Figure 2.

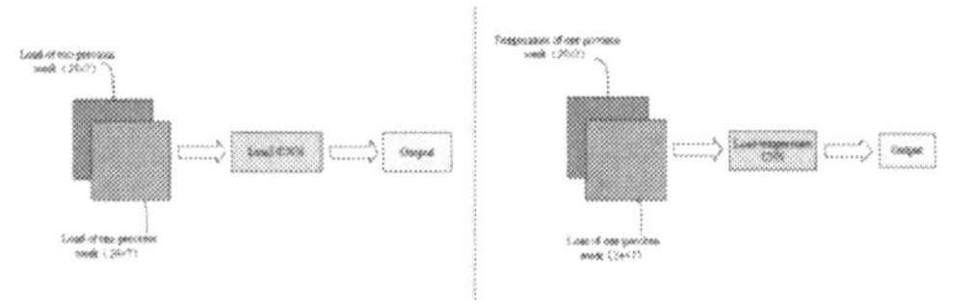

Fig.2 Load-CNN, which extracts features by inputting the load data of the first two weeks into the Load-CNN neural network.And Load-Temperature CNN, which inputs the load and temperature data from the previous week into the Load-temperature CNN neural network for feature extraction.

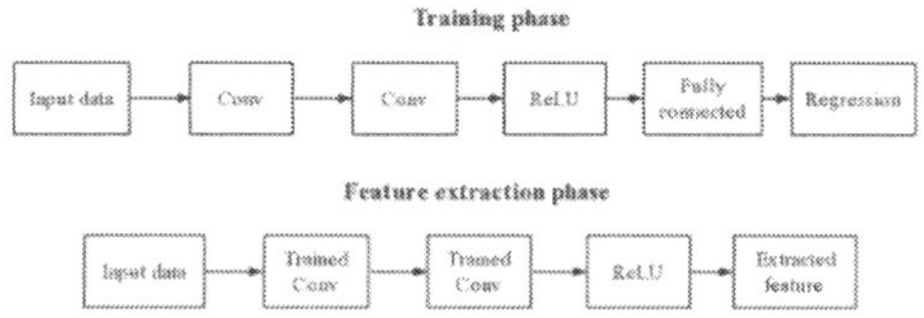

Fig.3 Phase diagram indicating the network architecture used during training stages and for feature extraction at the end.

The input of Load-CNN consists of 24 rows and 7 columns, corresponding to 24 hours and seven days within a week. The total input includes the load data from the first two weeks. The input of Load-temperature CNN includes the load data and temperature data from the previous week, also with 24 rows and 7 columns corresponding to 24 hours and seven days within a week. For two neural networks, training is required. During training, the output is the load value for the next hour, but in the end, the output becomes the load features to be extracted and the load-temperature features.

CNN has two main characteristics: weight sharing and local connections. A typical CNN consists of several parts, including convolutional layers, activation functions, pooling layers, fully connected layers, etc. Here, the max pooling layer is omitted in the design to avoid discarding some features. Additionally, the network used during the training stage and the final feature extraction stage may differ. The process is shown in Figure 3:

During training, two convolutional layers are used followed by an activation function using the ReLU function, then a fully connected layer and a regression layer. For feature extraction after training, two convolutional layers and ReLU activation function are used for feature extraction. For the convolutional layer, as shown in formula (5):

$$\mathrm{F}^{\mathrm{l}} = \mathrm{g}(\mathrm{F}^{\mathrm{l}-1} * \mathrm{W}^{\mathrm{l}} + \mathrm{b}^{\mathrm{l}}) \quad (5)$$

Where $\mathrm{F}^{\mathrm{l}-1}$ represents the feature map of the (l-1)th layer, W^{l} and b^{l} are learnable parameters representing the weights and biases of the lth layer. $*$ represents the linear convolution operator. $\mathrm{g}(\cdot)$ represents the ReLU activation function defined as formula (6):

$$\mathrm{g}(\mathrm{x}) = \max(0, \mathrm{x}) = \begin{cases} \mathrm{x}, \mathrm{x} > 0 \\ 0, \mathrm{x} \leq 0 \end{cases} \quad (6)$$

The initial input of the first layer has been mentioned previously. $\mathrm{F}^{\mathrm{l}-1}$ has a size of $W_d^{l-1} \times W_h^{l-1} \times W_n^{l-1}$, where W_d^{l-1} and W_h^{l-1} are the width (corresponding to the day) and height (corresponding to the hour) of the feature map of the (l-1)th layer, and W_n^{l-1} represents the number of outputs of the (l-1)th layer. The neural network uses filters composed of convolution kernels with a size of 3×3 and a stride of 1.

In order to perform feature extraction, neural networks need to be trained. Assuming there are M samples, and their next-hour load values are known. The actual values during training are set as y_i^a {i = 1, 2, ... M}, with predicted values as y_i^f {i = 1, 2, ... M}. Assuming there are L layers, the formula (7) is as follows:

$$y_i^f = \mathrm{W}^{\mathrm{L}}\mathrm{F}^{\mathrm{L}-1} + \mathrm{b}^{\mathrm{L}}; i = 1,2,\dots M \quad (7)$$

Therefore, the mean square error loss function is established, as shown in formula (8):

$$\psi = \frac{1}{2}\sum_{\mathrm{i}=1}^{\mathrm{M}} |y_i^a - y_i^f|^2 \quad (8)$$

The backpropagation algorithm is widely used to optimize weight W^{l} and bias b^{l} by minimizing the loss function. Prediction errors propagate from the higher layer to the previous layer, and the parameters of each layer are modified based on the propagation error. Stochastic gradient descent and Adam optimizer can be used to implement weight W^{l} and bias b^{l}. Finally, the trained CNN network can be used for nonlinear feature extraction.

2.3 LSTM-based power load prediction model

This model belongs to a single machine learning model, which uses a constructed LSTM neural network to perform electric load forecasting. LSTM stands for Long Short Term Memory, first proposed by Hochreiter S [30], it is a variation of Recurrent Neural Network (RNN) [31] with the ability to learn long-term dependencies. LSTM overcomes the impact of short-term memory, characterized by cell and "gate" structures.

The cell structure of LSTM is shown in Figure 4, which provides memory capability through the gating structure in the cell. Multiple cells are connected to form the hidden layer. The symbol "×" in the figure represents pointwise multiplication, and the symbol "+" represents pointwise addition. The cell structure includes a forget gate, an input gate, a cell state, and an output gate.

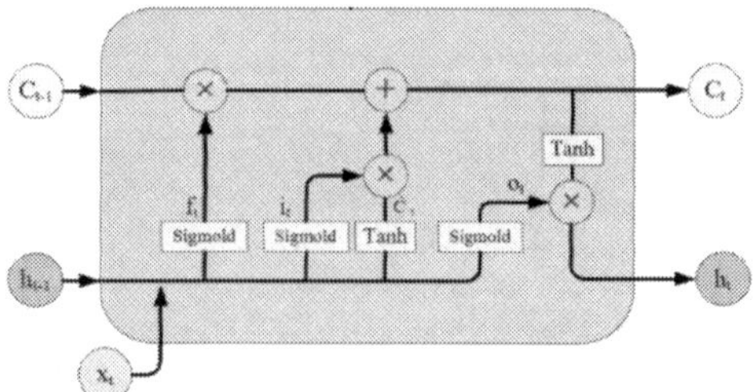

Fig.4 LSTM cell diagram, mainly composed of forget gate, input gate, output gate, and others.

The forget gate determines which information should be discarded or retained. Information from the previous hidden state h_{t-1} and the current input x_t are passed into a Sigmoid function, with the output value ranging between 0 and 1. The closer the output value is to 0, the more it should be discarded. This is expressed in formula (9):

$$f_t = \sigma(W_f \cdot [h_{t-1}, x_t] + b_f) \quad (9)$$

The input gate is used to update the cell state. First, the information of the previous hidden state h_{t-1} and the current input information x_t are passed to the Sigmoid function. Then, the information of the previous hidden state h_{t-1} and the current input information x_t are passed to the tanh function to create a new candidate value vector. Finally, the output value of the Sigmoid function and the output value of the tanh function are multiplied. This is shown in equations (10) and (11).

$$i_t = \sigma(W_i \cdot [h_{t-1}, x_t] + b_i) \quad (10)$$

$$\tilde{C}_t = tanh(W_C \cdot [h_{t-1}, x_t] + b_C) \quad (11)$$

The cell state C_t is obtained through the forget gate, input gate, and the previous layer's cell state. As shown in formula (12):

$$C_t = f_t \times C_{t-1} + i_t \times \tilde{C}_t \quad (12)$$

The output gate is used to determine the value of the next hidden state, which contains information from previous inputs. First, the information of the previous hidden state h_{t-1} and the current input information x_t are passed to the Sigmold function, and then the newly obtained cell state C_t is passed to the tanh function. Finally, the output of the tanh function is multiplied by the output of the Sigmoid function to determine the information the hidden state should carry, which is then passed to the next cell. Formulas (13) and (14):

$$o_t = \sigma(W_o \cdot [h_{t-1}, x_t] + b_o) \quad (13)$$

$$h_t = o_t \times tanh(c_t) \quad (14)$$

In order to provide appropriate training data, two vectors are generated [9] for each sequence: PredTrain and RespTrain, where PredTrain represents the predicted value and RespTrain represents the expected value. PredTrain is the input TS with the last 5 values removed, while RespTrain is the input TS with the first 5 values removed. The optimal lag chosen here is 5, meaning that each response depends on the previous 5 values. The training data for both input TS is generated in the same way. Predictive test vectors are also generated using only 13 months of data. "end" represents the end of the data, "12MonthsData" represents the first 12 months of data, and "13MonthData" represents the 13th month's data. The format for training and predictive testing data is shown in Equations (15), (16), (17), and (18):

$$\text{PredTrain} = \text{12MonthsData}(1{:}\,\text{end} - \text{lag}) \quad (15)$$

$$\text{RespTrain} = \text{12MonthsData}(\text{lag} + 1{:}\,\text{end}) \quad (16)$$

$$\text{PredTest} = \text{13MonthData}(1{:}\,\text{end} - \text{lag}) \quad (17)$$

$$\text{RespTest} = \text{13MonthData}(\text{lag} + 1{:}\,\text{end}) \quad (18)$$

2.4 Hybrid Power Load Forecasting Model Based on Resnet and LSTM.

2.4.1 Resnet

Resnet was first proposed in [32] to solve the degradation, vanishing, and exploding gradients problems that occur as neural networks become deeper. Vanishing gradients refer to the problem where the gradient approaches zero as the network gets deeper, resulting in difficulty in backpropagation when the error gradient of each layer is less than 1. On the other hand, exploding gradients happen when the error gradient of each layer is larger than 1, resulting in the gradient getting larger as the network gets deeper. The characteristics of this network are: (1) an ultra-deep network structure with over 1000 layers, (2) the introduction of residual blocks, and (3) the use of Batch Normalization [33] to accelerate training.

A. Batch Normalization

For addressing the issues of vanishing gradients and optimizing gradients, Batch Normalization is mainly employed. This operation is performed between each fully connected layer and activation function. The Batch Normalization algorithm normalizes the feature maps of a batch of data to follow a distribution pattern with a mean of 0 and variance of 1. This is achieved mainly through equations (19), (20), (21), and (22).

$$\mu_{\mathcal{B}} = \frac{1}{m}\sum_{i=1}^{m} x_i \quad (19)$$

$$\sigma_{\mathcal{B}}^2 = \frac{1}{m}\sum_{i=1}^{m}(x_i - \mu_{\mathcal{B}})^2 \quad (20)$$

$$\hat{x}_i = \frac{x_i - \mu_{\mathcal{B}}}{\sqrt{\sigma_{\mathcal{B}}^2 + \epsilon}} \quad (21)$$

$$y_i = \gamma \hat{x}_i + \beta \quad (22)$$

Here, $\mathcal{B} = \{x_{1...m}\}$ is the set of input values x_i , ϵ is a small constant used to prevent division by zero, y_i is the input, $\sigma_{\mathcal{B}}^2$ represents variance, with each element representing the variance in each dimension, and similarly $\mu_{\mathcal{B}}$ represents the mean, with each element representing the mean in each dimension. $\sigma_{\mathcal{B}}^2$ and $\mu_{\mathcal{B}}$ are calculated, while γ and β are parameters learned in backpropagation. γ adjusts the variance of the value distribution, with an initial value of 1, while β adjusts the position of the value mean, with an initial value of 0.

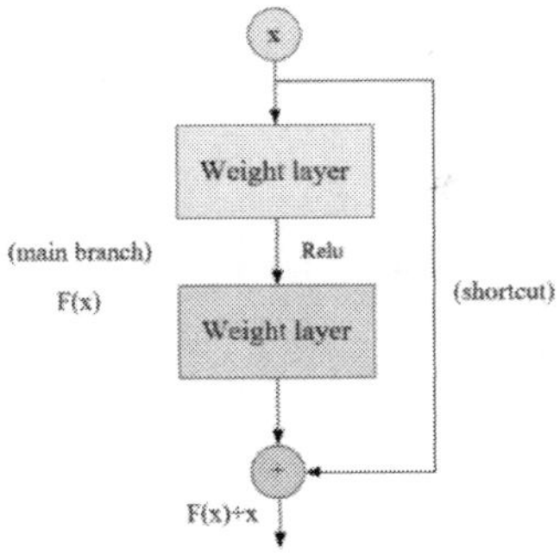

Fig.5 Residual block, where F(X) has two layers of weight layer and one layer of activation function.

B. residual block

To solve the degradation problem in deep neural networks, a residual block is applied to address it. The residual block adds a shortcut connection to weaken the strong correlation between each layer. Assuming the input is x and the expected output is H(x), which represents the actual mapping, the interpolation between the expected output and input values is F(x). F(x) is the residual mapping. Therefore, we can obtain formulas (23) and (24):

$$F(x) = H(x) - x \quad (23)$$

$$H(x) = F(x) + x \quad (24)$$

As shown in Figure 5, if the network tends to be optimal, continue to deepen the network. If the residual mapping F(x) becomes 0, H(x) = x; theoretically the performance of the network will not decrease with the increase of depth. If F(x) ≠ 0, but F(x) is close to 0, then x approximates the actual mapping H(x), and the problem of degradation is solved. However, it should be noted whether the shapes of the main branch and shortcut are the same, only when they are the same can they be added together. Therefore, in Resnet, the residual structure is divided into solid line residual structure and dotted line residual structure. When the shape of the main branch in the residual block is the same as the shape of the shortcut, the main branch and shortcut branches can be directly added together, which is the solid line residual structure. When the shapes are different, the shortcut branch needs to be processed by convolution operation to make the feature matrix shape of the shortcut branch consistent with that of the main branch, and then add them together.

2.4.2 Hybrid Model of Resnet and LSTM

The principle of LSTM has been mentioned in the previous text. Here we introduce a hybrid model of Resnet and LSTM based on literature [18]. This model uses Resnet as the pre-feature extraction unit and LSTM as the time series feature learning unit.

For the Resnet network part in the hybrid model, Resnet18 is adopted, consisting of five parts: Conv1, Conv2_x, Conv3_x, Conv4_x, and Conv5_x. Conv1 performs convolution and max pooling operations. The other four parts each consist of 2 residual blocks, with each residual block composed of two layers of a convolutional network. The feature matrices of the two branches of the residual block are added together and then activated by the ReLU function before being output. The data structure input to Resnet is (None, 56,1), where 56 is the number of feature parameters and the number of channels is 1. The number of convolutional kernels for each part is 32, 32, 64, 128, and 256, respectively. The size of the convolutional kernel is 3×3. In the max pooling layer, the pool size is also 3×3, and the step size is 2. After feature extraction by Resnet, a three-dimensional vector (None, 2, 256) is output to the LSTM network.

In this model, the residual block of the solid line residual structure performs a convolution operation with a stride of 1, while the dashed line residual structure performs a convolution operation with a stride of 2 on the shortcut branch, ensuring that the feature matrices of the short cut branch and the main branch have the same shape.

For the LSTM network, two layers are set here. The number of units in each layer is 1024 and 256. Dropout is used between the layers of the LSTM network to prevent overfitting of the model. Finally, two Dense layers are used to output the predicted load value. The first layer has 64 neuron nodes and the second layer has 1 neuron node.

3. Model Evaluation

In this section, we will select typical models from several categories mentioned earlier and compare their performance on different datasets to evaluate their strengths and weaknesses.

3.1 Evaluation Indicators.

Model evaluation metrics are diverse, and a single metric cannot fully reflect model performance. Therefore, this paper selects Mean Absolute Percentage Error (MAPE), Root Mean Square Error (RMSE), and Mean Absolute Error (MAE) [34] to compare and evaluate the models. The specific content and calculation formulas of each metric are shown in Table 1.

Table 1 Evaluation Indicators and Calculation Formulas.

Metric	Definition	Equation
MAPE	Average absolute percentage errors	$MAPE=\frac{1}{N}\sum_{n=1}^{N}\lvert\frac{\hat{y}_i-y_i}{y_i}\rvert\times100\%$
MAE	Average absolute error	$MAE=\frac{1}{N}\sum_{n=1}^{N}\lvert\hat{y}_i-y_i\rvert$
RMSE	Root Mean Square Error	$RMSE=\sqrt{\frac{1}{N}\sum_{n=1}^{N}(\hat{y}_i-y_i)^2}$

N represents the total number of samples, y_i represents the actual value of the sample, and $\hat{y}_i$ represents the predicted value of the sample. MAPE, MAE, and RMSE are all indicators where the smaller the value, the better.

3.2 Model Comparison

This article selected experimental results from four references [17,21,25,35] for model comparison. The source of each dataset is shown in Table 2.

Table 2 Dataset description and the dataset source

Dataset description	Dataset source
The publicly available online load data set published in Queensland, Australia from Jan. 1, 2006, to Dec.31, 2010	https://github.com/weiran4/AustraliaData
Real load data from the dataminer2 website of PJM in the US.	Data Miner 2 (pjm.com)
The building data set derived from hourly load data provided by the UTD from 1 January 2014 to 31 December 2015.	https://ieee-dataport.org/documents/short-term-load-forecasting-data-hierarchical-advanced-metering-infrastructure-and-weather
A pool of 69 customers out of thousands of customers' data, which were retrieved from the SGSC project initiated by the Australian Government	http://www.industry.gov.au/ENERGY/PROGRAMMES/SMARTGRIDSMARTCITY/Pages/default

3.2.1 Queensland Dataset in Australia

The publicly available online load dataset in Queensland, Australia from January 1, 2006, to December 31, 2010, contains 87649 rows of data. It has a sampling interval of 30 minutes with 48 sampling points per day. The dataset includes features such as load, dry-bulb temperature (drybt), dew point temperature (dewbt), wet-bulb temperature (wetbt), humidity, and price. The data from the first four years is used as the training set, data from January 1 to November 30, 2010, is used as the validation set, and the last month's data is used as the test set. Testing was conducted using data from December 1 to 7, 2010, and the results are shown in Table 3.

Table 3 Comparison of indicators of various models on the Australian dataset

Model	MAPE	MAE	RMSE
MLR	0.769	69.938	84.307
LSTM	0.822	65.982	89.396
CNN	0.815	67.747	86.494
Resnet	0.753	62.488	79.653
CNN-LSTM	0.830	69.501	89.264
Resnet-LSTM	**0.597**	**49.301**	**62.933**

*MLR, Mixed Logistic Regression, uses piecewise linear regression to fit nonlinear data.

It can be seen that Resnet-LSTM outperforms other models in terms of MAPE, MAE, and RMSE, demonstrating excellent performance. Among the six models compared, Resnet performs better than CNN and LSTM, as well as their hybrid network CNN-LSTM. However, the performance of Resnet is inferior to the Resnet-LSTM hybrid network. Resnet also outperforms the MLR algorithm.

3.2.2 Real load data from the dataminer2 website of PJM in the US

The data sets were collected from six load groups in the Mid-Atlantic, PJM_RTO, and Western regions of the United States, from November 22, 2020, to January 17, 2021, and from February 1, 2021, to March 28, 2021. They are represented as 2020 MID, 2020 PJM, 2020 Western, 2021 MID, 2021 PJM, and 2021 Western. These data sets were sampled 24 times a day, for a total of 1344 samples. The performance of models is shown in Table 4. Only the results of 2020 MID and 2021 MID are shown here.

Table 4 Model results based on the dataminer2 website dataset in PJM in the United States.

Distance	Model	Metrics		
		RMSE	MAE	MAPE(%)
2020 Mid	CEEMD+LSTM	430	335	1.1
	ARIMA	1015	725	2.3
	VMD+LSTM	**388**	**309**	**1**
	GRU+RNN	559	426	1.3
2021 Mid	CEEMD+LSTM	537	408	1.6
	ARIMA	939	690	2.7
	VMD+LSTM	**392**	**313**	**1.2**
	GRU+RNN	747	543	2.0

*CEEMD stands for Complementary Ensemble Empirical Mode Decomposition. VMD stands for Variational mode decomposition. GRU is a gated recurrent unit, and RNN stands for the recurrent neural network.

CEEMD and VMD are two different signal decomposition algorithms, while GRU and RNN are types of neural networks. Through comparison, it can be found that in these sections, the performance of the ARIMA model is significantly inferior to other neural network models. Among the three hybrid models involving neural networks, the VMD+LSTM model is the best.

3.2.3 The building data provided by the UTD

The dataset includes data from January 1st, 2014 to December 31st, 2015. The dataset was divided into training, validation, and test sets in a ratio of 0.6/0.2/0.2. The purpose of the test was to evaluate the performance of different neural networks with attention mechanisms. The results of the test are shown in Table 5.

Table 5 Results of the model on the building data provided by UTD.

Model	MAPE	MAE	RMSE
LSTM	0.070	2.991	3.904
LSTM-based SAM	0.067	2.884	3.826
CNN-GRU-based SAM	0.079	3.370	4.323
LSTM-CNN-based SAM	**0.027**	**1.144**	**1.589**

* SAM stands for self-attention mechanism, which is introduced to pay attention to the correlations among input data during the training process.

Comparing the results of the four models, the LSTM network fused with SAM outperformed the original LSTM network. However, the performance of the CNN-GRU-based SAM hybrid network was poor, performing worst among the four models. The LSTM-CNN-based SAM model was the best, with all numerical values significantly lower than the other models.

3.2.4 The data retrieved from the SGSC project

The dataset consists of 69 customers selected for the SGSC project initiated by the Australian government. These customers were selected based on the criteria of being energy users with a hot water system. The available data volume varies for each customer. Data were selected from June 1, 2013, to August 31, 2013. The data span for each customer is 92 days. The data is divided into training, validation, and testing data sets in a ratio of 0.7/0.2/0.1. At the same time, different time steps were set during the inspection, including 2, 6, and 12. However, the model metric used here is Average MAPE Individual Forecasts. The model metrics obtained by applying different models and time backstep lengths are shown in Table 6.

Table 6 Indicators using different models and time steps.

Model	Scenario(Lookback)	Average MAPE Individual Forecasts(%)
CNN-LSTM	2-time steps	**40.38**
	6-time steps	41.07
	12-time steps	42.85
LSTM	2-time steps	44.39
	6-time steps	44.31
	12-time steps	**44.06**
BPNN	2-time steps	49.62
	6-time steps	**49.04**
	12-time steps	49.49
KNN	2-time steps	74.83
	6-time steps	**71.18**
	12-time steps	81.13
ELM	2-time steps	**122.90**
	6-time steps	136.49
	12-time steps	123.45

*BPNN stands for Back Propagation Neural Network, KNN stands for k-nearest neighbor model, and ELM stands for Extreme Learning Machine, which is a method based on Feedforward Neuron Network (FNN).

Based on the above data, it can be seen that different parameters can lead to different results for the same model. For CNN-LSTM and ELM, 2-time steps are optimal; for LSTM, 12-time steps are optimal; for BPNN and KNN, 6-time steps are optimal. Overall, CNN-LSTM performs better than other single networks.

4. Conclusion and Outlook

Although it is not possible to compare the results obtained from the four datasets in Section 3 in a unified way, some characteristics related to the model can be summarized based on the results presented in each dataset. Therefore, the conclusions obtained in this article are summarized as follows:

1. Traditional model has a simple structure and performs well in special scenarios for fitting and prediction. However, overall, its performance is inferior to that of a single machine-learning model or a hybrid model.
2. A single machine learning model solves non-linear problems compared to traditional models. However, due to the use of only one approach, there are inevitably some shortcomings in some aspects.
3. Hybrid models combine multiple neural networks or other algorithms to optimize the model. Overall, hybrid models outperform single machine learning models, but hybrid models are more complex than single machine learning models. It cannot be said that the performance of a model after combining multiple neural networks is always better than the original model. The appropriate model should be selected based on the actual situation. Improper hybridization can degrade model performance.
4. Parameters also affect the performance of the model. After determining the network type and algorithm type, fine adjustment of the model parameters should also be done.

According to the experience in this article, there are the following optimization directions for future research:

1. Neural networks still have great potential for development, and can improve the shortcomings of a single network to form a new single network with better performance.
2. New combinations can be attempted based on existing neural networks and algorithms. Improve model performance by using new combination methods to compensate for their respective weaknesses.
3. Optimize the internal parameters of the model to improve its generalization ability.

References

1. Almeshaiei E, Soltan H. A methodology for electric power load forecasting[J]. Alexandria Engineering Journal, 2011, 50(2): 137-144.
2. Jahan I S, Snasel V, Misak S. Intelligent systems for power load forecasting: A study review[J]. Energies, 2020, 13(22): 6105.
3. Jin M, Zhou X, Zhang Z M, et al. Short-term power load forecasting using grey correlation contest modeling[J]. Expert Systems with Applications, 2012, 39(1): 773-779.
4. Nepal B, Yamaha M, Yokoe A, et al. Electricity load forecasting using clustering and ARIMA model for energy management in buildings[J]. Japan Architectural Review, 2020, 3(1): 62-76.
5. Singh A K, Khatoon S, Muazzam M, et al. Load forecasting techniques and methodologies: A review[C]//2012 2nd International Conference on Power, Control and Embedded Systems. IEEE, 2012:
6. Yang A, Li W, Yang X. Short-term electricity load forecasting based on feature selection and Least Squares Support Vector Machines[J]. Knowledge-Based Systems, 2019, 163: 159-173.
7. Bakirtzis A G, Petridis V, Kiartzis S J, et al. A neural network short term load forecasting model for the Greek power system[J]. IEEE Transactions on power systems, 1996, 11(2): 858-863.
8. Imani M. Electrical load-temperature CNN for residential load forecasting[J]. Energy, 2021, 227: 120480.
9. Muzaffar S, Afshari A. Short-term load forecasts using LSTM networks[J]. Energy Procedia, 2019, 158: 2922-2927.
10. Zhu J, Yang Z, Guo Y, et al. Short-term load forecasting for electric vehicle charging stations

based on deep learning approaches[J]. Applied sciences, 2019, 9(9): 1723.

11. Tang X, Dai Y, Wang T, et al. Short-term power load forecasting based on multi-layer bidirectional recurrent neural network[J]. IET Generation, Transmission & Distribution, 2019, 13(17): 3847-3854.
12. Guo X, Zhao Q, Zheng D, et al. A short-term load forecasting model of multi-scale CNN-LSTM hybrid neural network considering the real-time electricity price[J]. Energy Reports, 2020, 6: 1046-1053.
13. Wu L, Kong C, Hao X, et al. A short-term load forecasting method based on GRU-CNN hybrid neural network model[J]. Mathematical Problems in Engineering, 2020, 2020.
14. Xie K, Yi H, Hu G, et al. Short-term power load forecasting based on Elman neural network with particle swarm optimization[J]. Neurocomputing, 2020, 416: 136-142.
15. Nie Y, Jiang P, Zhang H. A novel hybrid model based on combined preprocessing method and advanced optimization algorithm for power load forecasting[J]. Applied Soft Computing, 2020, 97: 106809.
16. Dai Y, Zhao P. A hybrid load forecasting model based on support vector machine with intelligent methods for feature selection and parameter optimization[J]. Applied energy, 2020, 279: 115332.
17. Lv L, Wu Z, Zhang J, et al. A VMD and LSTM based hybrid model of load forecasting for power grid security[J]. IEEE Transactions on Industrial Informatics, 2021, 18(9): 6474-6482.
18. Wang Y, Chen J, Chen X, et al. Short-term load forecasting for industrial customers based on TCN-LightGBM[J]. IEEE Transactions on Power Systems, 2020, 36(3): 1984-1997.
19. Chen Z, Jin T, Zheng X, et al. An innovative method-based CEEMDAN–IGWO–GRU hybrid algorithm for short-term load forecasting[J]. Electrical Engineering, 2022, 104(5): 3137-3156.
20. Huang S, Zhang J, He Y, et al. Short-Term Load Forecasting Based on the CEEMDAN-Sample Entropy-BPNN-Transformer[J]. Energies, 2022, 15(10): 3659.
21. Chen X, Chen W, Dinavahi V, et al. Short-Term Load Forecasting and Associated Weather Variables Prediction Using ResNet-LSTM Based Deep Learning[J]. IEEE Access, 2023.
22. Liu M, Qin H, Cao R, et al. Short-Term Load Forecasting Based on Improved TCN and DenseNet[J]. IEEE Access, 2022, 10: 115945-115957.
23. Su J, Han X, Hong Y. Short Term Power Load Forecasting Based on PSVMD-CGA Model[J]. Sustainability, 2023, 15(4): 2941.
24. Cai C, Li Y, Su Z, et al. Short-Term Electrical Load Forecasting Based on VMD and GRU-TCN Hybrid Network[J]. Applied Sciences, 2022, 12(13): 6647.
25. Yi S, Liu H, Chen T, et al. A deep LSTM-CNN based on self-attention mechanism with input data reduction for short-term load forecasting[J]. IET Generation, Transmission & Distribution, 2023.
26. Ahmed M S, Cook A R. Analysis of freeway traffic time-series data by using Box-Jenkins techniques[M]. 1979.
27. MacQueen J. Classification and analysis of multivariate observations[C]//5th Berkeley Symp. Math. Statist. Probability. 1967: 281-297.
28. LeCun Y, Boser B, Denker J S, et al. Backpropagation applied to handwritten zip code recognition[J]. Neural computation, 1989, 1(4): 541-551.
29. Rafi S H, Deeba S R, Hossain E. A short-term load forecasting method using integrated CNN and LSTM network[J]. IEEE Access, 2021, 9: 32436-32448.
30. Hochreiter S, Schmidhuber J. Long short-term memory[J]. Neural computation, 1997, 9(8): 1735-1780.
31. Elman J L. Finding structure in time[J]. Cognitive science, 1990, 14(2): 179-211.
32. He K, Zhang X, Ren S, et al. Deep residual learning for image recognition[C]//Proceedings of the IEEE conference on computer vision and pattern recognition. 2016: 770-778.
33. Ioffe S, Szegedy C. Batch normalization: Accelerating deep network training by reducing internal covariate shift[C]//International conference on machine learning. pmlr, 2015: 448-456.
34. Jin Y, Guo H, Wang J, et al. A hybrid system based on LSTM for short-term power load forecasting[J]. Energies, 2020, 13(23): 6241.
35. Alhussein M, Aurangzeb K, Haider S I. Hybrid CNN-LSTM model for short-term individual household load forecasting[J]. Ieee Access, 2020, 8: 180544-180557.

Effect of pH on the Removal of Landfill Leachate Pollutant by DTRO

Tao Long*

Development and Planning Office, Kunming University, 650214, Kunming, China

Abstract. The operating pressure of the first-stage DTRO and the removal efficiency of COD, BOD, TP observed within 180 min at pH 6.5, pH 7, and pH 7.5, respectively, then the mechanism analysed and optimization methods recommended. The results show that the operating pressure reached its fastest increasing rate at pH 7.5, which grew from 32.13 to 32.66 Bar rapidly. The removal efficiency of all major pollutants increased steadily with uptime before stability. The maximum removal rates of COD, BOD, TP were 97.9%, 97.36%, and 97.7% at pH 7.5, respectively. The system performance can be improved by adjusting the pH value of the leachate to 6 ~ 7 during operation.

Keywords. landfill leachate; DTRO; pH; pollutant removal

1. Introduction

The effectiveness and operating conditions of the DTRO system for treating landfill leachate have been published in journals and conference papers [1-7]. However, few studies are available on how the pH value alter the working characteristics of RO membranes and thus affecting the removal efficiency of major pollutants. This article observes the removal efficiency of COD, BOD_5, and SS in a two-stage DTRO system under different pH values. The influence causes are analyzed, and the corresponding optimization methods are explored, for reference to other landfill leachate treatment projects.

2. Experiment and analysis

2.1 Materials and Methods

The landfill leachate treatment station is located in a plateau of Southwest China where belongs to the subtropical monsoon climate zone at an altitude of 1910m. The average annual temperature is 14.8 °C. The average minimum and maximum temperatures are 9.7 °C and 21.6 °C, respectively. The extreme temperatures reached -6.2 °C and 31.6 °C in the record. The average annual precipitation is 900mm, and the precipitation of rainy season (May-October) achieves about 86.7% of the total. The experimental equipment includes a leachate storage tank, sand filter, precision filter, first stage DTRO, second stage DTRO, clean water tank, air compressor, pump, etc. The treatment scale is 100t/d, and the designed water production rate is 75%. The power for the system installed is 61.68kW, with an operating value of 37.35kW. The DTRO system test reagents mainly include concentrated sulfuric acid (H_2SO_4) and sodium hydroxide (NaOH) for pH adjustment, cleaning agent, and scale inhibitors for system cleaning.

2.2 Main pollutants of leachate

Table 1 listed seven major pollutants and five heavy metal ions detected in the landfill within a typical day.

Tab. 1 The main pollutants of influent (mg/L)

Pollutants	(mg/L)	Pollutants	(mg/L)
SS	121	Cr^{6+}	0.018
TP	36.5	Pb	0.264
COD	2.53×10^3	Cr	< 0.05
NH_3-N	203.3	Cd	0.008
BOD_5	434	As	0.118
TN	300.2		

2.3 Range of pH values

The main pollutants of leachate were detected from October to November considering stable climate and environment temperature in these two months. The sampling interval of each group was set to three or four days. Fourteen groups of data were collected during the testing period. Figure 1 significantly shows the pollutants removal efficiency such as COD correlated with the change of pH value.

The 13th and 14th records showed the removal efficiency changed greatly because of the hydraulic wash on landfill by rain showers. The operating pressure rose fast as a result of more SS (suspended solids) entered the DTRO. Higher osmotic pressure improves the pollutants removal. Besides the COD, and other main pollutants also show

* Corresponding author: 12508414@QQ.com

regular changes with the altering of pH value and operating pressure. The field experiments can immediately check the accurate effect of pH on the removal efficiency.

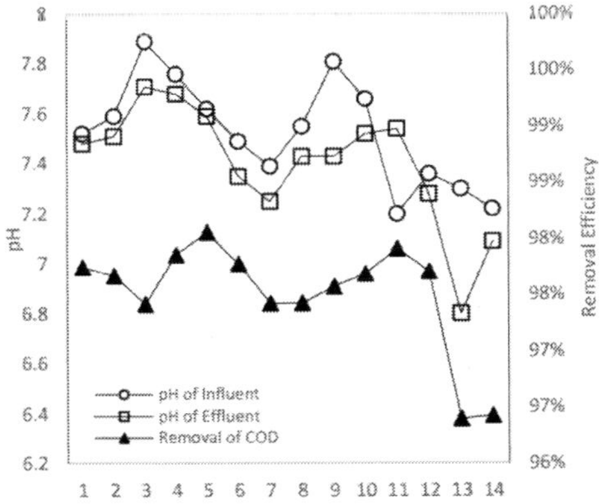

Figure 1. Relationship between pH and removal efficiency

The pH value of the influent recorded in 2 months was between 6.5 and 8. The pH drops after DTRO processing because of the RO membrane effects on acidic ions such as CO_3^{2-}, HCO^{3-} in liquid. As shown in formula 1, the CO_2 in the water produces a large amount of H^+ when it reacts to the right of the equation. Then it makes the effluent acidic as a result of $H^+CO_3^{2-}$ and HCO^{3-} cannot penetrate through the membrane and trapped in concentrated solution. The acidic effluent needs to be neutralized, considering the effluent would cause secondary pollution if the influent pH below 7.

$$CO_2 + H_2O \leftrightarrows HCO_3^- + H^+ \leftrightarrows CO_3^{2-} + 2H^+ \quad (1)$$

The carbonate, as a weak acid radical, is easy to combine with hydrogen ions to produce bicarbonate ions and hydroxide ions after ionization in water so that the concentration solution is alkaline. The operating pressure of DTRO increases rapidly as a result of the membrane fouling is aggravated when the pH value over 7. The economic and efficiency of DTRO would be reduced, considering more chemicals are required to deal with the high concentration pollutants and complicated components if the pH value is too high or too low. Therefore, it is appropriate to set the influent pH value to 6.5, 7, and 7.5 in the experiment, respectively. Then the effluent of DTRO was detected within 30, 60, 90, 120, 150, and 180 minutes.

3. Results and discussion

3.1 Effect of pH on operating pressure of the first-stage DTRO

Figure 2 illustrates the operating pressure rises steadily with the uptime. The highest-pressure was 32.66 bar at pH 7.5, 150 min. Additionally, the pressure reached 32.5 bar at pH 7 and 32.36 bar at pH 6.5, respectively.

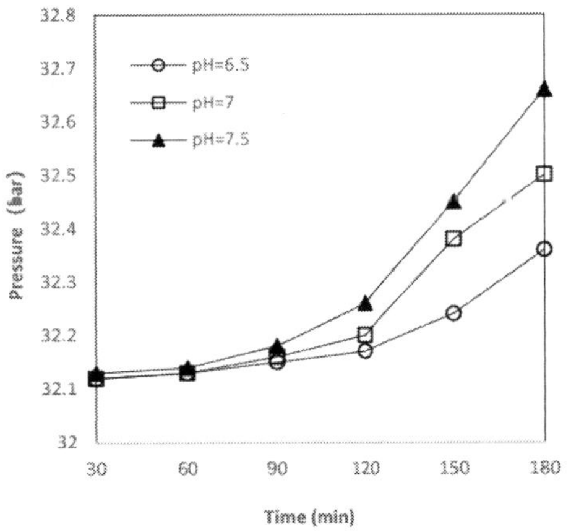

Figure. 2 Effects of pH on operating pressure

Alkaline leachate is likely to cause membrane fouling when pH over 7. The removal of COD, BOD, TP change regularly considering the operating pressure can directly affect the interception of pollutants. High pressure can improve the separation ability of RO to separate polymer materials, colloidal particles, and microorganism. But excessive pressure would make the salt transmission rate increases, then the removal efficiency gradually decreased.

3.2 Effect of pH on the removal efficiency of COD, BOD, and TP

Figures 3,4 and 5 show the trends of the removal of COD, BOD, and TP with the uptime. The change curves of these three pollutants seem similar under different pH value. The highest COD removal was 97.9% at pH 7.5, 180min. And the removal efficiency reaches 97.86% at pH 7 and 97.7% at pH 6.5, respectively.

The BOD_5 removal reached 97.36% at pH 7.5, 120 min, which was significantly higher than the value of 96.88% at pH 7, and 96.39% at pH 6.5, respectively. The removal efficiency showed a downward after 120 min.

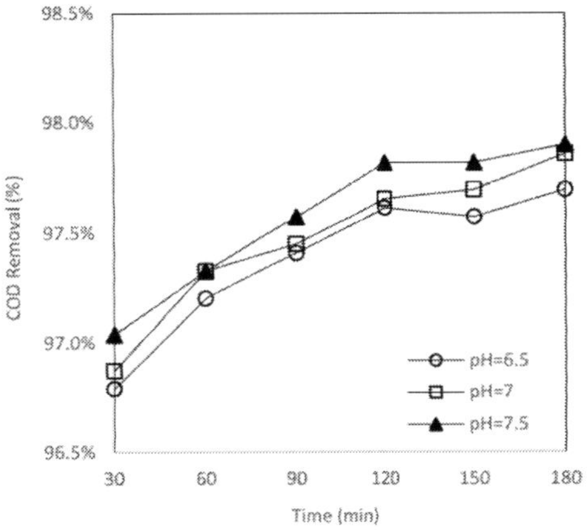

Figure. 3 Effects of pH on COD removal.

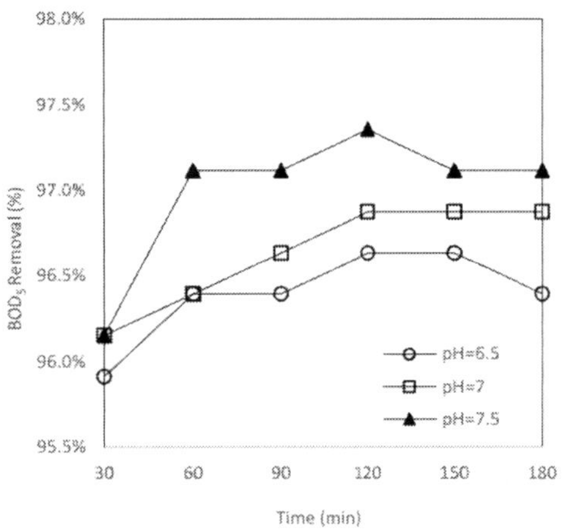

Figure. 4 Effects of pH on BOD5 removal

Figure 5 illustrates a more gradual change in TP removal. The removal efficiency at pH 7.5 is slightly higher than pH 7 and pH 6.5. The TP removal reaches the highest value of 97.7% at 150 min.

COD, BOD, and TP exist as various forms in the leachate. They may be suspended, colloidal, soluble, or be carried by organic matter and microorganisms. The leachate flows along the DT (Disk Tube) flow-path with a certain speed. The solvent, low molecular substances, and inorganic ions in the solution through the RO membrane enter the low-pressure side from the high-pressure side. RO can effectively intercept macromolecular substances and particles. Contaminants such as COD, BOD, and TP in the solution are trapped along with high molecular substances, colloidal particles, and microorganisms, which discharge as a concentrate.

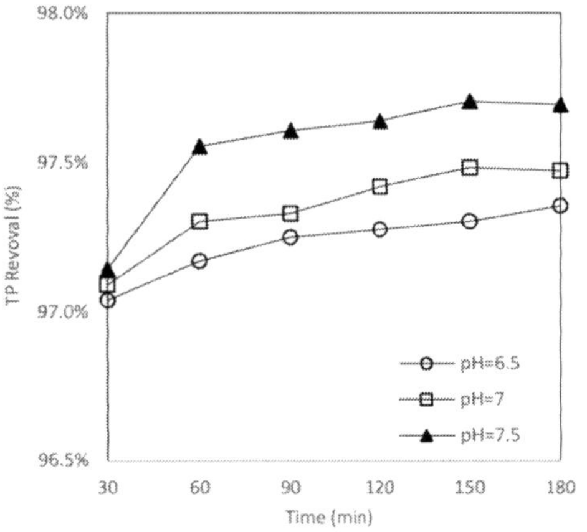

Figure. 5 Effects of pH on TP removal

The adsorption of polar organic on the surface of the RO membrane may be carried out by hydrogen bonding, dispersive force adsorption, and hydrophobic action [8]. Generally, there are three main influence factors of membrane effects on non-polar organics. First, the interaction between hydrophobic organics and water can enrich the slow-dissolving organics on the membrane surface. Second, the concentration polarization of polymeric organics also facilitates their adsorption on the membrane surface. Third, the interaction of ions in water (mainly Ca^{2+}) with organic functional groups alter the hydrophobicity and diffusivity of these organic molecules [9-11].

The pH affects the properties of the solute charge, and also the surface properties of the membrane. Thereby, the interaction between the solute and the membrane surface, the deposition amount of the solute and the membrane flux could be affected. Electro-adsorption on the surface of the RO membrane also effectively separates colloids and suspends solids from the leachate. The H^+ produced by the acidic leachate discharged through the RO membrane. Then a large amount of negatively charged ions are trapped by the membrane so that the membrane surface generates a repulsive force to the negatively charged colloid and macromolecular substance, which is not conducive to the adsorption of the surface [12]. Therefore, the adsorption performance changes and the removal efficiency increases when the pH value raised.

The Figures show that the removal efficiency generally increases or decreases slowly after 120 min when the concentration polarization reaches a certain value [13]. Because there still are a mass of vacancies on the surface for adsorption during the initial period. Differential concentration at the surface area caused by the rapid adsorption accelerates the diffusion of negatively charged particles and colloid. The vacant sites facilitate the rapid discharge of low molecular substances and inorganic ions through the RO membrane. However, the available adsorption sites gradually reduced with the uptime, and the passing ability decreased due to higher membrane surface coverage.

4. Conclusions

A large amount of charged ions were produced in the liquid when the pH value of the leachate changed. The accumulation of ions on the surface of the membrane contributes to the change of adsorption capacity. This is an important factor that affects the removal ability of the RO membrane.

Osmotic pressure also has a great influence on the separation efficiency of the RO membrane. Osmotic pressure accelerates the fouling phenomenon when pH over 7. The removal efficiency of the main pollutant is also improved when high osmotic pressure appears at the beginning. However, the pollutant removal efficiency does not rise but falls instead, due to concentrated polarization after the osmotic pressure reaches the critical point. Therefore, the pH value of leachate influent should be in the range of weak acidity. The weak acidic environment not only controls the net thrust of the system to a reasonable interval but also maintains stable pollutant removal efficiency.

The optimized pH value of the influent should between 6 and 7, which can decrease the dosage of chemicals to maintain better economics during operation, and obtain a better removal effect considering the proper operating pressure and adsorption capacity of the membrane surface.

References

1. G Chen, H Cai, XY Xiong, Z Peng. China Water & Wastewater, 2013, 27(16): 52 - 55
2. B Gao, JY Xiong, HB GU. China Water & Wastewater, 2013. 29(14): 38 - 40
3. LN Zhang. China Water & Wastewater, 2016 (16): 59 - 62
4. T Long, Y Sun, XJ Xun. China Water & Wastewater, 2017 (15): 71 - 74
5. J Zuo, Y Song, J Wang. Membrane Science & Technology, 2011, 31 (2): 110 - 115
6. L Li, F Su, X Zhang. Environmental Science & Management, 2014, 39 (9): 120 - 124
7. F Liu. Environmental Science & Technology. 2015(2): 25 - 29
8. HF Wang, B Xiang, YJ Li. Environmental Science & Technology, 2007, 30 (7): 106-108
9. SZ Wu, BZ Wang. Journal of Harbin University of Civil Engineering, 1999 (6): 53-58
10. Amy E. Childress, Menachem Elimelech. Environmental Science & Technology, 2000, 34 (17): 3710-3716
11. JA Nilson, FA Digiano. Journal, 1996, 88 (5): 53-66
12. AE Childress, M Elimelech. Journal of Membrane Science, 1996, 119 (2): 253-268
13. EM Vrijenhoek, S Hong, M Elimelech. Journal of Membrane Science, 2001, 188 (1): 115-128

Recovery process of waste ternary battery cathode material

Weihao Liu, Wenhua Zhang*, Yinbao Miao, Shuai Wang, Dongqi Kang, Xianbiao Chen, Jia Liu and Zhe Chen

School of Electrical Engineering, Nanchang Institute of Technology, Nanchang, China

Abstract. LIBs is a kind of energy storage device, which has the advantages of high energy density, no memory effect, good safety performance, and many cycles; It is widely used in many scientific and technological fields at home and abroad. With the large-scale increase in the use of lithium-ion batteries, the amount of scrap also increased. In order to better realize resource recovery, energy conservation and emission reduction, it is necessary to study a series of new technologies for waste battery recovery; This review mainly introduces the recovery process of the waste cathode material (LiNixCoyMn1-x-yO2) of the ternary battery, and carries out the resource recovery. The content describes the three major links of the recycling process. The first link introduces the pretreatment of waste ternary batteries. The second link analyzes the current methods of recycling the cathode materials of waste ternary batteries. After analyzing the advantages and disadvantages of each method, the process of wet recycling is introduced in detail. It is also the most commonly used way of recycling waste batteries at this stage. The last link describes the regeneration process of cathode materials.

1. Introduction

With the rapid development of modern science and technology, the shadow of electronic products such as mobile phones, cameras, laptops and other electronic products can be seen everywhere in our life. As the main energy supply battery, the demand and output of lithium-ion battery are also increasing year by year, and its output once increased by eight times from 2000 to 2010[1], its lifespans is about 1 to 3 years, when the service life of a certain number of years will ushered in the wave of battery retirement, so we face the problem of how to deal with the waste battery, because lithium ion batteries are rich in precious metals, the content of these metals is even higher than the content of natural ores. For example, waste lithium batteries contain a large number of strategically scarce metals such as cobalt, whose content is almost 850 times of the average cobalt content of minerals in China. If discarded randomly, it will cause a great waste of resources. However, if handled improperly, it will also bring huge pollution to the environment, such as groundwater and soil pollution, and affect the health of people and animals in real life. Therefore, by recovering precious metals from these wastes, not only can environmental pollution be reduced, but also resource sustainable development can be achieved.

2. Pretreatment of waste ternary battery

2.1 Pretreatment

2.1.1 Discharge treatment.

The internal structure of lithium-ion battery can be divided into positive pole, negative pole, diaphragm, electrolyte and shell as shown in Figure 1.In order to obtain the cathode material powder, the waste lithium-ion batteries should be pretreated. The first step of pretreatment is to discharge the waste lithium-ion batteries. The recovered waste lithium-ion batteries generally contain electricity with a residual voltage of more than 2.0 V. If these residual electricity cannot be effectively treated, it will inevitably lead to a series of potential safety hazards, which makes it difficult to carry out the experiment normally; The methods of discharging waste batteries are generally divided into physical methods and chemical methods.

The physical method is divided into external resistance method, short circuit method and low temperature environment treatment method. The external resistance method is to use a wire to connect the two poles of the waste battery with the resistance load for a period of time, and then use a multimeter to measure the open-circuit voltage. Generally, the residual voltage after discharge is less than 1V, which can be considered as complete discharge[2], This discharge mode is fast and thorough, but the disadvantage is that a lot of heat will accumulate in a short time, resulting in battery explosion; Short circuit

* Corresponding author: 2015994552@nit.edu.cn

method usually refers to the use of metal powder or graphite to short circuit the waste battery discharge[3], Compared with the discharge results of these two substances: Although the metal powder short circuit discharge speed, but the battery temperature rises rapidly, may lead to explosion; The graphite short-circuit discharge speed is moderate, will not cause the battery temperature rise too high, is a safe way of discharge[4].Low-temperature environmental treatment usually refers to discharge in a low-temperature environment of liquid nitrogen. The specific operation method is to store waste batteries at a low temperature in the early stage, and then store them in a vacuum for 1 hour. The principle is to freeze the electrolyte inside the battery to make it lose its conductive property, so as to make the battery inert. Finally, the diaphragm and fluid collection are removed by mechanical shock. This method will not make the temperature of the experiment too high, the disadvantage is that the experimental equipment requirements are very high, the investment of capital is relatively larger.

Most chemical methods use Na_2SO_4 solution[5]、 NaCl solution[6]、 $MnSO_4$ solution[7]、 Soak the waste battery for several hours, and the residual electricity is released naturally by soaking in the solution. Although the cost of using this method is relatively low, but the experiment time is long, and the use of a large number of chemical solvents will bring certain pollution to the environment, but also corrosion of the experimental equipment; For example, in the electrolytic system of NaCl solution, the chlorine gas and hydrogen produced will bring certain pollution to the environment. The discharge process also causes the discharge solution to be more alkaline. Serious corrosion will be caused to waste lithium ion batteries, resulting in electrolyte leakage, a large number of alkanes, olefin and dimethyl carbonate and other organic gases and flocculation precipitation. This method is an open system, which poses a great threat to the safety of working environment and the health of operators in industrial production practice.

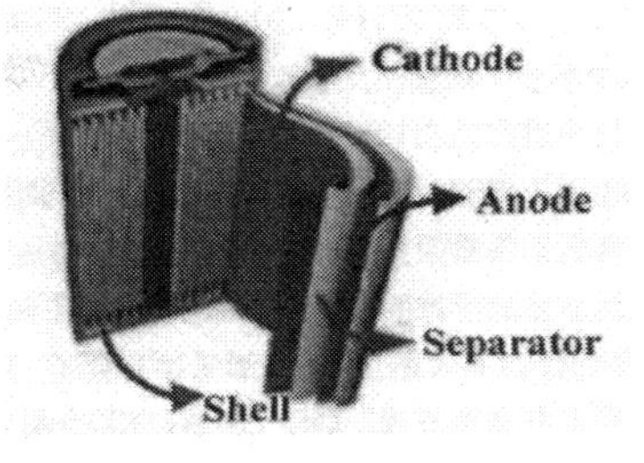

Fig. 1 Internal structure of lithium battery[8]

2.1.2 *Disassembly processing.*

The purpose of disassembly of waste batteries is to realize the separation of components, the main component of the negative electrode is copper foil and carbon black, the adhesive relay between the two is very small, direct percussion shock can be separated and recycled, polymer diaphragm can be concentrated for recycling, metal shell can also be used for secondary recycling; For the most valuable recycling positive electrode material disassembly methods mainly include manual disassembly method, mechanical disassembly method, organic dissolution method, alkaline leaching method, high temperature decomposition method, etc. The following methods will be introduced.

Manual disassembly method: Manual disassembly method is generally suitable for waste batteries with small volume, and the quantity is generally only suitable for laboratory scale recycling. Although the materials obtained from the treated batteries are of high purity, the disassembly efficiency is not high, and the disassembly process is easy to cause some pollution to the environment, so the whole process also needs to wear a good mask and corrosion resistant gloves.

Mechanical disassembly method: Mechanical disassembly method is mainly used in the industrial field, through the crusher of waste battery crushing treatment, get different particle size powder, after the physical screening to screen out the positive electrode material required for the experiment. The advantage of this method is that it can realize continuous automated production, withstand more resistance and accumulate greater production effectiveness. The disadvantage is that HF gas produced in the disassembly process may pollute the environment, so it is necessary to add a gas treatment device to avoid this harmful phenomenon.

Organic dissolution method: Because the binder is mostly poly vinylidene fluoride (PVDF) and other organic matter, the use of organic solvent can be dissolved, so as to achieve the separation of positive electrode material and aluminum foil[9].Commonly used organic solvents mainly include N-methyl-2-pyrrolidone (NMP), N,N-Dimethylformamide (DMF), Dimethylacetamide (DMAC), and sulfoxide (DMSO), etc[9].Compared with the above organic solvents, DMSO dissolving binder with low toxicity and low cost was the most suitable. The experimental results showed that the positive electrode material and aluminum foil could be completely separated under the conditions of 60 ℃, 85 min and ultrasonic assistance. NMP organic solvent has the advantages of good effect, short time and no damage to aluminum foil. Under the condition of temperature 80 ℃, time 30 min and ultrasonic assistance, NMP can remove the binder quickly, but the disadvantages are volatile, high toxicity and high price.

Alkaline leaching method: Since the positive electrode material is metal oxide, which only reacts with acid instead of alkali, alkali solution can be used to separate the positive electrode material and aluminum foil. The positive electrode material can be separated by dissolving aluminum foil with NaOH. The effective separation between the positive electrode material and aluminum foil can be achieved by normal temperature and moderate stirring. $2Al + 2NaOH + 2H_2O \rightarrow 2NaAlO_2 + 3H_2\uparrow$. The experimental results show that the dissolution effect of aluminum foil is obvious when NaOH 3 mol/L, temperature 50 ℃, solid-liquid ratio 100 g/L, time 1 h .The method of alkali leaching is simple and easy to operate. The disadvantage is that strong alkali is easy to corrode the equipment, and the H_2 gas produced by subsequent acid leaching process will be flammable and explosive, which will bring certain security risks. Therefore, ventilation treatment should be carried out.

High temperature decomposition method: according to the carbon material, binder and aluminum foil melting boiling point of different can also use high temperature decomposition method to achieve separation[10], It was found that the decomposition rate of PVDF gradually increased with the increase of temperature from 350℃, and reached 100% at 580℃. Therefore, in the process of the experiment, the mixture can be put in 700℃ and calcined in Muffle furnace for 2 h, then the binder can be removed, so that the positive electrode material and aluminum foil can be separated. In short, high-temperature decomposition method has the advantage of simple and efficient treatment, but the disadvantage is that the experimental process will consume huge energy, but also produce harmful gases such as HF, if not properly treated, it will bring secondary pollution to the environment.

To sum up, when testing in the laboratory, manual disassembly is often the first step in the experiment after the discharge of waste ternary battery. After that, alkaline leaching or organic dissolution method is often used to dissolve aluminum foil. In general, alkaline leaching method is generally used in the experiment in order to save costs without affecting the experimental results. Finally, in order to remove the high melting point binder and carbon black, high temperature decomposition method is often used to treat; In the industrial field, in order to improve production efficiency and reduce labor costs, mechanical disassembly has become the most widely applicable method in the first step after the discharge of waste ternary battery, and the subsequent treatment methods are roughly the same as those used in laboratory experiments. However, it should be noted that these treatment methods will produce a certain amount of harmful gases or liquids in the experimental process, which will cause secondary pollution to the environment, so it is necessary to install appropriate tail gas treatment devices and waste liquid discharge devices. At present, there is no relevant literature to report specific treatment devices.

3. Selective recovery of valuable metallic elements

Cathode material is an important part of ternary battery, and its recovery value is the highest relative to other components. Its cost accounts for about 1/3 of the total cost. There are many metal elements in cathode materials, and the metal elements with recycling value are usually selectively recycled. The common methods of recovering such metallic elements are pyrometallurgy, hydrometallurgy, biological metallurgy, etc.

3.1 Pyrometallurgy

The pyrometallurgy is mainly applied to the recovery and treatment of waste batteries in the industrial field. The main principle is to obtain metal oxides or compounds by melting the cathode materials at high temperature under the premise of adding reducing agents, and then conduct wet separation and recovery; The advantages of pyrometallurgy are that the process is relatively mature, the process flow is simple and clear, and this process does not require physical separation and mechanical treatment. The specific process is to put the cathode material, metal ore and reducing agent carbon powder of the three-element battery into the furnace and wash into the protective gas for heat treatment for 5 hours, during which the temperature is set at 1450 ℃. The experimental results make the binder and electrolyte achieve perfect decomposition, and also get the metal and metal compounds related to the cathode material. The defect of the experimental process is that it is easy to produce harmful gases and pollute the environment in the environment with high temperature greater than 1000 ℃, and the energy consumption is too high, and the requirements for the experimental equipment will also be higher. Therefore, the waste battery treatment often combines the pyrometallurgy and hydrometallurgy to achieve complementary advantages.

3.2 Hydrometallurgy

Among all the recycling methods of waste batteries, hydrometallurgy is the most widely used. Its principle is to transfer the active substances of positive electrode materials to the leaching solution in the form of ions, and finally realize the separation and recovery of valuable metals through precipitation, extraction or ion exchange. With low recovery cost, high recovery efficiency, low pollution and other advantages, hydrometallurgy can be divided into acid leaching treatment, alkali leaching treatment, biological treatment.

3.2.1 Acid leaching treatment.

Acid leaching treatment is divided into inorganic acid treatment and organic acid treatment[11], Common inorganic acids include: Common inorganic acids include: HCL,H_2SO_4,HNO_3, the results showed that hydrochloric acid had the best leaching efficiency and high leaching rate of valuable metal elements can be achieved without adding reducing agent. Factors affecting metal leaching rate include concentration of acid solution, time, temperature, amount of reducing agent, etc. Experimental results show that hydrochloric acid has the best leaching rate under optimal conditions. When the concentration of hydrochloric acid is $4 mol\cdot L^{-1}$, the ratio of solid to liquid is 1:20, and the reaction time is 18 h at 90℃, almost 100% of the metal elements including Li, Ni, Co and Al can be leach from the waste cathode material.

Although strong acid treatment makes the metal leaching rate relatively high, inorganic acid will generate many harmful gases during the experiment, which will bring secondary pollution to the environment. For example, chloric acid will generate chlorine gas in the process of acid leaching, sulfuric acid acid leaching will generate sulfide, nitric acid leaching will generate nitrogen oxide, and the equipment will be corroded due to the strong acidity of inorganic acid.

Compared with inorganic acid treatment, organic acid leaching has the advantages of degradable, short time, low temperature, high leaching rate and less pollution [12], The disadvantage is that the cost is high and proper methods should be adopted to realize the regeneration and recycling of organic acids[13], The organic acids citric acid, malic acid, oxalic acid and trichloroacetic acid used in the experiment have good leaching effect on valuable metals from waste batteries. Chen et al.[14]As for the recovery process of the mixture, citric acid was used for acid leaching. The experimental results showed that when the concentration of citric acid was 1.5mol·L^{-1}, the ratio of solid to liquid was 1:20, the reduction dose was 0.5:1, and the temperature was 80℃ for 2h, the leaching rates of Li,Ni,Co and Mn could reach 99%, 91%, 92% and 94%, respectively. X. H. Zhang[15]trichloroacetic acid was used as the leaching agent for acid leaching. The experimental conditions were as follows: the concentration of trichloroacetic acid was 3mol·L^{-1}, the ratio of solid to liquid was 50g·L^{-1}, the content of reducing agent was 4%, and the leaching rates of Li,Ni,Co and Mn were 99,7%, 93%, 91.8% and 89.8%, respectively, at 60℃ for 30min. But the leaching rate of Al was 7%. Malic acid has certain reducibility due to its own hydroxyl group, Zhou Tao et al [16]The acid leaching experiment shows that when malic acid is used as the leaching agent, under the conditions of malic acid concentration of 1.2mol·L^{-1}, volume fraction of H2O2 of 1.5%, solid-liquid ratio of 40g/L, reaction temperature of 80℃ and reaction time of 30min, The leaching rates of Li, Co, Ni and Mn in malic acid extract reached 98.9%, 94.3%, 95.1% and 96.4%. Zhang et al[17]After oxalic acid leaching, Co^{2+},Ni^{2+} and Li^{2+} ions will react with oxalic acid and precipitate directly from the leaching solution in the form of co-precipitation. After filtration, the filtrate mainly contains metal ions Li^{2+} ions, which can be recovered as Li_2CO_3. The generated NCM can be regenerated into NCM by forging and burning in a high temperature environment according to a certain proportion. First, 0.6 mol·L^{-1}oxalic acid and waste positive electrode powder were put into the reactor, the solid-liquid ratio was 20g·L^{-1}, and the water bath temperature was kept at 70 °C. With the increase of reaction time, lithium was dissolved, and the transition metal was deposited on the surface of the material in the form of oxalate, thus achieving the separation of lithium and transition metal. Through the above comparison, we can see that the appropriate use of organic acids for selective leaching can greatly shorten the experimental process and save various human and material resources, which is one of the most important factors for future industrialization.

To sum up, pyrometallurgy and hydrometallurgy are two inexpensive processes commonly used in industrialization to recover lithium batteries. Pyrometallurgy can deal with large scale waste lithium ion batteries of different models, while hydrometallurgy can quickly and effectively deal with specific models of waste lithium ion batteries. Therefore, pyrometallurgy and hydrometallurgy can take a place in the field of lithium battery recycling by virtue of their unique advantages.

3.2.2 *Alkaline leaching treatment.*

Although the acid leaching treatment can make the metal ions have a high leaching rate, the acid leaching treatment is not selective for the leaching of metal ions, which makes the metal separation and purification more complex, which will lead to excessive discharge of wastewater. Therefore, the method of ammonia leaching to recover the cathode active substances in waste lithium ion batteries can prevent the occurrence of such drawbacks. Ammonia, ammonium sulfite and ammonium carbonate were used as the leaching system to explore their influence on the leaching efficiency of active substances of ternary lithium battery positive material. The results showed that When the molar ratio of ammonia, ammonium sulfite and ammonium carbonate in the leaching solution is 1:0.5:1, the leaching effect of Ni,Co and Mn was better at 80°C and 1 h, and the total selectivity of Ni,Co and Li was greater than 98.60%.

3.3 Combined pyro-wet process

Some defects existing in single pyrometallurgical or hydrom roasting is etallurgical process can be solved by using pyro-wet combined process. Sulfuric often used in the stage of fire pretreatment[18]Nitrification roasting[19]And carbon reduction roasting[20]The waste lithium battery is treated by the process to convert Li, Ni, Co, Mn and other metals into the corresponding metal sulfate, metal nitrate or metal carbonate. In contrast, carbon reduction roasting does not require the introduction of inorganic acid, which is relatively environmentally friendly, so it is also the most commonly used recovery process in the fire-wet process. Zhang et al[20]Under the optimal conditions of temperature 650 ℃, time 3 h and carbon content 19.9%, the extraction rate of lithium can reach 84.7%, and the extracted slag can be recovered by H_2SO_4 acid leaching and extraction separation process with the recovery rate of Ni, Co and Mn above 90%.

3.4 Bio-metallurgy

Biometallurgy is a kind of mineral biooxidation process enhanced by microorganisms. In this process, insoluble metal oxides are converted into water-soluble metal sulfates, which can realize the leaching of waste battery materials. Common microbial leaching systems are eosinophilic sulfur oxidizing bacteria (SOB), iron oxidizing bacteria (IOB), mixed bacterial system (MS-MC), and for the passage of three,the MS-MC system can leach most metals from $LiNi_xCo_yN_{1-x-y}O_2$ with a leaching rate up to 95%. Although the bacterial leaching method has the advantages of high efficiency, low pollution and low cost, the culture conditions are strict and can only be realized in a specific environment, and the culture time period is long, which makes it difficult to implement in practical application. Compared with bacteria, fungi can survive in a wide range of PH without reducing metal leaching rate. Aspergillus Niger strain MM1 can dissolve a large amount of Co and Li, and the leaching rates can reach 82% and nearly 100%, respectively. Therefore, fungal bioleachation may be an environmentally friendly method for dissolving and recovering large amounts of metals from waste LIBs[21].

4. Regeneration of positive electrode materials

The regeneration of cathode material refers to the direct use of waste lithium battery cathode material leaching solution as raw material, repreparation of new three the process of the cathode material. The leaching solution after acid leaching can be used to resynthesize cathode materials, including coprecipitation method, sol-gel and hydrothermal method. The resynthesis process can realize a closed waste lithium ion battery recovery process.

Coprecipitation method: The most commonly used method for wet material regeneration is coprecipitation method. Precipitators include hydroxide, carbonate and oxalate precipitation. The main principle is that excessive metal ions such as Co, Ni and Mn are similar in nature and are not easy to separate[22], Therefore, it is particularly important to control the pH value of the solution to precipitate Ni, Co and Mn at the same time to the maximum extent. Different precipitating salts (carbonate, hydroxide and ammonium bicarbonate) are used, the pH of the solution to be controlled is also different. NaOH is usually used as the precipitating agent, use ammonia to adjust the pH to 11. Yanko Marinov Todorov et al[23]2mol·L^{-1} metal sulfate solution and 6 mol·L^{-1} NaOH aqueous solution were prepared, and the pH value was adjusted to 11 by ammonia water to obtain $Li(NiCoMn)_{1/3}(OH)_2$ precursor, which was mixed with Li_2CO_3 and then ball milling. Calcined at 900 °C for 20 h, $Li(NiCoMn)_{1/3}(OH)_2$ was obtained.

Compared with coprecipitation method, sol-gel method and hydrothermal method are less used in practical industrial applications because of their high cost and complex experimental instruments. The following is a brief introduction to their methods.

Sol-gel method: sol-gel method [24,25,26]It refers to a synthesis method in which part of the product prepared by metal carboxylate in water or non-water solvent or sol formed by complete hydrolysis is evaporated at high temperature to remove the solvent and finally form a gel, and then the gel formed is solidified and dried after heat treatment to prepare the product. The advantages of this method are: small product size, narrow particle size distribution, large specific surface area, easy to control the form, low heat treatment temperature and time. However, due to the relatively complex synthesis process, it is not suitable for industrial production.

Hydrothermal method: hydrothermal method[27]The synthesis of cathode material is to place the material and water into a hydrothermal reactor in a closed environment at a certain temperature and pressure. Considering the effect of synthesis temperature on the electrochemical properties of products, hydrothermal synthesis method has its unique advantages: the temperature of hydrothermal method is low, which avoids volatilization of lithium ions at high temperature and causes lithium deficiency compounds. In addition, hydrothermal method also has the advantages of homogeneous product, no inert gas protection, simple process, etc. However, this method has some difficulties in the synthesis of complex materials.

5. Conclusion

Compared with some foreign developed countries, our country's research on the new technology of waste battery recycling is still in the initial stage. Although the country has mentioned some treatment methods for some pollution sources in the environmental protection Law, there are also some recyclable metal elements in lithium ion batteries. At present, the country does not have a complete treatment system for the recycling and treatment of these metal elements. Most of the enterprises still have poor technology and imperfect system, and some illegal operators even take the immoral way of deliberately raising the recycling cost and earning profits.

Looking into the new era, some big cities, such as Shenzhen and Shanghai, have gradually implemented the industrial chain model of waste battery recycling, and the system is constantly improving. From the perspective of the future, the battery recycling industry is optimistic.

Acknowledgments

This work was supported by the Jiangxi Provincial Education Department Project(No. GJJ211913; No. GJJ211915), the National Natural Science Foundation of China (No. 22269014; 62041406) and Jiangxi science and technology support plan (No. 20171BAB206040).

References

1. Xianlai Zeng,Jinhui Li,Narendra Singh. Recycling of Spent Lithium-Ion Battery: A Critical Review[J]. Critical Reviews in Environmental Science and Technology,2014,44(10).
2. Chen Yisong, Zhao Junwei, Qiao Jie, Liu Yongtao. Analysis of Power Battery Recycling and Countermeasures for Electric Vehicles in China [J]. Journal of Automotive Engineering, 2018,8(02):97-103.
3. A New High-Efficiency Pulverizing Process for Waste Lithium ion Batteries (Wu Caibin et al., Patent No. CN201010510406.6)
4. Deepak Pant,Tenzin Dolker. Green and facile method for the recovery of spent Lithium Nickel Manganese Cobalt Oxide (NMC) based Lithium ion batteries[J]. Waste Management,2017,60.
5. Shi Hongcai. Recovery and reuse of Ni-cobalt-manganate lithium anode materials from waste lithium-ion power batteries [D]. Zhengzhou University,2017.
6. Chenxing Yi, Lijie Zhou, Xiqing Wu, Wei Sun, Longsheng Yi, Yue Yang. Technology for recycling

and regenerating graphite from spent lithium-ion batteries[J]. Chinese Journal of Chemical Engineering, 2021,39(11):37-50.

7. SONG Xiu-ling, DAI Shu-qi, XU Yong-sheng, et al. Experimental study on discharge of used lithium-ion batteries [J]. Applied Chemical Industry, 2015, 44(4): 594−597.
8. Hu Yafei. Study on Wet Recovery of Valuable Metals from Waste ternary Lithium ion Batteries (18650 Type) [D]. Beijing university of chemical industry, 2020. DOI: 10.26939 /, dc nki. Gbhgu. 2020.001344.
9. Hao Tao. Study on Recycling of LiNi_(0.6)Co_(0.2)Mn_(0.2)O_2 cathode Materials [D]. Kunming university of science and technology, 2019. DOI: 10.27200 /, dc nki. Gkmlu. 2019.001612.
10. Hao Tao; Zhang Yingjie; Dong Peng; Beam wind; Duan Jianguo; Meng Qi; Xu Bin; Research Progress of Cathode Material Recycling for Lithium ion Battery [J]; Journal of Anhui University of Technology; 2007 Silicate Bulletin; 08, 2018
11. Chen Jinyi, Wang Qi. Leaching valuable metals from waste lithium batteries by ascorbic acid [J]. Metallurgical Management,2019(15):17-19. (in Chinese)
12. Shen Rod, Gu Weixing, Yuan Haiping, Zhu Nanwen. Waste ternary lithium ion battery technology of extraction and purification research progress [J]. Journal of environmental science and technology, 2018, 9 (02) : 114-121. The DOI: 10.19672 / j.carol carroll nki. 1003-6504.2018.02.017.
13. He Lipo, Sun Shuying, Yu Jianguo. Research progress on recovery of valuable metals from decommissioned lithium ion Batteries [J]. Acta Chimica Sinica,2018,69(01):327-340. (in Chinese)
14. Chen X, Fan B, Xu L,Zhou T, Kong J.An atom-economic process for the recovery of highvalue-added metals from spent lithium-ion batteries. Journal of Cleaner Production, 2016,112:3562-3570.
15. Zhu Xianfeng, Zhao Ruirui, Chang Yi, Chen Hongyu. Research on Acid Leaching of ternary Cathode Material for waste lithium ion Batteries [J]. Batteries,2017,47(02):105-108. (in Chinese) DOI:10.19535/ J.1001-1579.2017.02.011.
16. Zhou Tao, Xu Liping, Fan Bailin, et al. A new process for recovering valuable metals from waste cobalt-nickel-manganate lithium batteries [J. Journal of Xuzhou Institute of Technology (Natural Science Edition),2017,32(1): 6-12.ZHOU Tao, XU Liping, FAN Bailin, et al. A new process forrecovering valuable metals from waste Nickel cobalt manganatelithium battery batteries[J].Journal of JiangInstitute of Technology(Natural Sciences Edition).2017,32(1):6-12.
17. ZHANG X, BIAN Y, XU S,et al. Innovative application of acidleachingto regenerate Li(Ni,Co,,Mn)O, cathodesfrom spentlithium-ion batteries[J]. ACS Sustainable Chemistry & Engineering, 2018,6(5):5959-5968.
18. Deepak Yadav,Rangan Banerjee. A comparative life cycle energy and carbon emission analysis of the solar carbothermal and hydrometallurgy routes for zinc production[J]. Applied Energy,2018,229.
19. Chao Peng,Fupeng Liu,Zulin Wang,Benjamin P. Wilson,Mari Lundström. Selective extraction of lithium (Li) and preparation of battery grade lithium carbonate (Li_2CO_3) from spent Li-ion batteries in nitrate system[J]. Journal of Power Sources,2019,415.
20. Juntao Hu,Jialiang Zhang,Hongxu Li,Yongqiang Chen,Chengyan Wang. A promising approach for the recovery of high value-added metals from spent lithium-ion batteries[J]. Journal of Power Sources,2017,351.
21. Zhang Yingjie, Ning Peichao, Yang Xuan, Dong Peng, Lin Yan, Meng Qi. Old ternary lithium ion battery recycling technology research progress [J]. Chemical engineering progress, 2020, 33 (7) 6:2828-2840. The DOI: 10.16085 / j.i SSN. 1000-6613.2019-1666.
22. Mu Deying, Liu Zhu, Jin Shan, Liu Yuanlong, Tian Shuang, Dai Changsong. Recovery and reuse of cathode material and electrolyte of waste lithium ion battery [J]. Progress in Chemistry,20,32(07):950-965.
23. Yanko Marinov Todorov,Koichi Numata. Effects of the Li:(Mn + Co + Ni) molar ratio on the electrochemical properties of $LiMn_{1/3}Co_{1/3}Ni_{1/3}O_2$ cathode material[J]. Electrochimica Acta,2004,50(2).
24. Peng Zhengshun. Synthesis and Electrochemical Studies on Spinel Phase LiMn_2O_4 Cathode Materials Prepared by Different Processes[J].Rare Metals,1999(02):64-69.
25. Patoux S, Doeff M M. Direct synthesis of $LiNi_{1/3}Co_{1/3}Mn_{1/3}O_2$ from nitrate precursors[J]. Electrochemistry Communications, 2004, 6(8): 767-772.
26. Kim J H, Park C W, Sun Y K. Kim J H, Park C W, Sun Y K. Synthesis and electrochemical behavior of $Li[Li_{0.1}Ni_{0.35-x/2}Co_xMn_{0.55-x}/O_2$ cathode materials. Solid State Ion, 2003, 164: 43-49
27. Huang B, Li X, Wang Z, et al. A novel carbamide-assistant hydrothermal process for coating Al_2O_3 onto $LiMn_{1.5}Ni_{0.5}O_4$ particles used for cathode material of lithium-ion batteries[J]. Journal of alloys and compounds, 2014, 583: 313-319.

Study on parameters limit of fracturing optimization process for thin and differential zone oil Wells

Hang Gao

Institute of Geology, No.2 Oil Production Plant, Daqing Oilfield Company, Heilongjiang Daqing 163000, China

Abstract: The potential target of water flooding measures has gradually changed to thin differential reservoir in three types of reservoirs. The fracturing proportion of thin differential reservoir in water flooding block oil well has reached more than 60%, and the existing fracturing design code can not fully meet the needs of thin differential reservoir fracturing. Through the effect evaluation of the technical parameters matching different types of thin differential reservoir, the field test is carried out to establish the limit of the optimal technological parameters for fracturing of thin differential reservoir oil well, which is of great significance for improving the fracturing effect and improving the technical level of fracturing in the future oilfield.

Key words: Thin differential zone, well fracturing, process parameter boundary.

1. There is a problem

In 2015, water flooding Wells contained 90.7% water before pressure, and the proportion of refracturing was 72%, and the proportion of fracturing in thin and differential reservoirs reached more than 60%. Due to the characteristics of poor development, small thickness, low permeability and longitudinal interaction distribution with high aquifer in thin and poor oil reservoir, it is difficult and ineffective to exploit potential. Meanwhile, the fracturing process parameters used in fracturing of water drive oil well are mainly for well-developed oil reservoir, and the existing fracturing design specifications cannot fully meet the needs of fracturing in thin and poor oil reservoir. According to statistics, the average oil increase of single well in thin and differential reservoir fracturing in recent three years can only reach 2.8t.

2. The main factors affecting the fracturing effect of thin and poor reservoir are studied

Through the analysis of 85 oil Wells previously fractured in the block, 40 oil Wells mainly fractured with thin and differential oil layers are selected as our research objects. By comparing the relationship between relevant influencing factors and fracturing effect, the influences of water cut before fracturing, strength of fluid produced before fracturing, thickness of target fracturing layer and amount of sand added in interval on fracturing effect are clarified:

Firstly, from the perspective of well selection conditions, the pre-pressure water cut and the strength of pre-pressure liquid production affect the fracturing effect. When the pre-pressure water cut is too high, there is a high-producing liquid layer in the well, and the contradictions between the layers are still prominent after fracturing. When the strength of pre-pressure fluid production is too low, the well is seriously short of liquid supply and the fracturing effect is poor due to the lack of formation energy. Therefore, the strength of pre-pressure water cut and pre-pressure liquid production are evaluated together. Based on the dispersion diagram of pre-pressure water cut and pre-pressure liquid production strength, it is clear that the fractured well should meet the requirement of pre-pressure water cut below 94% and pre-pressure liquid production strength greater than 2t/m at the same time when the oil increase is 3t as the standard.

Second, from the perspective of potential scale, the relationship between the effective thickness of the whole well fractured target layer and the oil increase volume is calculated. The greater the thickness of the whole well fractured target layer, the better the fracturing effect will be. With the gradual increase of the thickness, the oil increase range will be increased.

Thirdly, from the perspective of construction scale, the relationship between sand addition and oil increase in the fracturing interval of the whole well is calculated. The more sand addition in the fracturing interval, the better fracturing effect will be. However, with the increase of sand addition, the range of oil increase decreases.

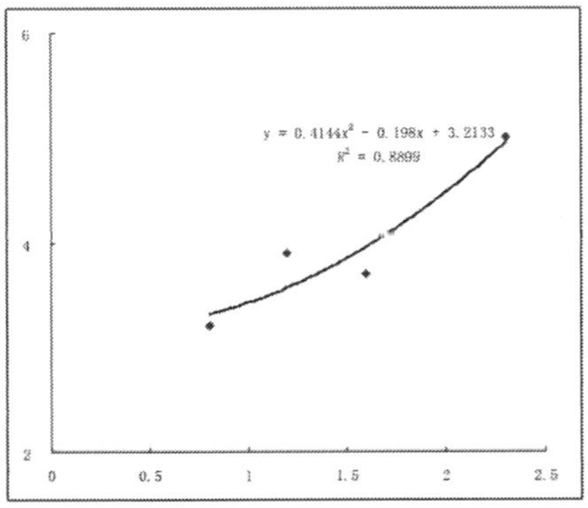

Fig. 1 Relation curve between effective thickness of target layer and oil increase

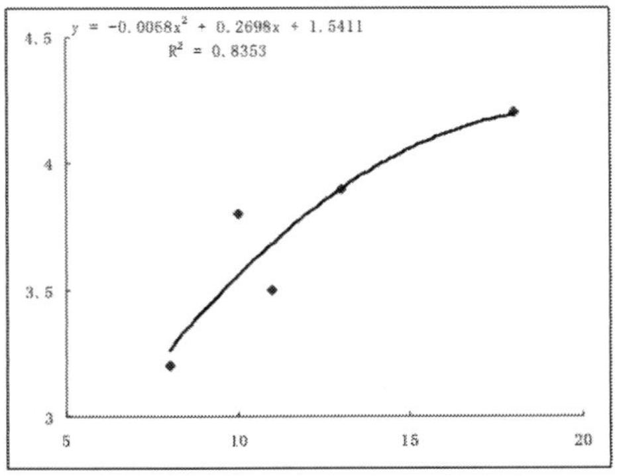

Fig. 2 Relation curve between sand addition and oil increase in fractured interval

3. Combined with the characteristics of reservoir development, the potential target of thin and differential reservoir is defined

According to the development of reservoir, the distribution form of sand body and the connection relationship between oil and water, the thin differential reservoir is subdivided into six types. Based on the evaluation of the effect of 106 thin differential intervals, it is concluded that the proportion of the three types of sand body in the implementation interval is 96.1%, and the oil increase intensity is 1.0t/m. The oil increase effect is obviously higher than that of the other three types, so three types of sand bodies, namely filling continuous piece type, edging bypass type (thin injection and thick production) and clumping distribution type, are regarded as the main potential objects for fracturing thin and differential Wells at present.

4. Establish fracturing technical specification of thin and differential reservoir

On the basis of the previous fracturing effect of thin and differential oil well, through the comparison of various parameters of sand body types in thin and differential oil reservoir, the technical limits of various fracturing parameters were studied with the target of 3.5t fracturing oil increase.

4.1 Establish the optimal selection criteria of measure well formation

Under the premise that the pre-pressure water cut is lower than 94%, the influence of formation liquid supply capacity on the oil increase effect of fractured well is considered. The low strength of pre-pressure liquid production indicates insufficient liquid supply capacity, while the high strength of pre-pressure liquid production indicates that multiple small layers of the well have been well used. Therefore, the relationship between the strength of pre-pressure liquid production and the amount of oil increase of three types of sand bodies is calculated respectively, and the limit of the strength of pre-pressure liquid production of classified sand bodies is clearly defined. The filling contiguous sand body is between 2-7t/m, the edging bridge sand body is between 2-6t/m, and the turd distribution sand body is between 2-4t/m.

Combined with the analysis results of factors affecting the fracturing effect of thin and differential formations, the larger the fracturing thickness of the whole well is, the better the oil increase effect will be. By establishing the matching relationship between the converted thickness of the fractured interval and the oil increase amount of the whole well, it is concluded that the converted thickness of the fractured interval aiming at the measured oil increase of 3.5t needs to be more than 3.0m. Meanwhile, the fracture thickness limit is further subdivided into three potential objects. The filling contiguous sand body shall be at least 3.0m, the edging bridge sand body shall be at least 3.5m, and the turd distribution sand body shall be at least 3.3m.

https://doi.org/10.1051/e3sconf/202339401005

4.2 Define the boundary of process parameters

According to the theoretical chart of fracture parameter optimization in thin differential zone, the penetration ratio requirements of different types of sand bodies at the highest recovery degree after fracturing are defined. Under the condition of 175m injection-production well spacing in thin differential zone, the penetration ratio formula is as follows: The penetration ratio = fracture half-length/well spacing *100%, the minimum fracture half-length is calculated, and then the maximum fracturing sand addition amount of classified sand body is obtained according to the relationship curve between fracture half-length and sand addition amount. According to the regression relationship between the sand addition amount and the effect of different sand body types, the minimum sand addition amount is obtained. Thus, the sand addition limit range of classified sand body is obtained.

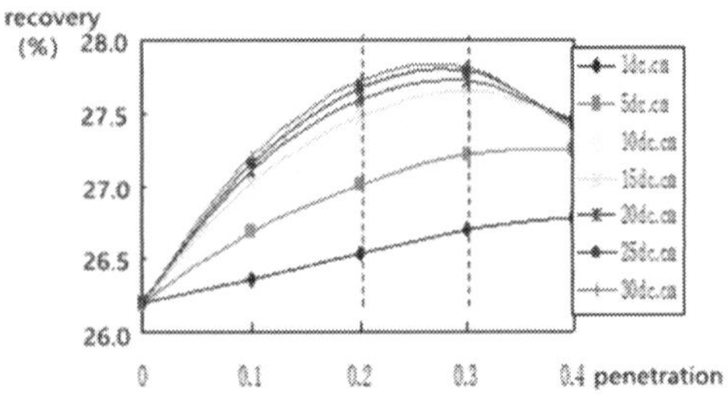

Fig. 3 Fracture parameter optimization diagram of thin differential layer

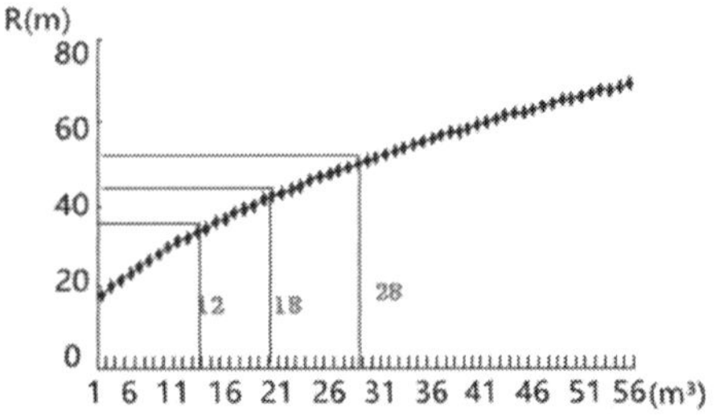

Fig. 4 Relation curve between fracture half-length and sand addition amount

Based on the above analysis, considering the factors affecting the fracturing of classified sand bodies, the fracturing well selection standard of thin differential reservoir is clearly defined, the construction parameter design is optimized to increase the reconstruction scale of thin differential reservoir, and the technical parameter specification that best matches the classified sand bodies of thin differential reservoir is established.

Table 1 Fracturing technical parameters boundary table of classified thin differential reservoir

Sand body type	Preferred well zone criteria			Optimization parameter design
	Water cut before pressure (%)	Strength of liquid production before pressing (t/d·m)	Fracturing is converted to thickness (m)	Amount of sand added in interval (m^3)
Filling continuous type	< 94	2-7	≥3.0	9-12
Border bypass type	< 94	2-6	≥ 3.5	14-18
Turd distribution type	< 94	2-4	≥ 3.3	16-28

The fracturing technology specification of thin and differential zone oil well was used to select 30 Wells in the block for verification. After the implementation, the oil increase per well was 4.2t per day. Compared with the previous fracturing effect of thin and differential zone oil well, the fracturing effect of thin and differential zone oil well was significantly improved.

5. Conclusion

1. After water flooding development enters the ultra-high water cut stage, the remaining oil distribution is scattered, and well fracturing has become an important means to improve the utilization degree of sand body and explore potential of local remaining oil.

2. The difficulty of oil well fracturing is gradually increasing, and the existing standards of oil well fracturing measures cannot meet the requirements of the current situation. Therefore, it is of great significance to establish the parameter boundary for thin and differential oil reservoir fracturing. 3. Establish fracturing technical specifications for different types of thin differential reservoirs. On the basis of fine remaining oil analysis and combined with the matching adjustment of Wells, the fracturing technical specifications for thin differential reservoirs have seen a better oil increase effect.

References

1. Development theory and technology progress of low permeability reservoirs in China [J]. Hu Wenrui, Wei Yi, BAO Jingwei. Petroleum Exploration and Development. 2018(04)
2. Application of integrated technology of large-scale fracturing water injection development in ultra-low

permeability reservoir [J]. Wang Hongwei. Oil Drilling and Production Technology. 2018(01)

3. JIANG Hongfu, WANG Yunzeng, LIU Qiuhong, JIANG Dagui, Rao Hua. Petroleum Geology and Development in Daqing. 2016(02)
4. Experimental study on optimization of rectangular well pattern in ultra-low permeability fractured reservoir [J]. Tian Wenbo, Yang Zhengming, XU Xuan, XIAO Qianhua, Teng Qi. 2013(02)
5. Reunderstanding of fluid percolation in low permeability reservoirs [J]. Petroleum Exploration and Development. 2012(05)

 https://doi.org/10.1051/e3sconf/202339401006

Influence of groundwater level change on deep foundation pit and its control technology

Xuze Ge[1], Yang Gao[1], Yanbo Jiang[1], Xinyue Zhao[1], Kun Wu[1], Jie Zhang[1], Cheng yuan[1], Xue Lin[1] and Qingbiao Wang[2,3,4,*]

[1]College of Civil Engineering and Architecture, Shandong University of Science and Technology, Qingdao, China
[2]College of Resources, Shandong University of Science and Technology, Tai'an 271019, China College of Civil Engineering and Architecture, Shandong University of Science and Technology, Qingdao 266590, China;
[3]College of Safety and Environmental Engineering (College of Safety and Emergency Management), Shandong University of Science and Technology, Qingdao 266590, China;State Key Laboratory of Mining Disaster Prevention and Control Co-Founded by Shandong Province and the Ministry of Science and Technology, Shandong University of Science and Technology, Qingdao 266590, China;
[4]National Engineering Laboratory for Coalmine Backfilling Mining, Shandong University of Science and Technology, Tai'an 271019, China

Abstract: During the construction of deep foundation pits with rich water straight grooves, instability of deep foundation pits due to groundwater frequently occurs, posing a huge challenge to construction safety and economy. Based on the deep foundation pit of the intercity connecting line of Beijing New Airport, the technical method of combining theoretical analysis and on-site monitoring is adopted to analyze the influencing factors of groundwater on the retaining structure of deep foundation pit, and control and analyze the influencing factors. The deformation control technology for deep foundation pit is proposed. The following conclusions are obtained: (1) Based on the on-site geological conditions and the mechanism of groundwater level change, the deformation and failure forms and precipitation forms of deep foundation pits are obtained through theoretical analysis. (2) Based on the Deformation monitoring technology of deep foundation pit, the change of groundwater level and Deformation monitoring of retaining structure are carried out. (3) Combining on-site monitoring with theoretical calculation, conducting comparative analysis of data, summarizing the influencing factors of deep foundation pit damage, obtaining targeted deformation control technology for deep foundation pit, guiding the construction of practical projects, and providing reference for similar projects.

Key words: deep foundation pit; Field monitoring; Pipe surge; A surge; Control technique.

1. Preface

With the large-scale investment in infrastructure construction in China, the construction of deep foundation pits has become a trend. However, due to the complex excavation conditions of deep foundation pits and the ever-changing environment, groundwater poses challenges and hazards to deep foundation pit excavation. Therefore, it is necessary to systematically study the impact of groundwater level changes on deep foundation pit construction, and develop detailed and targeted control techniques to reduce the impact of groundwater on deep foundation pit construction.

Many scholars at home and abroad have conducted a series of discussions on the impact of groundwater level changes on deep foundation pits and control technologies. Zhou Chuanshui [1] obtained the deformation law of the foundation pit diaphragm wall by studying the groundwater level, and optimized the method of groundwater level control and the supporting structure of the diaphragm wall; Xu Zhongtao [2] analyzed the deformation laws of deep foundation pits and supporting structures caused by dewatering excavation and the impact of different supporting structure design parameters on foundation pit deformation through combining theory with on-site monitoring and using numerical simulation methods; Li Yuzi [3] has conducted research on the dynamic change law of groundwater level, prediction and early warning of water level drawdown, and optimization of monitoring well network through dynamic monitoring and numerical simulation methods; Based on on-site monitoring and finite element numerical simulation, Wang Hao [4] analyzed and studied the deformation of deep foundation pit, surrounding surface settlement, and the impact on building foundation settlement caused by dewatering and excavation process; Long Zhang [5] optimized the monitoring method of groundwater level in deep foundation pits through theoretical research on existing projects; Xia Wen, Li Pan, Qiu Wei, Li Xudong[6] analyzed the impact of underground water level in deep

* Corresponding author: skd990748@sdust.edu.cn

foundation pit excavation on surrounding building settlement under existing support methods through on-site real-time monitoring and numerical simulation; Wu Peng [7] analyzed the deformation law of the retaining structure of the deep foundation pit during the process of groundwater precipitation, as well as the surrounding structure settlement and groundwater level changes, and conducted a conditional analysis of the foundation pit construction method; Tao Hongliang [8] preliminarily deduced the theory of computation of the water gate through theoretical calculation and numerical calculation analysis based on the analysis of the factors affecting the embedding depth of the water stop curtain; Gao Dong [9] conducted finite element simulation on the excavation, support, and precipitation methods of deep foundation pits, and the obtained data was compared and analyzed using the software Origin; Zhang Dongdong [10] used a combination of indoor tests and on-site experiments to conduct numerical simulation of experimental data, and conducted research on vacuum drainage and precipitation methods, precipitation optimization, and intelligent methods for deep foundation pit engineering precipitation; Wang Chuanlin [11] systematically studied the influence of precipitation depth on the stability of foundation pits; Zhao Min et al. [12] studied the variation of groundwater level and optimized the dewatering method of deep foundation pit based on the example of deep foundation pit engineering in the loess area of Xi'an City; Li Lin [13] used finite Meta computing and FLAC program to study the influence of dewatering mode on the retaining structure of deep foundation pit and surrounding ground settlement.

According to the current research status of the impact of groundwater on deep foundation pit excavation, most scholars use theoretical analysis to analyze the impact of groundwater on deep foundation pit excavation, while few use the deformation law of deep foundation pit to find accurate control methods, which are closer to engineering practice. Some use numerical analysis and simulation for simulation experiments, but the on-site conditions are complex, there are many uncertainties in on-site construction, and there are certain deviations in the control technology used.

On the basis of theoretical calculation, combined with on-site monitoring data, this paper compares and analyzes the change of underground water level of deep foundation pit with the Deformation monitoring data of retaining structure, discusses the relationship between the deformation law of retaining structure and the change law of underground water level in detail, and adopts stage analysis, by comparing the monitoring data with theoretical calculations, the deformation control technology for deep foundation pits is obtained to verify the feasibility of the excavation and monitoring plans, which can guide actual construction, avoid engineering hazards caused by improper construction, and reduce engineering risks and economic losses.

2. Project Overview

2.1 Overview of tunnel route

The line is located in Da Xing District, Beijing, and Lang Fang City, Hebei Province. The intercity railway connecting line runs from north to south, with a starting and ending mileage of DK42+038~DK46+115, all tunnels are constructed by open excavation method.

2.2 Overview of geological conditions

The sources of natural groundwater include shallow groundwater, runoff water, drainage water, atmospheric precipitation, etc. The tunnel passes through the Yong Ding River and the Tian Tang River, where the water resources are affected by seasonal changes and the water level significantly increases during the rainy season.

The groundwater level in the site is Quaternary pore phreatic water, and the groundwater in local silt or silt layers is slightly pressurized. The sand layer has abundant water content, and the groundwater depth along the line varies greatly. It is mainly supplied by atmospheric precipitation and underground runoff, and the discharge is mainly runoff and manual mining. The sand layer in the tunnel body is distributed intermittently due to its thin layer, and the water content is not large.

3. Analysis of groundwater level change mechanism

3.1 Impact forms of groundwater changes on foundation pits

Taking the pile anchor retaining structure in the retaining structure of straight open cut deep foundation pit as the research object, the deformation and damage forms of deep foundation pit caused by changes in groundwater level include the following aspects:

(1) When the water pressure is too high, seepage phenomenon will inevitably occur. When the seepage force of groundwater flows in the gap, it will exert a force on the retaining structure, causing deformation of the retaining pile and instability of the foundation pit.

(2) Groundwater will reduce the integrity of the soil behind the pile, resulting in uneven overall stress on the retaining structure. The water and soil move, causing piping.

(3) When the water level in the foundation pit is high, with the continuous excavation of the soil mass, the overlying soil mass is jacked up by the groundwater level, causing cracks at the bottom of the pit, causing sediment to surge from the cracks, resulting in a surge phenomenon.

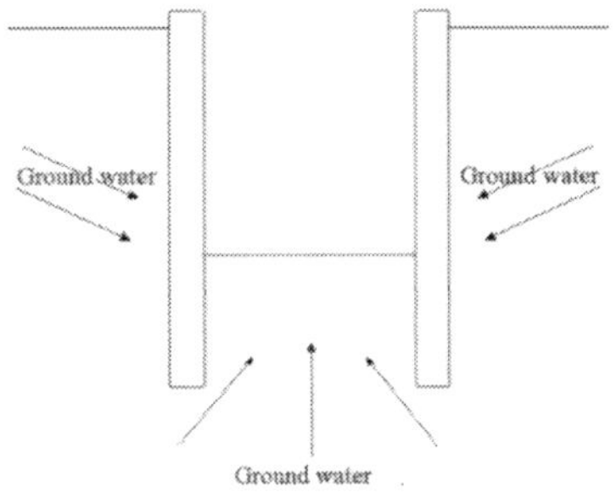

Figure 1. groundwater sketch effect on deep foundation pit

3.2 Dewatering mechanism of deep foundation pit

During excavation of the foundation pit, the unloading of the soil in the pit will increase the water level difference inside and outside the retaining structure, and the groundwater from the outside of the foundation pit will flow into the inside of the foundation pit, inevitably causing a decrease in the water level and pore water pressure outside the foundation pit, resulting in a large uplift of the foundation base.

After excavation of the foundation pit, various parameters also change, with significant changes in pore water pressure. According to the Skempton formula:

$$\Delta u = B[\Delta\sigma_3 + A(\Delta\sigma_1 - \Delta\sigma_3)] \quad (1)$$

Where, A and B are pore pressure coefficient.

To facilitate observation, formula (1) can be simplified as follows:

$$\Delta u = B[A\Delta\sigma_1 + \Delta\sigma_3(1-A)] \quad (2)$$

The magnitude of water content changes the soil structure and causes consolidation of the foundation pit soil. Generally, measures to reduce the impact of precipitation on the foundation pit are taken to reduce the water level.

According to the Terzaghi effective stress principle, After the groundwater level drops, the seepage pressure of water increases. In the construction of foundation pits, it is often used to install water stop curtains to control water seepage, but the effect is not ideal. Seepage phenomena often occur in foundation pits. Under normal circumstances, groundwater moves in the soil, and formula (3) is as follows:

$$J = \gamma_w \cdot i \quad (3)$$

Where: J ——Osmotic pressure; γ_w ——Water density; i ——Hydraulic gradient.

When the groundwater drops too fast, it can be seen from the formula that the seepage pressure will increase rapidly, and the seepage pressure will act on the soil particles, causing "flowing soil" phenomenon and deformation of the soil behind the structure.

In the well point dewatering foundation pit, a water stop curtain is used to form a vacuum near the well point pipe, causing the foundation pit to become in a negative pressure state, and the soil mass moves towards the negative pressure direction. This is due to the compression effect caused by the accumulation of soil particles, causing partial soil mass to settle.

4. Research on Deformation monitoring of deep foundation pit

4.1 Groundwater level monitoring

(1) Monitoring instruments

The deep foundation pit project uses a steel ruler water level gauge to measure the underground water level, and the water level pipe is made of PVC engineering plastic. As shown in Figure 2.

Figure 2. ruler gauge

Figure 3. PVC water level tube

(2) Embedded measurement points

Underground water level observation holes shall be respectively arranged within the precipitation range and the precipitation impact range. Water level pipes are buried separately according to different water layer depths. When monitoring the lower water level, the upper water level is sealed around the water level pipe, and perforated filter pipes are used at the monitoring water level position.

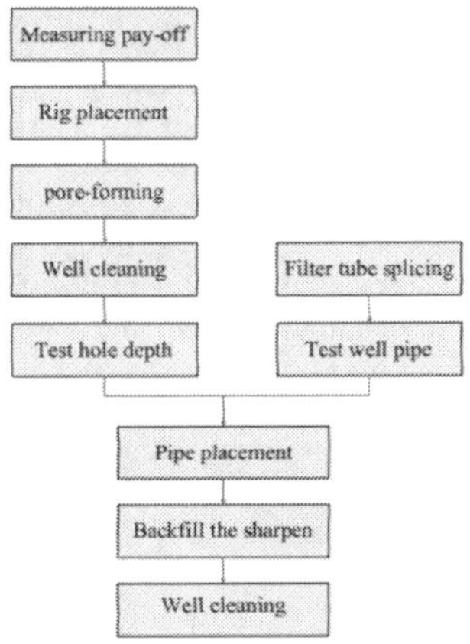

Figure 4. Water level pipe embedding flow chart

 https://doi.org/10.1051/e3sconf/202339401006

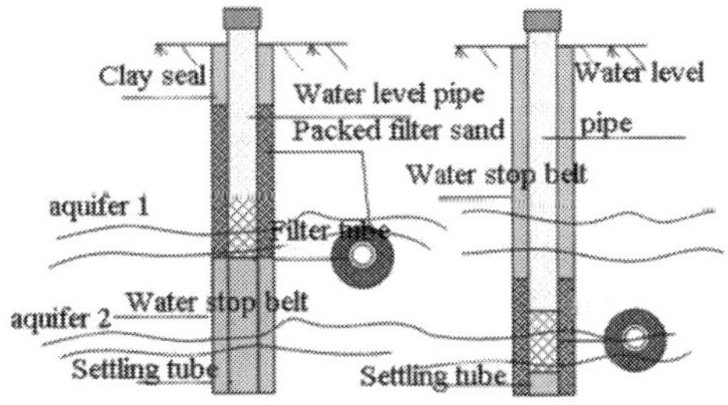

Figure 5. Schematic diagram of water level pipe installation

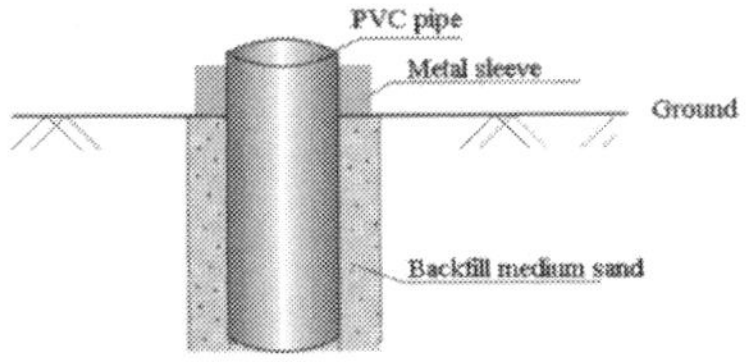

Figure 6. Schematic diagram of protection measures for water level holes

(3) Water level monitoring method

Place the probe of the water level gauge into the water level pipe. When the probe touches the groundwater, the water level gauge sends an alarm signal. At this time, the depth of the groundwater level can be obtained by waiting for the instrument to stably read the readings.

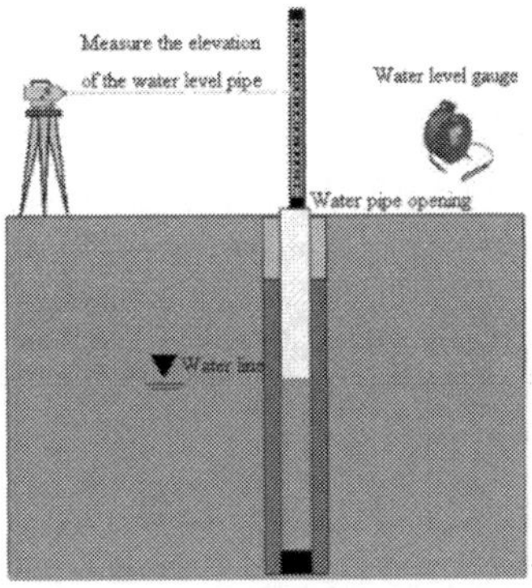

Figure 7. Schematic diagram of water level measurement

According to the pipe top elevation and elevation difference, the elevation and buried depth of the groundwater level can be calculated. During observation, conduct independent observation for each observation hole for 3 consecutive times, and take the average value. The formula is as follows:

$$H_{stage} = H_{Pipe\ top} - h_{Pipe\ top-surface\ of\ water} \quad (4)$$

(4) Site water level monitoring

There are 21 monitoring points for groundwater level, numbered DSW1~DSW21.

4.2 Deformation monitoring of enclosure structure

(1) Monitoring instruments

The monitoring instrument is a CX806D inclinometer, and the inclinometer tube is made of PVC material, as shown in Figure 8 and Figure 9.

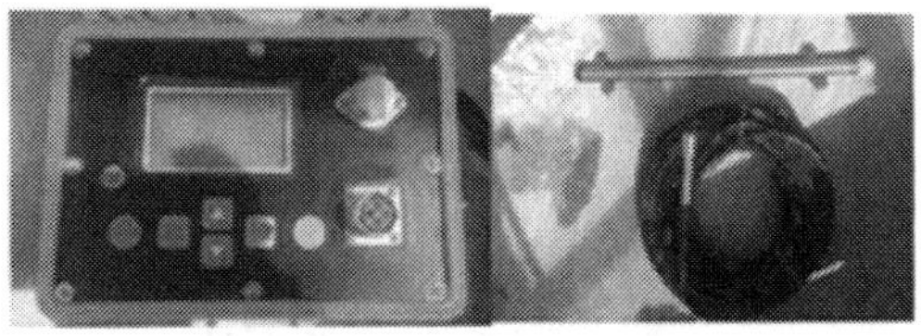

Figure 8. CX806D type foundation pit slopemeasuring instrument

Figure 9. PVC inclinometer

(2) Installation method

Before the excavation of the foundation pit, the deep horizontal displacement monitoring points shall conduct initial data collection on the inclinometer pipe position monitoring points for three consecutive times, and take the average value of the three data as the initial value. At the interface of each section of the inclinometer pipe, slurry is prevented from entering the inclinometer pipe. Tape is used to seal the pipe, and an anchor hoop is set every 1m to tightly bind the inclinometer pipe to the steel mesh to ensure the perpendicularity of the inclinometer pipe.

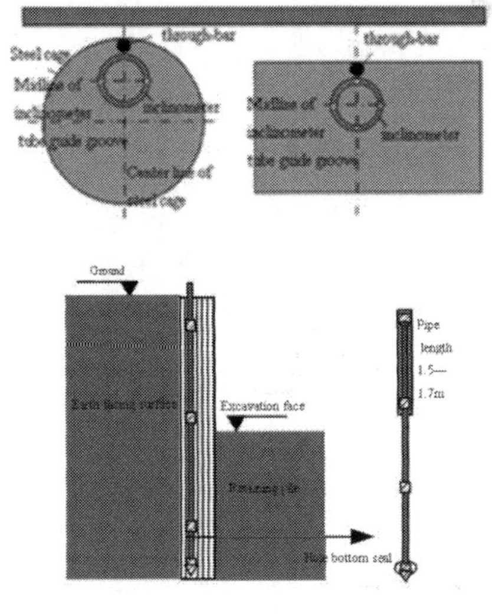

Figure 10. Installation of the inclinometer

Figure 11. Protection measures for the inclined tube

(3) Monitoring methods
During observation, it is necessary to correct the change in the coordinate port. place the inclinometer vertically into the inclinometer tube until it reaches the bottom of the tube. After the instrument data is stable, read and store it, and measure it every 0.5 meters. After the front side is completed, perform a reverse measurement, and rotate the inclinometer head for 180° to insert it again.
After the measurement is completed, turn the probe over for 90°, and perform the measurement again according to the above procedure. The inclination measurement principle is shown in Figure 12.

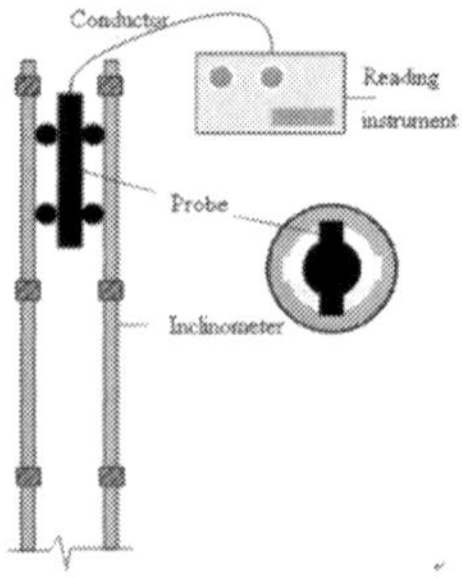

Figure 12. Schematic diagram of the tilting operation

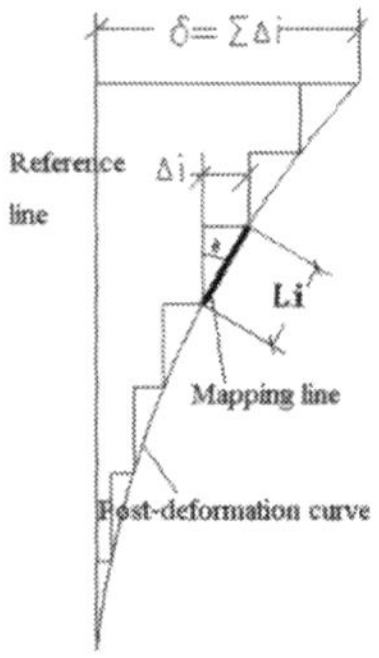

Figure 13. Principle of inclination

(4) On-site monitoring
The data of the civil engineering reserve project CJLLXTJYL DK48+300~DK48+700 section of the Beijing New Airport section of the new intercity railway connecting line was selected for research. The total length of the foundation pit in this section is 400m, the width of the foundation pit is 14m, the excavation depth of the foundation pit is 16m, the length of the retaining pile is 26m, and the deep horizontal displacement monitoring points are arranged at a spacing of 30m. There are a total of 34 monitoring points, numbered ZQT1~ZQT34.

5. Monitoring numerical analysis and research

5.1 Research on deformation laws of enclosure structures

(1) Analysis of monitoring results of deep horizontal displacement

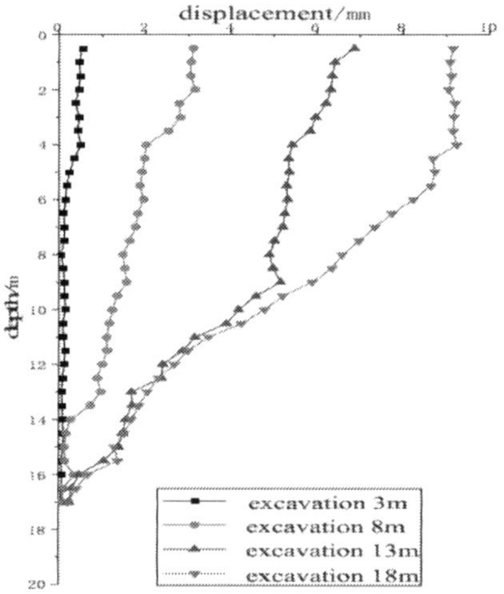

Figure 14. Deep horizontal displacement of ZQT1 hole

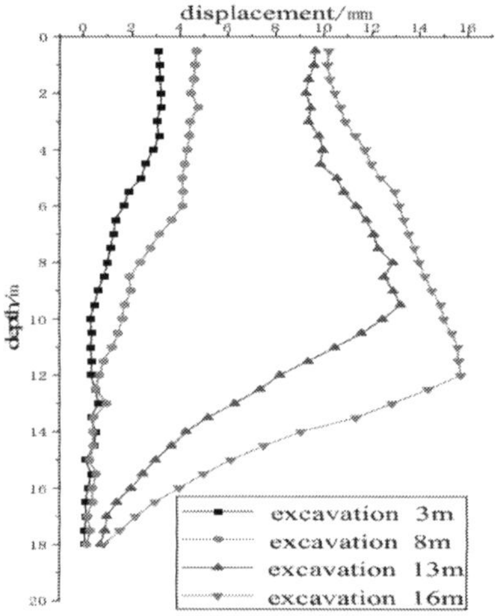

Figure 15. Deep horizontal displacement of ZQT8 hole

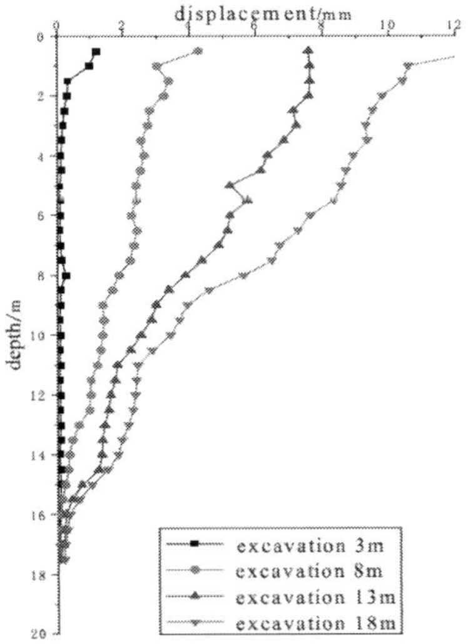

Figure 16. Deep horizontal displacement of ZQT16 hole

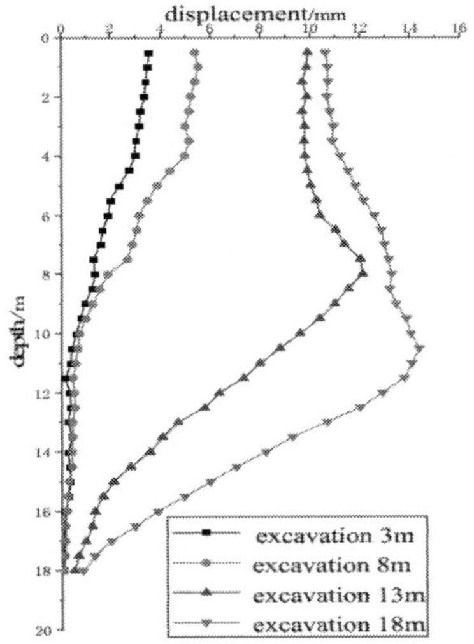

Figure 17. Deep horizontal displacement of ZQT24hole

The monitoring data of deep horizontal displacement inclinometer holes ZQT1, ZQT8, ZQT16, and ZQT24 were selected for study. The ZQT1, ZQT16 points are located at both ends of the foundation pit, and ZQT8 and ZQT24 are on the same monitoring section. The groundwater level in this section is high. The monitoring work has collected initial data values before excavation. The pouring of the retaining structure and the embedding of the inclinometer pipe were completed in June 2017, and the soil in the foundation pit was excavated in September 2017. The strength of the retaining structure meets the excavation construction conditions. During the excavation process, the monitoring of the deep horizontal displacement of the retaining structure shall always be maintained.

(2) Analysis of deformation law of deep horizontal displacement

According to the data and rules in the figure, when the soil excavation within the retaining structure is within 4 meters, the deep horizontal displacement of the retaining pile has little change, with the maximum displacement of 3.61 mm and the displacement controlled within 4 mm. Moreover, the bottom displacement of the retaining structure has basically not changed, and the retaining pile is in a stable state as a whole.

When the excavation reaches 8m, the deformation of the retaining pile is obvious, and the horizontal displacement of the pile top increases. When the excavation reaches 13m, the maximum displacement of the retaining pile transitions from the pile top to the middle of the pile, with the maximum deformation reaching 12.51mm, when the excavation reaches the bottom of the 16m foundation pit, the deformation of the retaining piles at points ZQT8 and ZQT24 reaches 16.11mm. At this time, the foundation pit is in the excavation exposure period. If the bottom plate and side wall are not poured in a timely manner, the deformation of the retaining piles will continue to increase. Point ZQT8 and point ZQT24 are on the same section, which clearly shows that the deformation amount and deformation law of the retaining piles on both sides of the foundation pit are very similar.

(3) Comparative Analysis of Deformation Laws of Enclosure Structures

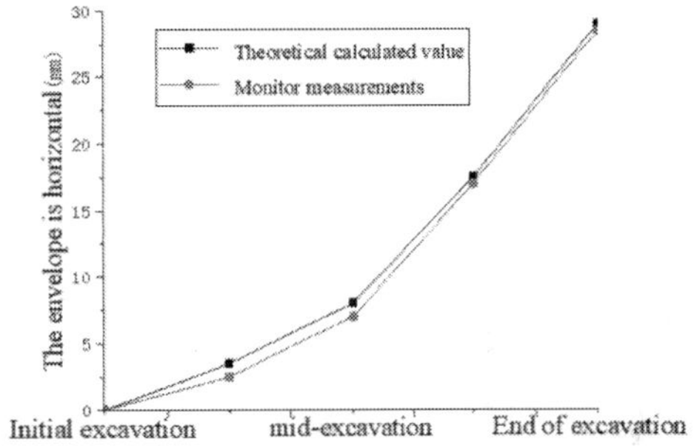

Figure 18. Comparison of calculated values of deformation of retaining structure

The deformation of the retaining structure can better reflect its changes based on the changes in the deep horizontal displacement of the retaining structure. Therefore, the deep horizontal displacement deformation of the retaining structure is selected to replace the deformation of the retaining structure for comparative analysis of the calculated deformation values of the retaining structure. It can be seen that the deformation of the retaining structure gradually increases with the excavation of the deep foundation pit. There is a certain difference between the actual monitoring measurement value and the theoretical calculation value, but the difference between the two values is not significant, indicating that the theoretical calculation is correct and the monitoring data is accurate.

5.2 Research on the variation law of groundwater level in deep foundation pit

(1) Groundwater level monitoring results
Select the water level observation holes DSW5, DSW6, DSW15, DSW16 in the foundation pit, and the JSJ10, JSJ11 of the dewatering wells outside the foundation pit to analyze the changes of groundwater level during the dewatering of the foundation pit inside the retaining structure and the dewatering of the outside of the retaining structure. The water level changes are shown in Figure 19 and Figure 20.

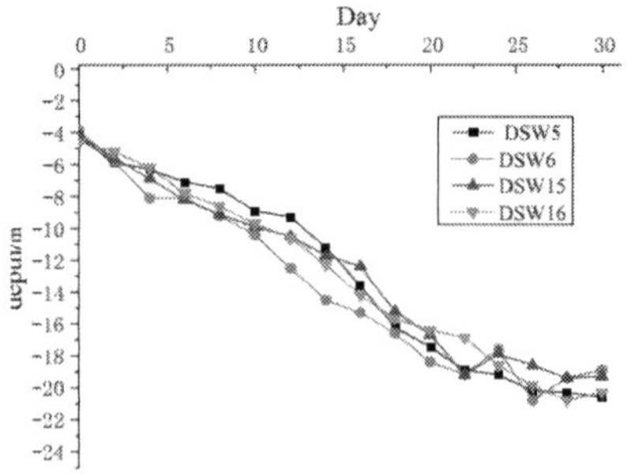

Figure 19. Water level change of the precipitation well inside the foundation pit

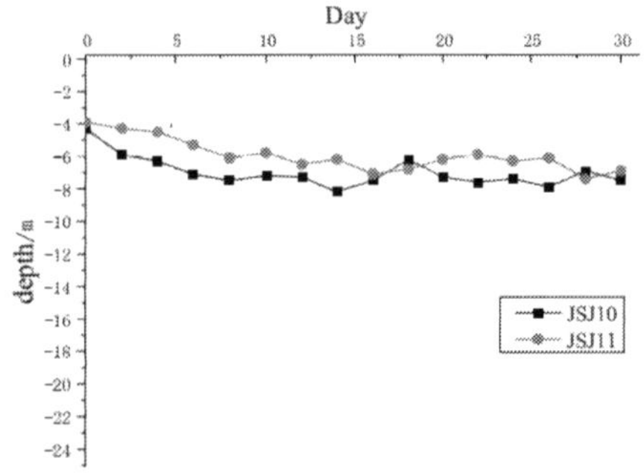

Figure 20. The water level change of the observation well outside the foundation pit

(2) Analysis on deformation law of groundwater level monitoring
Due to the preliminary dewatering of the inner side of the foundation pit before construction, it can be seen from the data and rules in the figure that the groundwater level in the foundation pit is continuously decreasing, and the water level changes tend to stabilize within 25 days. The water level outside the foundation pit decreased somewhat during the initial precipitation, but after 10 days, the water level remained stable. But at the beginning of precipitation, the water level on the outer side of the foundation pit is affected by the precipitation on the inner side of the foundation pit. However, based on subsequent analysis of the precipitation effect, its impact is limited.
(3) Comparative analysis of groundwater level changes in deep foundation pits

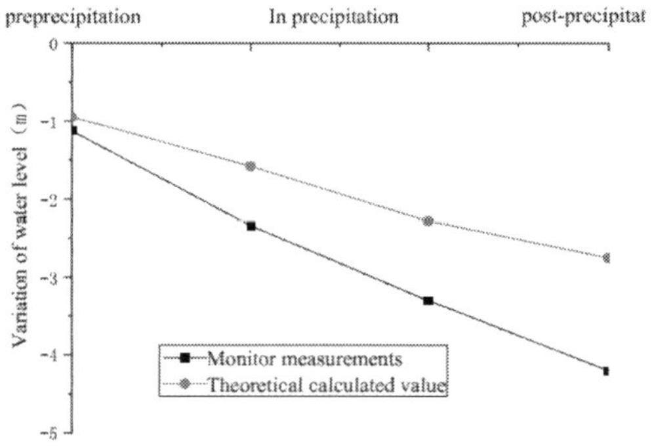

Figure 21. Comparison of calculated groundwater level

The monitoring measurement value in the figure is the variation of the groundwater level outside the foundation pit, and the actual monitoring water level variation is significantly higher than the theoretical calculation value by up to 4m. In combination with the civil engineering reserved line project of the Beijing New Airport Section, due to construction reasons such as untimely anchoring and shot creting of the anchor network, and untimely drainage and blocking of the groundwater level, some groundwater outside the foundation pit flows into the inside of the foundation pit, causing the groundwater level outside the foundation pit to exceed the theoretical calculation value. However, the overall numerical variation trend is not significantly different from theoretical calculations.

6. Deformation control technology for deep foundation pit

(1) When the groundwater level seeps into the pit due to improper construction reasons, grouting fluid should be used to seal the formation, stop excavation or adjust the one-time excavation depth to reduce the exposure time of the excavation surface.
(2) When the water level around the foundation pit is high, first ensure that there are no new leakage points on the retaining wall; Subsequently, clean up the debris at the leakage point, use a drill bit to chisel out a surface with a width of 15 cm to 20 cm around the leakage point of the diaphragm wall, and use quick-drying cement or fast-setting concrete to seal the leakage point until there is no leakage.
(3) When there is a large amount of water leakage, high water flow pressure, and the phenomenon of mud and sand seepage at the leakage point, using quick drying cement or instant setting concrete is insufficient to seal it. The method of inner sealing can be considered, including water glass cement double liquid grouting to seal the inner pores of the soil, and the treatment method of setting a diversion pipe to guide water and reduce pressure; The periphery of the leaking water site shall be sealed with quick-drying cement or concrete with good instantaneous setting effect. When the strength of the surrounding cement is sufficient, a diversion device shall be installed at the leaking point and the leaking point shall be completely sealed, forming a method of combining a

diversion and drainage system with sealing outside the pit. This method can use cotton to plug the leaking point to prevent the mud and sand from seeping out with the water flow.

Figure 22. Anti-leakage measures on the inside of the foundation pit

7. Conclusion

Based on the geological conditions of the open cut straight trench deep foundation pit and the deep foundation pit of the intercity connecting line of Beijing New Airport as the engineering support, the technical method of combining theoretical analysis and on-site monitoring is adopted to analyze the influencing factors of groundwater on the deep foundation pit retaining structure, and control and analyze the influencing factors. The deformation control technology for the deep foundation pit is proposed. The main conclusions are as follows:

(1) Based on the on-site geological conditions and the mechanism of groundwater level change, the deformation and failure forms and precipitation forms of deep foundation pits are obtained through theoretical analysis.

(2) Based on the Deformation monitoring technology of deep foundation pit, the change of groundwater level and Deformation monitoring of retaining structure are carried out.

(3) Combining on-site monitoring with theoretical calculation, conducting comparative analysis of data, summarizing the influencing factors of deep foundation pit damage, obtaining targeted deformation control technology for deep foundation pit, guiding the construction of practical projects, and providing reference for similar projects.

References

1. Zhou Chuanshui. Study on the influence of groundwater level changes on the horizontal displacement of diaphragm walls in deep foundation pits [J]. Juye, 2022 (09): 85-87.
2. Xu Zhongtao Dewatering excavation and numerical simulation analysis of deep foundation pit of subway station in water rich soft soil [D]. Nanchang University, 2021. DOI: 10.27232/d.cnki.gnchu.2021.002850.
3. Li Yuzi Research on the Change Law and Numerical Simulation of Groundwater under Deep Foundation Pit Dewatering Conditions [D]. Yangtze River Academy of Sciences, 2021. DOI: 10.26982/dcnki.gcjky.2021.000010.
4. Wang Hao Research on Deformation Characteristics of Deep and Large Foundation Pit Dewatering Excavation in Subway Stations [D]. Shandong University of Science and Technology, 2020. DOI: 10.27275/d.cnki.gsdku.2020.001783.
5. Long Zhang. The impact of groundwater level on surrounding buildings in deep foundation pit monitoring [J]. Zhonghua Construction, 2020 (05): 92-93.
6. Xia Wen, Li Pan, Qiu Wei, Li Xudong. Analysis of the impact of changes in groundwater level during deep foundation pit excavation on surrounding settlement [J]. Construction Technology, 2019,48 (S1): 138-141.
7. Wu Peng Deformation monitoring and numerical simulation analysis of dewatering and excavation of deep foundation pit in water rich sandy pebble stratum [D]. Xiangtan University, 2019. DOI: 10.27426/d.cnki.gxtdu.2019.001548.
8. Tao Hongliang Research on Groundwater Control Methods for Deep Foundation Pit Engineering of the First Class Terrace of the Yangtze River in Wuhan [D]. China University of Geosciences, 2021. DOI: 10.27492/d. cnki. gzdzu.2021.000193.
9. Gao Dong Analysis of the Impact of Foundation Pit Precipitation on Ground Settlement [D]. Anhui University of Architecture, 2018.
10. Zhang Dongdong Research on Dewatering and Decompression Dewatering Technology for Deep Soft Soil Foundation Pit [D]. Shanghai Jiao Tong University, 2017.
11. Wang Chuanlin. Research on the influence of groundwater level height on the stability of deep foundation pits [J]. Engineering and Construction, 2016,30 (02): 236-238.
12. Zhao Min, Li Guangguang, Meng Lingdong. Monitoring and Analysis of Groundwater Level in Deep Foundation Pit in Loess Area [J]. Journal of Yancheng Institute of Technology (Natural Science Edition), 2016, 29 (01): 70-73. DOI: 10.16018/j.cnki.cn32-1650/n.201601015.
13. Li Lin Research on the Influence of Engineering Precipitation on the Properties and Surrounding Environment of Deep Excavation [D]. Tongji University, 2007.

Research on Mesoscopic Characteristics of Limestone Grouting Diffusion Based on Discrete Fracture Network

Jiawen Li [1], Shugao Zhan [1], Zuliang Zhong [2], Yi Wang [2,*]

[1] China Construction Tunnel Co., Ltd., Chongqing 401320, China
[2] School of Civil Engineering, Chongqing University, Chongqing 400045, China.

Abstract. The limestone stratum has well-developed rock fissures and abundant groundwater. Under extreme rainstorm conditions, tunnel surrounding rock seepage is serious especially. In order to reveal the grouting diffusion mechanism of karst fractured rock mass and realize effective water plugging, in this paper, the basic mechanical parameters of limestone are obtained by laboratory triaxial test, and the porosity distribution of rock mass is obtained by nuclear magnetic resonance test. The PFC2D discrete element software is used to simulate the mechanical characteristics of compressive deformation and fracture propagation pattern of the limestone and compared with the triaxial test results. Discrete fracture network theory is used to establish PFC2D numerical simulation of grouting diffusion model of fractured rock mass. And the filling diffusion mechanism and stress variation rule of cement slurry in fractured limestone under different grouting pressure are studied. The results show that (1) During the grouting process, the micro-cracks expand along the tip of the existing cracks, and extend to the surrounding cracks to finally form through cracks, so as to achieve the effect of grouting reinforcement. (2) With the increase of grouting pressure, the maximum displacement of particles increases continuously, but with the prolongation of grouting time, the maximum displacement of particles reaches a stable state and no longer increases. (3) During the grouting process of the fractured rock mass, with the increase of grouting time, the stress of the fractured rock mass finally reaches a stable state. This study provides an important reference for the setting of grouting parameters of limestone rock mass.

Keywords: Tunnel engineering, fractured rock mass, PFC2D, limestone, grouting diffusion.

1. Introduction

Subjected to complex geological structure, various structural planes (faults, joints and cracks, etc.) are formed inside the rock mass, which will cut the rock mass into rock blocks of different sizes and shapes, and finally form the fractured rock mass. Due to the complex structure and poor mechanical properties of fractured rock mass, grouting is often used to improve the mechanical properties of fractured rock mass in engineering practice. For the grouting process of fractured rock mass, materials with smaller particle diameter and certain cohesiveness are often used to fill the fractured rock mass with certain pressure. Wang[1] analyzed the strength of fractured rock mass filled by grouting materials based on uniaxial compression test, and the results showed that grouting materials could improve the strength of specimens, delay crack initiation, and reduce the strain at the crack tip. Because the permeability of cracks is much greater than that of pores [2], the grout in the grouting process is mainly spread by cracks, and only a few of them have pore seepage [3]. In actual engineering, the distribution of cracks in rock mass is difficult to identify clearly. So that in engineering construction, empirical design is often used for grouting reinforcement of fractured rock mass, which makes the grouting reinforcement theory of fractured rock mass lag behind the engineering practice.

In recent years, scholars have conducted a large number of grouting tests on fractured rock mass [4-6]. Du[7] discussed the diffusion characteristics of cement slurry in rock under the action of dynamic water through laboratory tests. Ding[8] conducted a grouting diffusion test through a visual test device, and analyzed the spatial-temporal distribution characteristics of the velocity field and pressure field in the process of grouting diffusion, and the results showed that the fracture structure of rock mass had a great influence on the grouting diffusion. Jiang[9] concluded that slurry filling could effectively improve the strength of fractured rock mass through laboratory tests, and analyzed the variation law of grouting pressure, slurry flow and slurry diffusion surface based on three-dimensional grouting test system. Zhang[10] developed a simulation test device for dynamic grouting of fractured rock mass, and determined the law of grouting diffusion

* Corresponding author: Yi Wang

https://doi.org/10.1051/e3sconf/202339401007

through the grouting test. It can be seen that the grouting test explained the grout diffusion filling process in the crack from the macro aspect, but for the meso aspect, the grout diffusion direction and stress distribution and other problems could not be effectively solved. Based on the rapid development of computer, some scholars used finite element method, finite difference method and discrete element method to build the rock mass grouting model, analyzed the changes of grout velocity field, displacement field and stress field, and further improved the development of grout diffusion filling law. But most of the studies were based on single fissure or orthogonal fissure rock mass [11-13]. There is a certain gap between it and the actual fractured rock mass. In order to objectively describe the grout diffusion and the change law of displacement and stress field of fractured rock mass in the process of grouting, this paper obtained reliable meso-parameters through the comparison and verification of laboratory test and numerical calculation. The random fracture network model in PFC2D was adopted to construct limestone fractured rock mass, and the evolution process of grouting diffusion and deformation law of limestone fractured rock mass are studied. The results are of great significance for the study of grouting parameter setting in limestone fractured rock mass.

2. Project Overview

The Tuzhu tunnel in Chongqing crosses the Guanyin gorge anticline, with small folds, faults, joints and fissures, interlayer sliding scrapes and fracture zones in and around the anticline axis. The underlying bedrock of the tunnel area is mainly Jialingjiang group, partly Leikoupo group, which is hard and mostly fragmented, with joints and fissures between undeveloped and more developed, laminated and generally combined between layers; the stratigraphy is mainly limestone, with occasional muddy rocks, as shown in Figure 1.

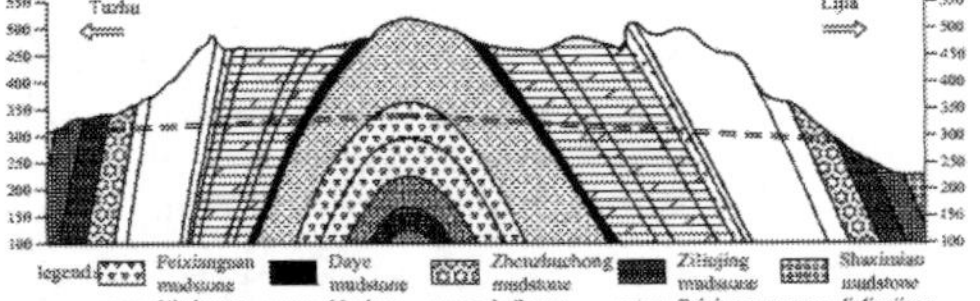

Figure 1: Schematic diagram of the engineering geology of the earth main tunnel

According to the survey, there are more cavities and caves in the soluble rock formations, more tension fissures in the Feixianguan formation in the core of the anticline, and the phenomenon of leakage of groundwater exists in various depth ranges. The presence of fissures and dissolution pores leads to different permeability and water-richness of the rock mass. Engineering disasters such as roof collapse, collapse and large water gushing will occur when the tunnel is driven into the densely fractured section, fault fracture zone and near the boundary of various aquifers, as shown in Figure 2. In the process of tunnel excavation, water leakage from the lining will jeopardize the safety of tunnel construction if it is not blocked in time [14, 15] and cause the surface water level to drop. So the construction process takes targeted grouting and water plugging to achieve the purpose of "mainly blocking and limited discharge" [16], to ensure the safety of tunnel construction and operation, and reduce the impact on the residents' living water.

Figure 2: Tunnel lining water seepage

This paper relies on the curtain grouting treatment section of Chongqing Tuzhu tunnel, as shown in Figure 3. The PFC2D discrete element numerical software is used to simulate the slurry diffusion process based on hydraulic splitting, and the optimal slurry injection pressure and slurry filling diffusion law are obtained by analyzing the evolution of displacement field and stress field during the fractured rock grouting process.

Figure 3: Curtain grouting

3. Parameter Calibration and Model Building

3.1 Parameter Calibration

3.1.1 Particle Mesoscopic Parameter Calibration

The macroscopic properties of rock mass are specifically expressed as the mesoscopic parameters of particles in the particle flow numerical model. After the modeling is

 https://doi.org/10.1051/e3sconf/202339401007

completed, the accuracy of the simulation results largely depends on the precise selection of mesoscopic parameters. Parallel bonding model is adopted between particles. In this test, 12 limestone specimens are selected for calibration of rock mass physical and mechanical parameters, and a total of 6 crack specimens meeting the test requirements are obtained. After the test, 6 samples with obvious macroscopic cracks and good integrity are selected and divided into 2 groups, group A and group B, with 3 samples in each group. In this test, the French TOP INDUSTRLE rock full stress multi-field coupling triaxial tester is used for triaxial test, as shown in Figure 4. The confining pressure is 5 MPa and the vertical loading control is 0.1 mm/min. Numerical software PFC2D is used to simulate limestone specimens with a diameter of 50 mm and a height of 100 mm, as shown in Figure 5.

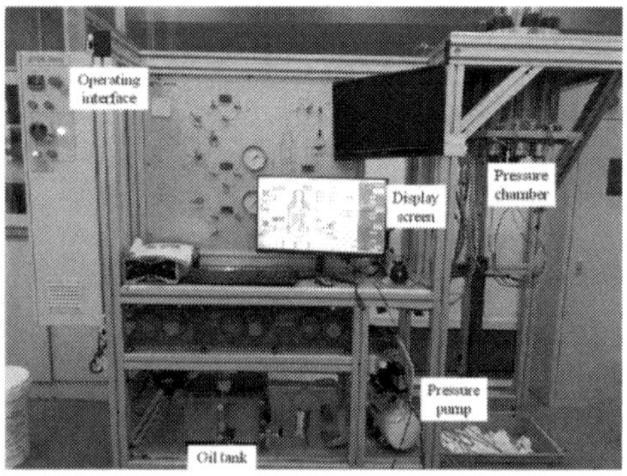

Figure 4: French TOP INDUSTRLE rock total stress multi-field coupling triaxial test instrument

From Figure 5, it can be seen that the limestone is a hard rock, and the average peak strength reaches 185 MPa under the action of the surrounding pressure of 5 MPa. The peak strength of the particle flow simulated limestone specimen matches the test results well and meets the macroscopic consistency. Because this specimen has more internal microcracks, the A2 stress-strain curve exhibits an obvious compression phenomenon in the initial section, then enters the linear elastic phase, and eventually suffers brittle damage due to reaching the peak strength, resulting in a sharp drop in the curve.

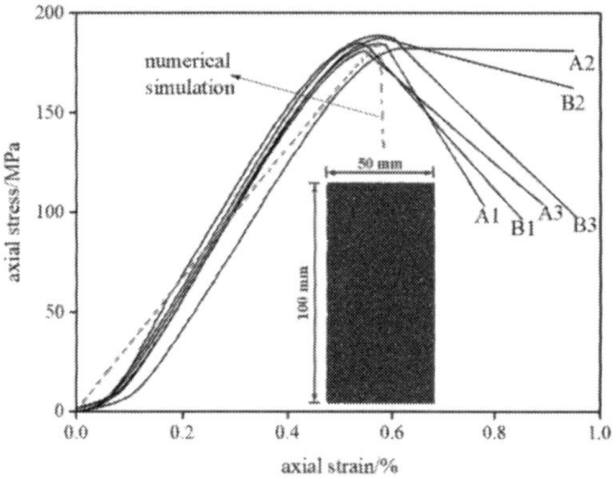

Figure 5: Stress-strain curve of triaxial compression

Under triaxial compression, the fracture morphology of limestone is mainly shear failure (A2, A3, B1, B2, B3), and occasionally V-shaped failure (A1), as shown in Figure 6. By comparing the peak strength and damage morphology of stress-strain curve of the indoor tests and numerical simulations, the granular mesoscopic parameters are determined as shown in Table 1.

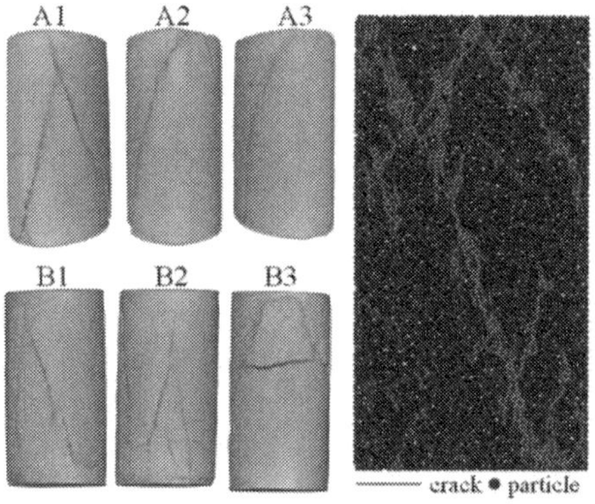

Figure 6: Comparison of failure morphology of limestone specimen between test and numerical simulation

Table 1: Mesoscopic parameters of limestone particles calibrated by triaxial compression test

Granular mesoscopic parameters	Maximum radius (mm)	Minimum radius (mm)	Particle contact modulus (GPa)	Normal tangential stiffness ratio of particles	Particle friction coefficient
	1.5	1	16.6	1.2	0.6
Mesoscopic parameters of parallel bonding	Minimum distance between parallel bonds (mm)	Parallel bonding modulus (GPa)	Parallel bond stiffness ratio	Parallel bond tensile strength (MPa)	Internal cohesion of parallel bond(MPa)
	0.05	16.6	1.2	50	60

3.1.2 *Fluid Mesoscopic Parameters*

Model porosity is a key factor in determining the permeability of cement slurry. The porosity magnitude of intact and fractured rocks are obtained through laboratory tests, so the histogram of porosity distribution of rock distribution is obtained by nuclear magnetic resonance analysis, as shown in Figure 7. The porosity of intact rock ranges from 1.14% to 1.47%, and that of fractured rock ranges from 2.94% to 3.53%. Therefore, to ensure that the numerical simulation is more consistent with the actual situation, the porosity of 3.5% is chosen to simulate the fractured rocks.

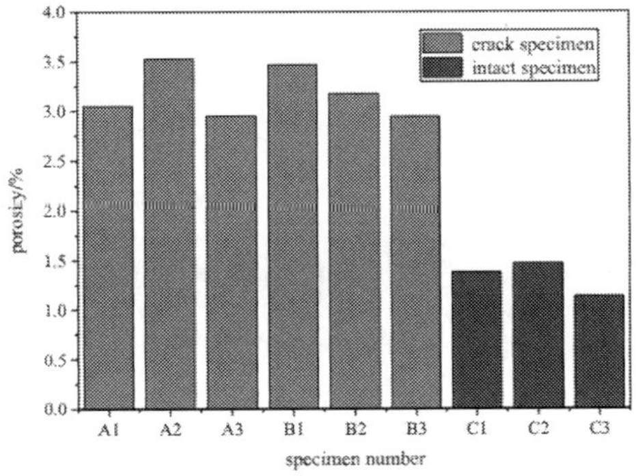

Figure 7: Histogram of porosity distribution

The model permeability coefficient is obtained by reference [17]. Cement slurry is a typical bingham fluid, and the permeability coefficient of water in limestone strata can be obtained by nuclear magnetic resonance. However, it is difficult to obtain the permeability coefficient of cement slurry in limestone strata. The water-cement ratio of cement slurry selected in the test is 1:1. In order to simulate the permeability process of cement slurry in limestone strata, the permeability coefficient of cement slurry in limestone strata can be obtained from formula (1) :

$$k_c = k_0 \frac{\eta_w \rho_c}{\eta_c \rho_w} \quad (1)$$

Where k_c is the permeability coefficient of cement slurry in limestone formation; k_0 is the permeability coefficient of water in limestone; η_w is the dynamic viscosity coefficient of water. η_c is the dynamic viscosity coefficient of cement slurry. ρc is the density of cement slurry; ρ_w is the density of water. The final permeability coefficient of cement slurry in limestone can be determined by the formula (1) as 2.3×10^{-5} cm/s. The mesoscopic parameters of the fluid are finally determined through laboratory test and numerical simulation results, as shown in Table 2:

Table 2: Mesoscopic parameters of fluid

Permeability coefficient(cm/s)	time step	Initial aperture (mm)	Porosity(%)
2.3×10^{-5}	1×10^{-3}	1	3.5

3.2 Modeling of Fractured Rock Mass

In order to describe the distribution of fractures in a rock mass, DFN method is selected, which is to characterize its size, location and occurrence [18]. The fracture distribution function can be obtained from equation (2) :

$$n(l,\theta,\varphi,V) = \alpha(\theta,\varphi)\cdot l^{-r}\cdot V^{\frac{D}{3}} \quad (2)$$

In the formula, l is the fracture length, 35~40 mm; φ is the fracture angle, which is 20~70° in simulation. V is the region where the crack is located, and x and y are -70~70 mm, respectively. r is the scale index of the power law function; n is the power law function of crack distribution; D is the topological dimension (less than 3 in actual calculation). After the generation of fractured rock mass network, although it follows certain mathematical statistics law, the discrete fractured rock network samples can not be accurately expressed. For the crack density calculation method, p10 can accurately calculate the crack frequency value in the sample [19]. Finally, a discrete fracture network model with length and width of 150 mm×150 mm is obtained, as shown in Figure 8. To take into account DFN's conductivity, FISH language is used to break the bond between the particles at the fracture so that the cement slurry fills quickly in the fracture, simulating the realistic characteristics of rapid fissure infusion.

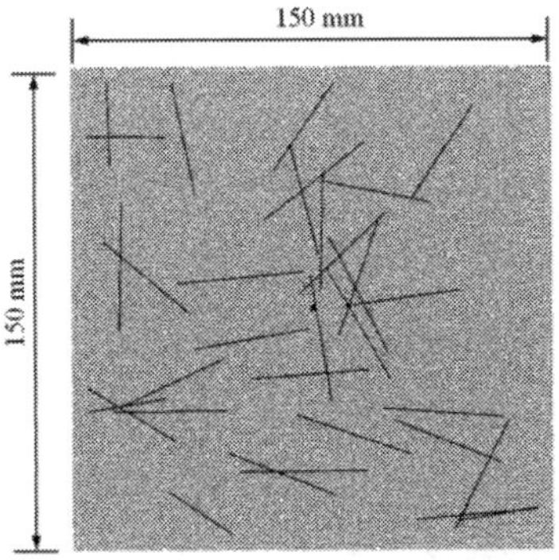

Figure 8: Discrete fracture network model

4. Analysis of Numerical Simulation Results

By using particle flow to simulate the grouting process, the crack propagation and stress distribution of fractured rock mass under the action of grouting pressure can be directly observed, and the grouting process can be better discussed on the mesoscopic level. After the grouting is started, the slurry enters the fluid domain between particles along the pipe from the borehole, as shown in Figure 9. The fluid pressure in the fluid domain and the pipe changes, affecting the contact between particles. As the grouting time progresses, the slurry spreads outwards in a radial pattern. And the energy is consumed in the process of slurry diffusion. When the diffusion reaches a certain distance, the slurry pressure is not sufficient to overcome the resistance within the rock, the slurry stops diffusing and the grouting reinforcement process ends.

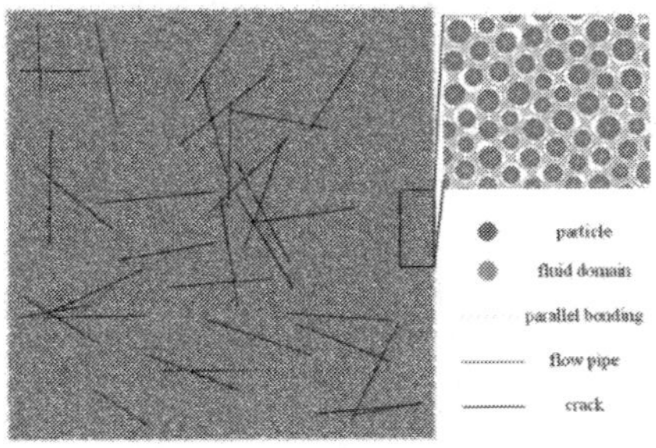

Figure 9: Fluid domain and pipeline model diagram

4.1 Filling and Diffusion Law of Cement Slurry under Different Grouting Pressures

The grout filling diffusion is an important index to test the grouting effect of fractured rock mass. The effects of micro-crack propagation and grout filling under four different grouting pressures at different time steps were simulated numerically. As shown in Figure 10, with the increase of grouting pressure, the number of micro-cracks gradually increases. Under 1 MPa grouting pressure, no new micro-cracks are generated in the fractured rock mass, and only existing cracks are filled. Under 2 MPa grouting pressure, fractured rock mass produces 1 micro-crack. Under the action of 3 MPa and 4 MPa, it produces 8 micro-cracks. However the crack development velocity under 4 MPa grouting pressure is faster than 3 MPa. With the increase of grouting pressure, the crack propagation speed increases gradually, and the first micro-crack is produced under 4 MPa grouting pressure at 2000 time step earlier than 3 MPa and 4000 time step earlier than 2 MPa. It can be seen that with the increase of grouting pressure, the grout diffusion speed is faster and the filling efficiency is higher. Among them, the filling diffusion range under 3 and 4 MPa grouting pressure is consistent and better than 1 and 2 MPa.

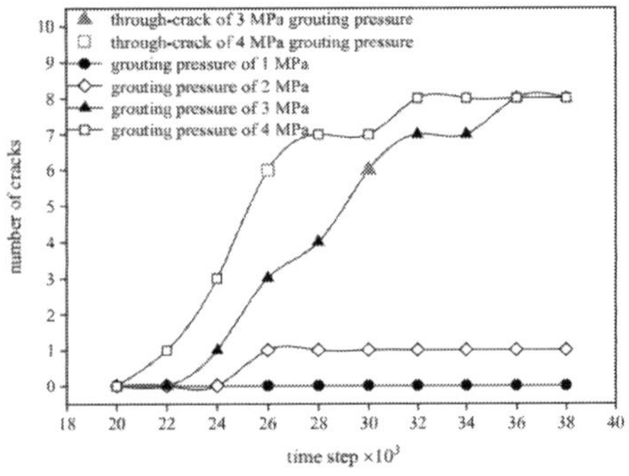

Figure 10: Development of the number of micro-cracks under different grouting pressures

The red micro-cracks in Figure 11 represent the cracks that occur when the grouting pressure exceeds the bonding force between rock particles. The cement slurry is first filled along the existing crack, and then diffuses outward along the grouting hole. Due to the stress concentration at the tip of the existing crack, micro-cracks appear. It can be seen from Figure 11 that the micro-crack develops from the long axis of the existing crack to another crack, and finally forms a through crack. As can be seen from Figure 10, two through cracks formed in the fractured rock mass at the time step of 30,000 and 26000 respectively under the grouting pressure of 3 MPa and 4 MPa, while no through-cracks formed at 2 MPa and 1 MPa. Therefore, crack development can be summarized into three stages: the first stage is crack sprouting, the second stage is crack development, and the third stage is crack penetration. In the first stage, the cement slurry only fills the existing cracks around the grouting holes, such as cement slurry expansion of 1 MPa grouting pressure. In the second stage, after the cement slurry is filled with the existing cracks, the stress of the grout at the crack tip increases with the increase of time, which promotes the occurrence of micro-cracks at the existing crack tip, and then makes the slurry split rock mass increase the reinforcement range. For example, after 24000 time step of 2 MPa grouting pressure, 22000~30000 time step of 3 MPa grouting pressure and 20000~26000 time step of 4 MPa grouting pressure, cracks are always in the developing stage. In the third stage, cracks are continuously developed and connected with other existing cracks, which further enlarges the range of reinforcement, mainly at the time step of 30,000 and 26000 of 3 MPa and 4 MPa grouting pressure.

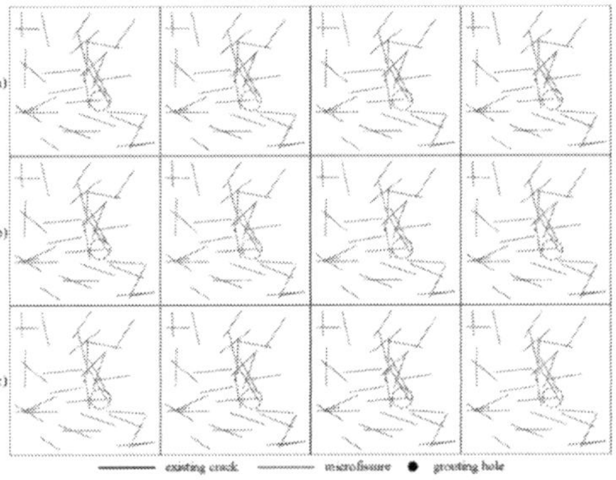

Figure 11: Micro-crack propagation at 26000, 30000, 34000 and 38000 steps under different grouting pressures (a) 2 MPa grouting pressure (b) 3 MPa grouting pressure (c) 4 MPa grouting pressure

4.2 Variation Law of Particle Displacement of Fractured Rock Mass under Different Grouting Pressures

As can be seen from Figure 12 (a), the rock mass is cut by existing cracks, and the rock mass particles will move along the developed part of the cracks under the action of grouting pressure. As shown in Figure 12(b), under the action of four grouting pressures, the maximum displacement of particles reaches 0.078mm, 0.225mm, 0.796mm and 0.980mm respectively. With the extension of grouting time, the maximum displacement of particles reaches stability and no longer increases. Among them, under 3 MPa and 4 MPa grouting pressure, the particle displacements have a sudden increase trend at 30000 and 26000 time step, respectively. Because at this stage, the micro-cracks are connected with the existing cracks, which leads to the sudden increase of particle displacement.

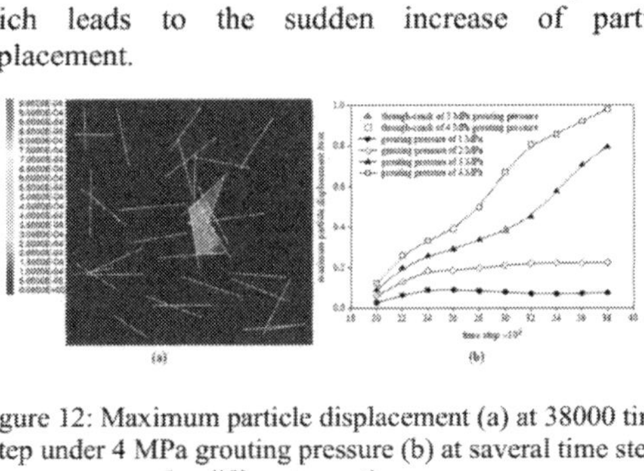

Figure 12: Maximum particle displacement (a) at 38000 time step under 4 MPa grouting pressure (b) at saveral time step under different grouting pressures

https://doi.org/10.1051/e3sconf/202339401007

4.3 Variation Law of Stress Field of Fractured Rock Mass under Different Grouting Pressures

As shown in Figure 13, under different grouting pressures, the horizontal stress of fractured rock mass continuously radiates outward along the fracture development area. Combined with Figure 8, it can be seen that the maximum horizontal stress is concentrated at the developing part of micro-crack, and it also verifies the stress concentration phenomenon of horizontal stress at the existing crack tip during the grouting process. As can be seen from Figure 14, the horizontal stress tends to be stable with the increase of grouting time step, where under 1 MPa grouting pressure fractured rock mass reaches the maximum value of 8.08 MPa at 26000 time step. The fractured rock mass reaches a maximum of 17.36 MPa at 38000 time step under 2 MPa grouting pressure. Besides under 3 MPa it reaches the maximum horizontal stress of 20.15 MPa at 24000 time step. And the fractured rock mass reached 20.91 MPa at 24000 time step under 4MPa grouting pressure. Under the grouting pressure of 1 and 2 MPa, the horizontal stress increases with the increase of time step, and finally reaches stability. However, 3 MPa and 4 MPa are in the crack penetration stage at 30000 and 26000 time step respectively, which results in the rapid decrease of stress in rock mass. Under the action of 2 MPa grouting pressure, the crack penetration threshold of the third stage is not 20.15 MPa. Therefore, with the increase of grouting pressure, the grout fills the existing cracks around the rock, and the horizontal stress in the fractured rock keeps a high level and constant.

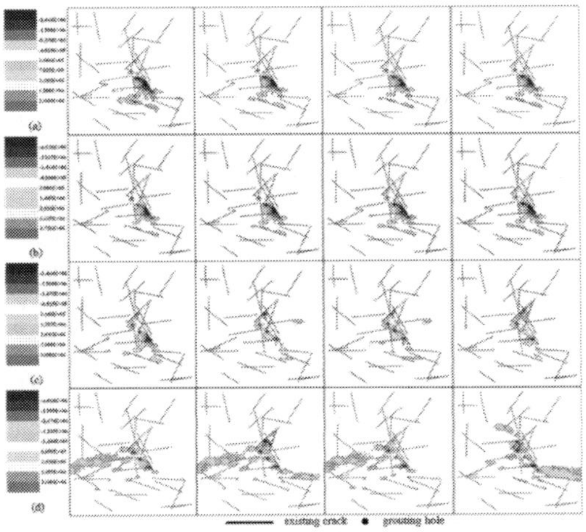

Figure 13: Horizontal stress at 26000, 30000, 34000 and 38000 under different grouting pressures (a) 1 MPa grouting pressure (b) 2 MPa grouting pressure (c) 3 MPa grouting pressure (d) 4 MPa grouting pressure

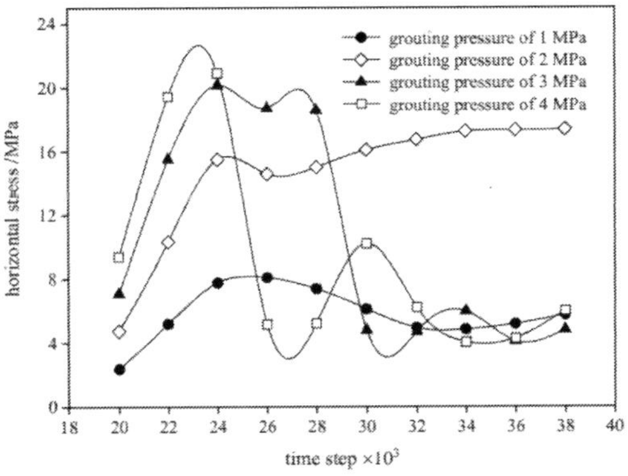

Figure 14: Horizontal stress distribution at each time step under different grouting pressures

5. Conclusion

In this paper, particle discrete element software is used to study the filling and diffusion effects of cement slurry in cracks, particle displacement and stress variation rules under different grouting pressures on limestone fractured rock mass based on random fracture network model. The main conclusions are as follows:

(1) Since the slurry diffusion filling effect of fractured rock body under different grouting pressure is different, based on the different states of slurry diffusion filling, it is divided into 3 stages, which are crack sprouting stage, crack development stage and crack penetration stage.

(2) Under the action of different grouting pressure, with the increase of grouting time, the micro-cracks expand along the tip of existing cracks and have the tendency to extend to the surrounding cracks, and finally form through cracks when the grouting pressure reaches the threshold, so that the existing cracks are connected to each other to achieve the effect of grouting reinforcement.

(3) With the increase of grouting pressure, the maximum particle displacement increases continuously. However, when the fractured rock mass produces through cracks, the maximum particle displacement increases abruptly. However, with the extension of grouting time, the maximum particle displacement eventually reaches a stable state and does not increase.

(4) In the process of grouting, with the increase of grouting time, the horizontal stress of the fractured rock mass finally reaches stability. In the process of grouting, due to the stress concentration of the crack tip, the micro crack expands along the existing crack tip and extends to the surrounding crack. When the stress threshold of the fractured rock mass is exceeded, the through crack is formed, and the stress of the fractured rock mass decreases rapidly after the formation of the through crack. With the increase of grouting time, it finally keeps constant at lower stress level.

Acknowledgements

The work was supported by Natural Science Foundation and Frontier Research Program of Chongqing Science and Technology Commission of China (award number: cstc2013jcyjA30005).

References

1. Y.X. Wang, H. Zhang, H. Lin, et al. Mechanical Behavior and Failure Analysis of Fracture-Filled Gneissic Granite, Theoretical and Applied Fracture Mechanics, vol. 108 (2020), p. 102674.
2. Z.D. Zhu, Z.H. Niu, X.C. Que, et al. Study on Permeability Characteristics of Rocks with Filling Fractures Under Coupled Stress and Seepage Fields, Water, vol. 12 (2020), no. 10, p. 2782.
3. Y.L. Zhao, Q. Liu, C.S. Zhang, et al. Coupled Seepage-Damage Effect in Fractured Rock Masses: Model Development and a Case Study, International Journal of Rock Mechanics and Mining Sciences, vol. 144 (2021), p. 104822.
4. W.Q. Zhang, H.T. Zhou, W.H. Guo: Experimental Study on Seepage Characteristics of Fractured Rock Mass and its Electrical Response, Journal of Hydrologic Engineering, vol. 24 (2019), no. 7, p. 04019017.
5. X.M. Jiang, G.S. Zheng, W.H. Sui, et al. Anisotropic Propagation of Chemical Grouting in Fracture Network with Flowing Water, ACS Omega, vol. 6 (2021), no. 7, p. 4672-4679.
6. X.L. Li, M.M. Hao, Y.H. Zhong, et al. Experimental Study on the Diffusion Characteristics of Polyurethane Grout in a Fracture, Construction & Building Materials, vol. 273 (2021), p. 121711.
7. X.M. Du, C. Liu, C.J. Wang, et al. Diffusion Characteristics and Reinforcement Effect of Cement Slurry on Porous Medium Under Dynamic Water Condition Considering Infiltration, Tunnelling and Underground Space Technology, vol. 130 (2022), p. 104766.
8. W.Q. Ding, C. Duan, Q.Z. Zhang: Experimental and Numerical Study on a Grouting Diffusion Model of a Single Rough Fracture in Rock Mass, Applied Sciences, vol. 10 (2020), no. 20, p. 1-23.
9. D.H. Jiang, X.Z. Cheng, H.J. Luan, et al. Experimental Investigation on the Law of Grout Diffusion in Fractured Porous Rock Mass and its Application, Processes, vol. 6 (2018), no. 10, p. 191.
10. E.M. Zhang, Y.C. Xu, Y. Fei, et al. Influence of the Dominant Fracture and Slurry Viscosity on the Slurry Diffusion Law in Fractured Aquifers, International Journal of Rock Mechanics and Mining Sciences, vol. 141 (2021), p. 104731.
11. W.T. Liu, Y.B. Liu, Y.H. Du, et al. Study on the Development and Distribution Rule of Tensile Shear Fracture Propagation Angle of Fractured Rock Mass Under Seepage Pressure, Geotechnical and Geological Engineering, vol. 37 (2019), no. 5, p. 4151-4161.
12. C. Liu, X.S. Huang, W.P. Yue, et al. Extension of Grouting-Induced Splitting Fractures in Materials Similar to Coal Rocks Containing Prefabricated Fractures, Journal of Geophysics and Engineering, vol. 17 (2020), no. 4, p. 670-685.
13. F. Xiao, Z.Y. Zhao: Influence of Fracture Deformation on Grout Penetrability in Fractured Rock Masses, Tunnelling and Underground Space Technology, vol. 102 (2020), p. 103431.
14. C.W. Zhu, W. Wu, H.W. Ying, et al. Drainage-Induced Ground Response in a Twin-Tunnel System through Analytical Prediction Over the Seepage Field, Underground Space, vol. 7 (2022), no. 3, p. 408-418.
15. Z.G. Zhang, M.D. Mao, Y.T. Pan, et al. Experimental Study for Joint Leakage Process of Tunnel Lining and Particle Flow Numerical Simulation, Engineering Failure Analysis, vol. 138 (2022), p. 106348.
16. K.K. Panthi, B. Nilsen: Uncertainty Analysis for Assessing Leakage through Water Tunnels: A Case from Nepal Himalaya, Rock Mechanics and Rock Engineering, vol. 43 (2010), no. 5, p. 629-639.
17. H. Su: Application of Discrete Element Particle Flow in Hydraulic and Geotechnical Engineering (Science Publications, China 2017), p.129.
18. P.T. Wang, F.H. Ren, W.H. Tan, et al. Model of Roughness Discrete Fractures Network for Uniaxial Compressive Test and its Mechanical Properties, Yan Tu Li Xue, vol. 38 (2017), p. 70-78.
19. X.Q. Wang: Study on Model Reconstruction and Mechanical response Characteristics of jointed coal (Ph.D., China University of Mining and Technology (Beijing), China 2017), p. 25.

Biomass-sourced polymers for pressure-sensitive adhesive applications

Chen Xu, Yong Shen*

State Key Laboratory Base of Eco-Chemical Engineering; College of Chemical Engineering, Qingdao University of Science and Technology, Qingdao 266042, China.

Abstract. Pressure-sensitive adhesives are self-adhesive chemical products that have gradually supplanted traditional glues in recent years. Most of the commercially available pressure-sensitive adhesives are derived from petrochemical resources. The increasing attention to environmental protection, coupled with the introduction of stringent regulations and rising oil prices, has led to widespread attention being paid to the development of sustainable pressure-sensitive adhesives from biomass sources. This paper summarizes the biomass-sourced pressure-sensitive adhesives and their modification methods that have been reported in recent years as promising alternatives to petrochemical resources.

1. Introduction

Pressure-sensitive adhesive is a self-adhesive material that can adhere to the surface of objects with only slight pressure, and is widely used in automotive film, daily office, medical, packaging and other fields [1]. The formation of pressure-sensitive adhesion does not require a chemical reaction with the surface of the object in contact, which is achieved by the van der Waals forces between the pressure-sensitive adhesive and the substrate. Therefore, it is necessary that the pressure-sensitive adhesive material has a certain degree of fluidity to infiltrate the surface of the object to be adhered to, as well as the elasticity of the solid and the creep resistance required for the separation process [2]. Pressure-sensitive adhesives can be divided into six types: rubber, polyester, polyether, silicone, polyurethane and acrylate [3]. The use of pressure-sensitive adhesives has increased dramatically in recent years due to the rise of the courier industry. Most of the pressure-sensitive adhesives today come from petrochemical resources, and with the increasing demand and the contradiction between non-renewable raw material reserves, the development of new, recycled pressure-sensitive adhesives has become an inevitable trend for the future. The use of naturally occurring macromolecules, such as natural rubber, starch, egg whites and animal glue, glutinous rice, etc., for the production of adhesives began long ago, and these traditional adhesives are still in use and still exist in many ancient buildings today. Although these natural pressure-sensitive adhesives are renewable, they often contain a variety of impurities and are costly, limiting their large-scale industrial production. Natural pressure-sensitive adhesives have significant limitations, such as poor UV resistance and lack of water resistance, and are not suitable for today's complex applications, but it is still possible to extract raw materials from biomass and use polymerization and chemical modification to obtain new renewable pressure-sensitive adhesives that are comparable to today's commercial pressure-sensitive adhesives [4].

2. Polyester pressure-sensitive adhesives of bio-sourced

Polyester pressure-sensitive adhesives are becoming a hot topic of research due to their good biocompatibility and availability of raw materials from a wide range of biomass sources, as well as their rapid degradability in nature [5]. Kim and Shin et al. used diethylene glycol (DEG) as initiator and tin (II) ethylhexanoate $Sn(Oct)_2$ as catalyst to achieve sequential ring-opening polymerization of the bio-sourced monomers ε-decalactone (DL) and L-LA to prepare ABA-type tri-block polymers (PLLA-PDL-PLLA). The soft segment PDL has a molecular weight of up to 100 kg/mol, which effectively enhances the energy storage modulus and creep resistance of the polymer. The authors noted that tackifiers and other additives are necessary components of most polyester-based pressure-sensitive adhesives, and the addition of bio-derived rosin ester tackifiers and epoxidized soybean oil plasticizers blended with PLLA-PDL-PLLA yielded renewable pressure-sensitive adhesives [6]. The optimal formulation has a peel strength of 2.6 N/cm, a probe tack of 3.0 N, and hold tack test results greater than 2000 min, which is comparable to commercially available pressure-sensitive adhesives.

* Corresponding author: shenyong@qust.edu.cn

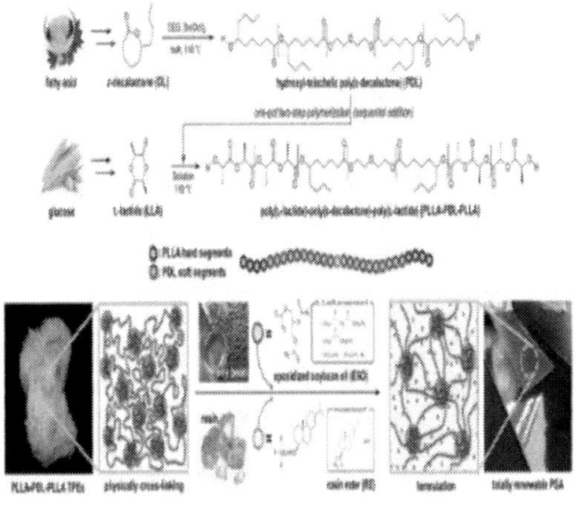

Figure 1. Synthesis of PLLA-PDL-PLLA from ε-decalactone (DL) and L-LA and preparation of renewable pressure-sensitive adhesives [6]. Reprinted with permission from ref. 6. Copyright (2015) American Chemical Society.

The cashew nut industry produces more than 2.2 million tons of cashew nuts per year, while cashew nut shells go unnoticed. these cashew nut shells can be produced by a simple extraction process to produce cashew nut shell liquid and are available in more than 450,000 tons per year. Hillmyer and Ellison et al. used cashew nut shell liquid as a raw material to prepare a new seven-membered ring polyester with long alkyl substituents, poly(pentadecyl caprolactone) (PPDCL) and polymerized it with L-LA blocks to obtain the tri-block polymer LPL (Figure 2) [7]. By blending with renewable rosin resin tackifiers, a series of pressure-sensitive adhesives with different adhesion properties can be obtained by changing the ratio of tackifiers. Among them, the 180° peel strength is 10.2 N/cm at 60 wt.% of rosin resin tackifier addition, which is higher than most commercial pressure-sensitive adhesives, and the ring adhesion is 10.0 N/cm and hangs vertically for 500 min without failure, with good creep resistance. The pressure-sensitive adhesive can be degraded in acidic aqueous solutions, which can effectively solve the problem of "adhesive residue" in pulping facilities.

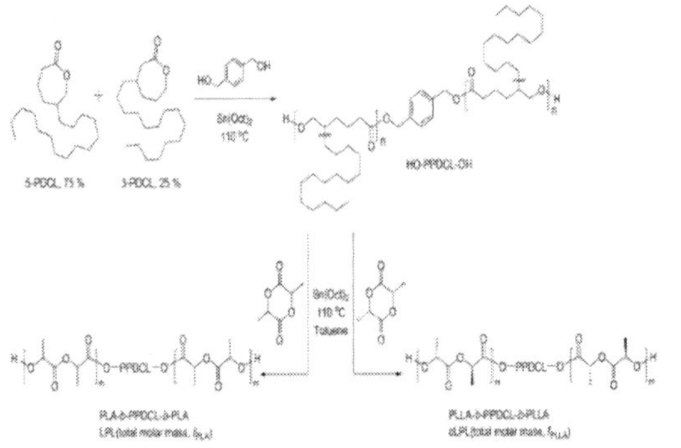

Figure 2. Synthesis of tri-block copolyesters from PDCL and LLA [7]. Reprinted with permission from ref. 7. Copyright (2020) American Chemical Society.

Williams et al. in 2020 reported a class of fully degradable bio-based pressure-sensitive adhesives. Hard segments were obtained by alternating copolymerization of biomass-derived monomers limonene oxide (LO), tricyclic anhydride (TCA), which had better properties than PLLA hard segments, where OL was derived from waste citrus peel [8]. Seven ABA-type polyesters were synthesized by a "one-pot" method through the selective polymerization of OL, TCA (1-4) and ε-decalactone using a single catalyst. These polyesters exhibited tunable peel strength (4-13 N/cm) and outperformed commercial adhesive or bio-based polyester formulations, and more importantly, did not require tackifiers or various additives. The authors further investigated the hydrolysis reaction of these polyesters and found that they degrade rapidly within 5 h of immersion in organic acid solutions at 60°C, with a mass loss of more than 98% of the polymer. Since these block polyesters can be used as pressure-sensitive adhesives without the addition of tackifiers and various additives, they are expected to be recyclable for reprocessing.

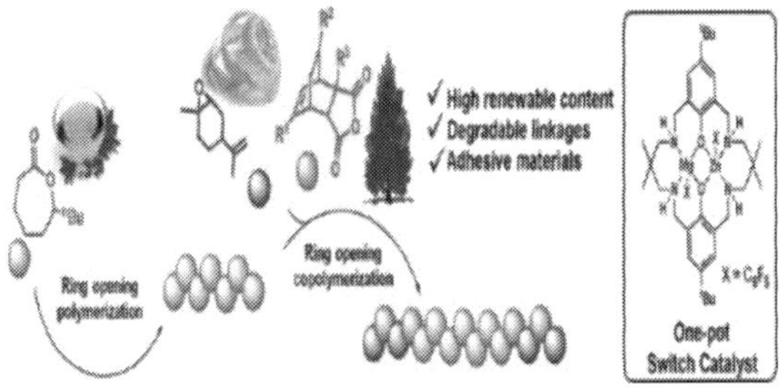

Figure 3. Fully renewable ABA type polyester synthesis [8]. Reprinted with permission from ref. 8. Copyright (2020) American Chemical Society.

3. Degradable pressure-sensitive adhesives derived from lignin

Lignin is the most abundant biopolymer in nature after cellulose, and it is attractive to use lignin as a raw material for pressure-sensitive adhesives to replace existing petroleum-based pressure-sensitive adhesives. Wang and Epps et al. reported a method to develop pressure-sensitive adhesives with high performance by obtaining chemicals directly from biomass [9]. The poplar was depolymerized in methanol solution using commercial Ru/C catalysts, and two aromatic compounds, 4-propylsyringol and 4-propylguaiacol, were obtained in a few purification steps (vacuum filtration and extraction), using the two aromatic compounds as hard segments, which allowed the synthesis of tri-block polymers with biologically derived poly(n-butyl acrylate) as intermediate segments. The polymers exhibit excellent adhesive properties with a 180° peel force of 4 N/cm and an adhesion force of 2.5 N/cm, comparable to many commercial pressure-sensitive adhesives and without the addition of any tackifiers or other additives.

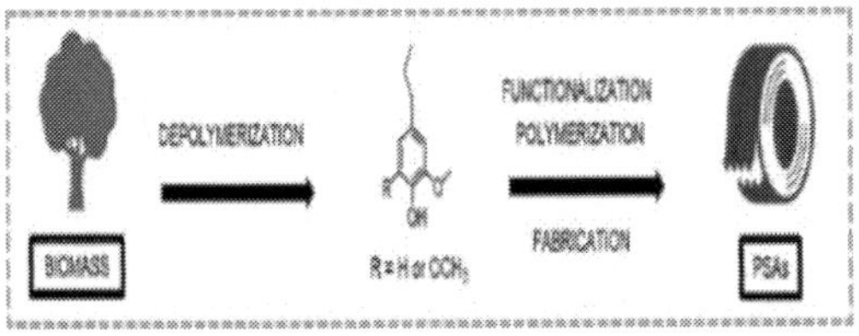

Figure 4. Monomers obtained from real biomass, not model compounds or lignin mimics [9]. Reprinted with permission from ref. 9. Copyright (2018) American Chemical Society.

The use of lignin, a non-toxic renewable material, as an additive to modify the performance of pressure-sensitive adhesives is an option to achieve sustainability, and the addition of new materials to the formulation usually results in significant changes in performance compared to conventional products, Nasiri et al. used acetyl bromide and acetic acid, methacryloyl chloride and maleic anhydride for modification to reduce the viscosity of latex [10]. Moreover, both the acrylic lignin and the maleated lignin obtained through the modification were effective in simultaneously improving the adhesion and shear strength in acrylic adhesive formulations while maintaining a constant peel strength.

4. Conclusion

Research on biomass-sourced pressure-sensitive adhesives has been one of the hot topics in recent years, with the main objective of developing a green material that can replace existing petroleum-based pressure-sensitive adhesives. With the global emphasis on environmental protection and sustainable development, the research and development of bio-sourced pressure-sensitive adhesives has become a trend. In this field, researchers have successfully developed a series of bio-sourced pressure-sensitive adhesives with excellent properties through the extraction and modification of various bio-based raw materials. This paper discusses the recent advances in biomass-derived pressure-sensitive adhesives and summarizes the methods reported in the literature in recent years for the preparation of pressure-sensitive adhesives using bio-based feedstocks. The performance of these pressure-sensitive adhesives is comparable to or even exceeds that of current commercial pressure-sensitive adhesives, making it possible to replace existing petroleum-based pressure-sensitive adhesives. Such substitution can reduce the dependence on petroleum resources and is also in line with environmental protection concepts.

References

1. S. Sun, M. Li, and A. Liu, A review on mechanical properties of pressure sensitive adhesives, International Journal of Adhesion and Adhesives 41 (2013), 98-106.
2. E.P. O'Brien, L.T. Germinario, G.R. Robe, T. Williams, D.G. Atkins, D.A. Moroney, and M.A. Peters, Fundamentals of hot-melt pressure-sensitive adhesive tapes: the effect of tackifier aromaticity, Journal of Adhesion Science and Technology 21 (2007), 637-661.
3. I. Márquez, N. Paredes, F. Alarcia, and J.I. Velasco, Adhesive Performance of Acrylic Pressure-Sensitive Adhesives from Different Preparation Processes, in: Polymers, 2021.
4. R. Vendamme, N. Schüwer, and W. Eevers, Recent synthetic approaches and emerging bio-inspired strategies for the development of sustainable pressure-sensitive adhesives derived from renewable building blocks, Journal of Applied Polymer Science 131 (2014).
5. Y. Zhu, C. Romain, and C.K. Williams, Sustainable polymers from renewable resources, Nature 540 (2016), 354-362.
6. S. Lee, K. Lee, Y.-W. Kim, and J. Shin, Preparation and Characterization of a Renewable Pressure-Sensitive Adhesive System Derived from ε-Decalactone, l-Lactide, Epoxidized Soybean Oil, and Rosin Ester, ACS Sustainable Chemistry & Engineering 3 (2015), 2309-2320.
7. H.J. Kim, K. Jin, J. Shim, W. Dean, M.A. Hillmyer, and C.J. Ellison, Sustainable Triblock Copolymers as Tunable and Degradable Pressure Sensitive

Adhesives, ACS Sustainable Chemistry & Engineering 8 (2020), 12036-12044.

8. T.T.D. Chen, L.P. Carrodeguas, G.S. Sulley, G.L. Gregory, and C.K. Williams, Bio-based and Degradable Block Polyester Pressure-Sensitive Adhesives, Angewandte Chemie International Edition 59 (2020), 23450-23455.
9. S. Wang, L. Shuai, B. Saha, D.G. Vlachos, and T.H. Epps, III, From Tree to Tape: Direct Synthesis of Pressure Sensitive Adhesives from Depolymerized Raw Lignocellulosic Biomass, ACS Central Science 4 (2018), 701-708.
10. A. Nasiri, J. Wearing, and M.A. Dubé, The use of lignin in emulsion-based pressure-sensitive adhesives, International Journal of Adhesion and Adhesives 100 (2020), 102598.

Temporal and Spatial Distribution Characteristics of Flood Disasters with Different Intensities in Arid-Semiarid Region in Northern Xinjiang, China

Xi Wang[1, *], Tao Zhang [1], Yun Wang [2], Xiuquan Huang [1], Hanxiang Gong[1,3], Baoxin Chen [1]

[1] Faculty of Humanities and Social Sciences, Macao Polytechnic University, Macao, 999078, PR China
[2] Gansu Weather Modification Office, Lanzhou 730000, PR China
[3]The Second Affiliated Hospital of Guangzhou Medical University, Guangzhou 510180, Guangdong, PR China

Abstract. Floods have devastating environmental and socioeconomic consequences. Flood disaster management is critical for the long-term management of water resources and agriculture in arid areas. Using the data on the number of deaths, destroyed homes, collapsed sheds, livestock deaths, and the crops affected area caused` by 1394 regional floods in Northern Xinjiang (As a typical arid-semiarid region of NorthWest China) from 1981 to 2019, the ratio-weight method and the dimensionless linear summation method was used to calculate the actual disaster damage exponent of flood disaster events. The damage exponent is further used to categorize the severity of disaster events into four grades: normal, moderate, severe, and extremely severe. The analysis results indicated that Bortala Mongolian Autonomous Prefecture experienced the highest frequency of catastrophic occurrences, while Ili River Valley experienced the highest intensity. The flood happened most often in Wenquan County. And Yining County suffered the worst calamity. The occurrence frequency and intensity of disasters from Grades 1 to 3 exhibited a unimodal distribution, with the majority and most vital occurrences in July, whereas the occurrence frequency and intensity of Grade 4 disasters followed a bimodal distribution, with the most occurrences in July and the strongest in April. The annual frequency of heavy rain and flood disasters in Northern Xinjiang increased by 7.7 times every 10 years, indicating a significant linear increase trend. The yearly occurrences of Grades 1 and 2 also increased linearly, by 5.2 and 2.0 times per decade, respectively. There was no linear tendency to increase or decrease flood disasters in Grades 3 and 4. The analysis and division of actual disaster damage are conducive to flood risk management, efficient prevention, and reducing disaster losses.

Keywords: Northern Xinjiang, flood, disaster exponent, disaster intensity, temporal and spatial distribution

1. Introduction

Climate change, solar activity, and the El Nio-Southern Oscillation all impact precipitation extremes and flooding events [1]. Extreme precipitation events have grown dramatically in size and frequency on both a global and regional scale [2], but it is seasonal, regional, and influenced by human activities. During the Paleocene-Eocene Thermal Maximum, extreme precipitation became more severe in tropical and subtropical Africa and parts of South America, while average annual precipitation and extreme rates increased at high latitudes. The authors concluded that rainfall in tropical East Africa became more episodic and extreme, possibly more seasonally distributed. And extreme precipitation rate is particularly sensitive over much of tropical Africa, especially during the rainy season. Extreme precipitation is exacerbated by increasing soil erosion, particularly in North Africa and the Tethys Ocean. Because of its extreme continentality, the continental United States, like Northwest China, experiences the smallest increase in global specific humidity in the western United States, which is arid but has extreme precipitation [3,4]. The number of heavy precipitation events in the United States increased by 20% in some densely populated areas [5].

The most developed region in China, with a dense population and large agricultural area, often experiences flash floods and river floods caused by extreme rainfall events. The frequency of floods in China significantly shows an increasing trend. In the recent 30 years, 8 flooding events have occurred on average every year (EM-DAT, http://www.emdat.be/) [6]. Changes in the frequency of extreme precipitation events are therefore highly relevant, especially where the resilience of existing infrastructure is already exceeded. Flood disasters have destructive solid power, often causing severe disasters such as death and injury of humans and animals, collapse

* Corresponding author: xwang@mpu.edu.mo

of houses, destruction of farmland, and collapse of bridges [7].

According to official data released by China government, from 2000 to 2019, the average number of people affected by floods in the country reached 114 million yearly, and the average annual death and missing persons were 1,195. The disaster area is 9.88 million hectares, and the direct economic loss accounts for 0.48% of GDP [8]. According to data from the China Ministry of Water Resources, 148 million people were affected by the flooding in 2020, and the direct economic losses from these floods were 178.96 billion Yuan, with these numbers being respectively 12.7% and 15.5% higher than the average values of the previous five years [9]. Heavy rainfall events frequently occur in Xinjiang, causing severe disasters [10,11]

Statistics show that, from 1961 to 2019, 3,220 flood disasters devastated 86 counties (cities) in Xinjiang virtually annually, resulting in a total of 1,365 fatalities, more than 51.7×10^4 destroyed homes, 16.0×10^4 collapsed sheds, 156.3×10^4 dead livestock, and covering a damaged farmland area of 376.9×10^4 hm^2 [12].

Some scholars have conducted preliminary research on flood disasters in Xinjiang. Zhang Q. et al. showed that higher probabilities of flood disasters are found in Northern Xinjiang than in Southern Xinjiang [13]. Sun G. et al. used district data from 1901 to 2010 to investigate the spatial distribution characteristics of extreme hydrological occurrences in Xinjiang, China. The frequency distribution was symmetric along the Tianshan Mountains, with an even distribution in Junggar Basin and Tarim Basin. Compared to the southeast, the northwest had a higher frequency of flood disasters. The incidence was highest in the west Tianshan Mountains and gradually declined south-eastward [14]. Based on the daily precipitation data of 44 national meteorological stations in Southern Xinjiang from 2010 to 2019, Yanying, W. et al. evaluated flood disasters and economic data, as well as the spatial and temporal changes in flooding situations in Southern Xinjiang [15]. According to Wang H. et al., the high precipitation is located mainly in Northern Xinjiang, the western part of the Tianshan Mountain Area. High precipitation values (300-530 mm) range from 15.1 to 505.8 mm annually. Recently the importance of changes in temporal patterns of heavy precipitation events on agriculturally dominated areas flooding has been highlighted [17,18]. Northern Xinjiang has a relatively humid climate, abundant water sources, relatively fertile land, a relatively dense population, and highly concentrated areas of productivity.

This article aims to provide a different perspective to measure the hazard of flood disasters and reveal the temporal and spatial distribution of actual disaster damage caused by flood disasters. The characteristics are conducive to formulating corresponding defense measures and are significant to protecting people's lives and disaster prevention and mitigation.

This research goal considers:

(1) To calculate the actual disaster damage exponent by statistical methods, based on the data of the number of deaths, destroyed homes, collapsed sheds, dead livestock, and the crops affected area caused by flood disasters in Northern Xinjiang from 1981 to 2019;

(2) To classify the intensity of the disaster events.;

(3) To analyze the temporal and spatial variation characteristics of storm and flood disasters with different intensities in Northern Xinjiang;

(4) And to explain the reasons for the interannual variation of disasters, using the number of days above heavy rain and rainstorms.

2. Experimental

2.1 Overview of the Study Area

Northern Xinjiang refers to the area north of the Tianshan Mountains in Xinjiang(79°57' - 91°32'E , 43°23' - 49°10'N), which is mainly mountainous, basin and Gobi, reaching the north slope of the Tianshan Mountains and the Yili Valley in the south, the Altai Mountains in the north, and the Junggar Basin in the middle, with a total area of about 5.95×10^5 km^2 (Figure 1). Flood disasters can be triggered by intense thunderstorms, tropical cyclones, large low-pressure systems, monsoons, ice plugs, or snowmelt [19]. In Northern Xinjiang, due to the interception and uplift of water vapor by tall mountains, the precipitation mainly occurs in the mountains, and when the precipitation reaches, it will lead to a flood of a certain magnitude [20]. In addition, the underlying gobi saline and alkaline land are in the majority, causing severe desertification and a fragile ecological environment, which can easily lead to flood disasters [21]. As a typical arid and semi-arid area in northwest China, Northern Xinjiang has a temperate continental climate, with considerable precipitation variation and uneven spatial distribution. The average annual precipitation of nearly 60a is 195mm, and 300-1000mm in mountainous and surrounding areas [22]. Floods in Xinjiang mainly originate from mountain areas and are induced by ice-snow melt and precipitation. Due to the increase in heavy rain in summer, floods have occurred more frequently since the late 1980s.

Generally, Northern Xinjiang is more prone to drought than Southern Xinjiang, with more floods due to increased summer torrential rain [23]. Since the 1980s, the global climate has shown a trend of warming and wetting. The increase of declining water in the warming background is conducive to alleviating the drought in Northern Xinjiang, but it also brings more challenges to disaster risk control.

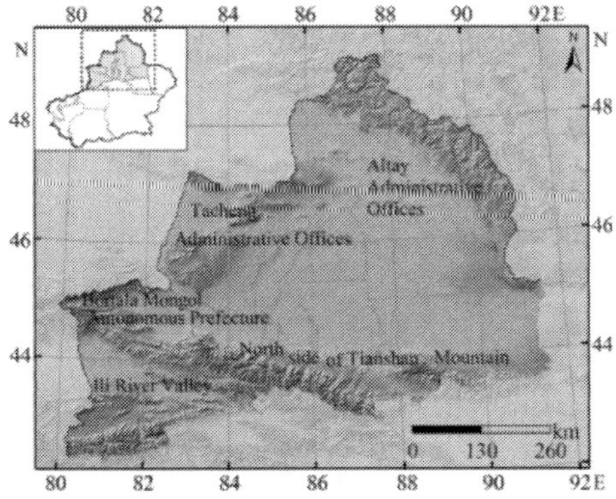

Fig. 1. Topography and administrative map of Northern Xinjiang.

2.2 Data

Data were based on the Xinjiang Uygur autonomous region department of flood disasters information from 1981 to 2019. Thirty-eight counties (city), 1394 disaster records in Northern Xinjiang contain occurrence time (month), area (county/city), deaths (persons), destroyed homes (houses), collapsed shed (seat), livestock deaths (head), crops affected area (hm^2). If there is a rainstorm and flood disaster in a county (city) area, the number of flood disasters in the county (city) is recorded as 1. According to the daily precipitation data of 33 meteorological stations in Northern Xinjiang from 1981 to 2019 provided by the Meteorological Bureau of Xinjiang Uygur Autonomous Region, the number of heavy rain days (12.1- 24.0mm) and the number of annual rainstorm days (≥24.1mm) was counted.

2.3 Methods

Because the disaster elements that constitute disaster events include 5, they have different units, and the strength of different disasters cannot be compared. In order to compare the strength of disaster events, it is necessary to construct an actual disaster damage exponent (Z_i) that can comprehensively express the five disaster factors. The weight of each disaster element was determined by the ratio method, and then the dimensionless linear combination method was used to obtain Z_i [24,25].

Suppose that each disaster factor consists of N sample, N=1394, so the disaster factor evaluation matrix $X_{N\times5}$ can be obtained. The calculation formula of the actual disaster damage exponent Z_i is:

$$Z_i = \sum_{j=1}^{5} a_j \frac{X_{i,j}}{\bar{X}_j} \quad \text{“(1)}$$

In the above equation, i=1,2, ···, N, j =1,2, ···, 5, X_{ja}, $\bar{X}_j$, a_j represent the maximum value of j, average, and weight of the first disaster element. The calculation formula of a_j:

$$a_j = \frac{\sum_{i=1}^{n} X_{i,j}}{\sum_{j=1}^{5}\sum_{i=1}^{n} \frac{X_{i,j}}{X_{ja}}} \quad (2)$$

The aforementioned calculation formula (2) reveals that the maximum value is utilized to establish the weight, which ensures the equivalence of the five disaster elements. The average value is used to calculate the actual disaster damage exponent so that the actual disaster damage exponent has a discrete type.

After the calculation of formula (1), the weight coefficient, the average, and the maximum value of five disaster factors can be noted in Table 1.

The closeness of the two can be determined by the size of the correlation coefficient between the actual disaster damage exponent and the disaster elements. Disaster damage exponent and the number of deaths, destroyed homes, collapsed sheds, livestock deaths, and crops affected area between the correlation coefficient of 0.23, 0.90, 0.89, 0.46, and 0.29, respectively, all passed the significance level of 0.01, which shows that the disaster damage exponent can not only express comprehensive five disaster elements, and the size of the damage exponent reflects the strength of the disaster event.

Table 1. Weight coefficient, average and maximum value of disaster factors.

	Deaths (persons)	Destroyed homes (houses)	Collapsed sheds (seats)	Livestock deaths (heads)	Crops affected area (hm^2)
Weight coefficient	0.18	0.33	0.25	0.09	0.16
Average value	0.3	148.7	31.3	470.1	940.9
Maximum value	24	6010	1650	70000	80000

The percentile method determines the damage exponent grade [26]. Percentile is a kind of position exponent. According to the range of threshold change, the disaster is divided into four grades (specific results are listed in Table 2).

Table 2. Grading criteria of rainstorm and flood disasters in Northern Xinjiang.

Percentile r (%)	Disaster exponent Z_i	Disaster grade
r≤50	$Z_i \le 0.13173$	Mild (Grade 1)
50.1≤r≤75	$0.13174 \le Z_i \le 0.56716$	Moderate (Grade 2)
75.1≤r≤90	$0.56717 \le Z_i \le 1.72884$	Severe (Grade 3)
r≥90.1	$Z_i \ge 1.72885$	Extremely severe (Grade 4)

3. Results

3.1 Spatial distribution of flood disaster in Northern Xinjiang

In 38 counties (cities) in Northern Xinjiang, from 1981 to 2019, the frequency and intensity of flood disasters showed obvious regional differences. And the cumulative number of disasters that occurred on average over these 39 years in each region was as follows: Boltala Mongolian Autonomous Prefecture (70), Altay region (47), Yili Valley (45), northern Tianshan Mountains (25), Tacheng Area (21), which showed that the disaster events occurred most in Bortala Mongolian Autonomous Prefecture, and the least occurred in Tacheng Area; among them,

https://doi.org/10.1051/e3sconf/202339401009

Wenquan County had the most flood disasters, with a total of 103 occurrences (Figure 2a). Regional average annual damage exponents were as follows: Yili Valley (2.354), Altay (0.819), Tcheng (0.450), northern Tianshan (0.384), Bortala Mongolian Autonomous Prefecture (0.379). The intensity of disaster events in Yili Valley is the strongest, and Bortala Mongolian Autonomous Prefecture is the weakest. With the damage exponent of 3.195, Yining County has the highest disaster intensity out of the 38 counties (cities) (Figure 2b).

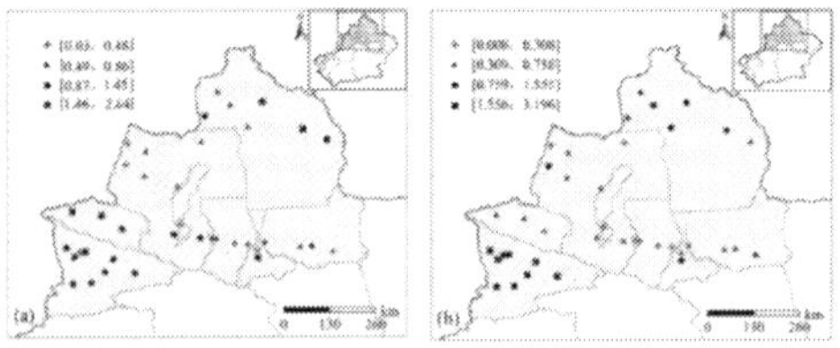

Fig. 2. Spatial distribution of annual average occurrence frequency and annual average disaster exponent of rainstorm and flood disasters in Northern Xinjiang. a) average annual occurrences (times), b) average annual damage exponent.

The number of flood disasters ranging from Grades 1 to 4 also demonstrated apparent regional variances. Grades 1 and 2 occur most frequently in Bortala Mongolia Autonomous Prefecture and least frequently in Tacheng Area; Grades 3 and 4 occur most frequently in Yili Valley; Flood disasters of Grade 3 occur least frequently in the northern side of Tianshan Mountains; Grade 4 occur least frequently in Bortala Mongolia Autonomous Prefecture. The county(cities) with the most occurrences of Grades 1 to 4 are Wenquan County (66 cumulative occurrences in 39 years), Wenquan County (25), Altay City (17), and Gongliu County (12).

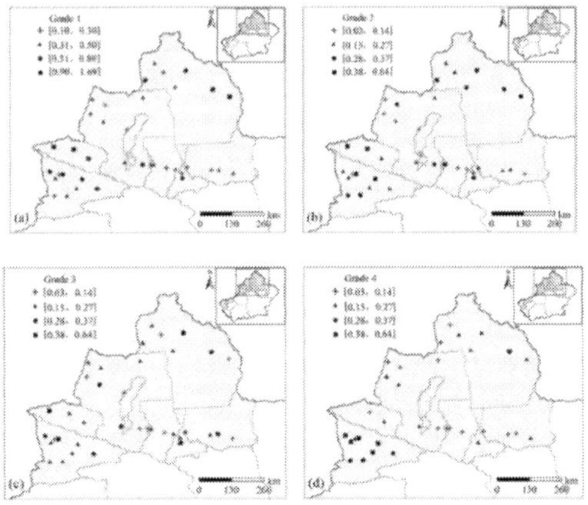

Fig. 3. Spatial distribution of average annual occurrence frequency of Grades1-4 rainstorm and flood disasters in Northern Xinjiang.

3.2 Seasonal and monthly distribution of flood disasters in Northern Xinjiang

The occurrence frequency and intensity of flood disasters show diversity. Grade 1 disasters occur mainly from May to August, and Grade 2 and 3 disasters mainly occur from May to July. Grades 1 to 3 disasters show a unimodal distribution, with the peak month occurring in July, while Grade 4 was bimodal, with the peak month in July and the second peak month in April (Figure 4a, b). The intensity of disasters from Grades 1 to 3 showed a unimodal distribution, with the peak month appearing in July; while the intensity of Grade 4 disasters showed a bimodal distribution, with the peak month appearing in April and the second peak month appearing in June (Figure 4c, d). The monthly distribution of the occurrences of Grades 1-4 flood disasters is very consistent with the monthly distribution of heavy rain days and rainstorm days, indicating the occurrence of flood disaster events in Northern Xinjiang may cause by continuous heavy rain and rainstorms.

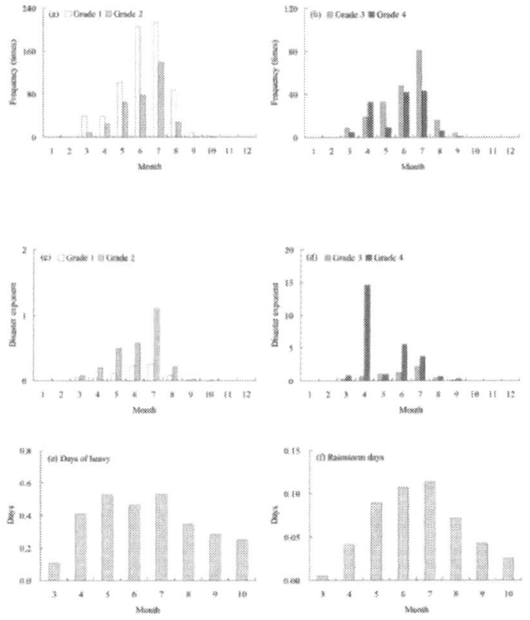

Fig. 4. Seasonal variation of occurrence frequency (a, b) and disaster exponent (c, d) of rainstorm and flood disasters in Northern Xinjiang. Seasonal variation of days of heavy (e) and rainstorm days in Northern Xinjiang.

3.3 Inter-annual occurrence distribution change of flood disaster

According to Figure 5, the annual occurrence of flood disasters in Grade 1 and 2 grew more dramatically from 1981 to 2019, by 5.2 times for Grade 1 and 2.0 times for Grade 2 every 10 years. Flood disasters in Grades 3 and 4 occurred on average 5.4 and 3.6 times yearly, respectively. However, the number fluctuated around the climate average.

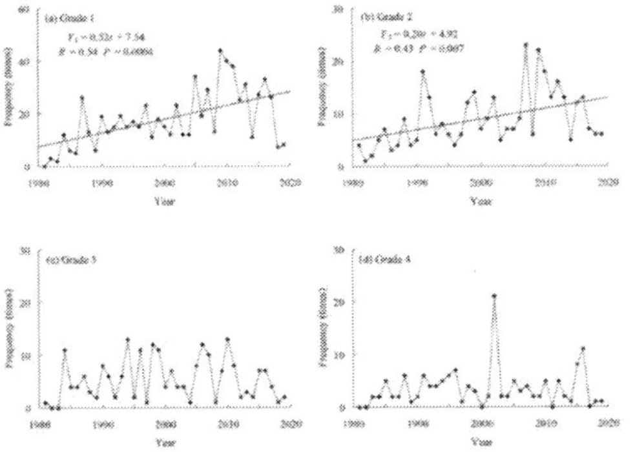

Fig. 5. Inter-annual variations in the number of occurrences of rainstorms and floods of Grades 1-4 in Northern Xinjiang. R indicates the correlation coefficient, and P indicates the Grade of reliability. The straight line indicates a linear trend in the figure.

As illustrated in Figure 6a, from 1981 to 2019, the annual occurrence of flood disasters (Y) in Northern Xinjiang increased by 7.7 times every 10 years, indicating a significant linear increase. The inter-annual variation of the occurrence number of heavy rain days (Rd) and the number of rainstorm days (Bd) from March to October both showed a significant linear increase trend (Fig. 6b, 6c); the correlation coefficient between Y and Rd was 0.39 (P=0.01), and the correlation coefficient between Y and Bd was 0.55 (P=0.0003). Therefore, the annual increase in the number of heavy rain days (Rd) and rainstorm days (Bd) from March to October results in an annual increase in the occurrence of flood disasters.

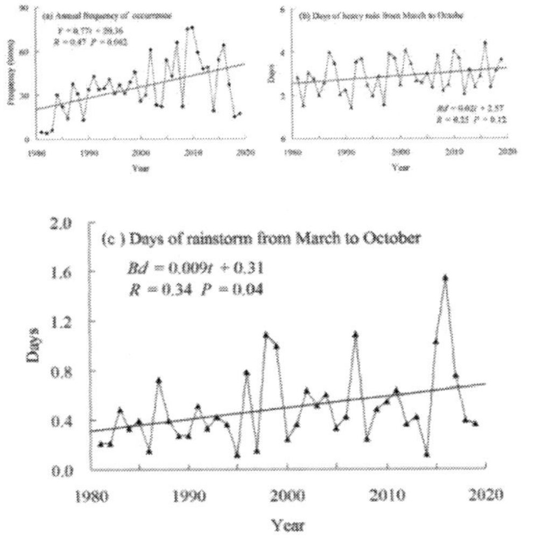

Fig. 6. Annual frequency of rainstorm and flood disasters in Northern Xinjiang (a), interannual variation of heavy rain days (b), and rainstorm days (c) from March to October. R indicates the correlation coefficient, and P indicates the Grade of reliability. The straight line indicates a linear trend in the figure.

4. Discussion and conclusions

This extreme rainstorm caused severe floods and seriously adverse socioeconomic impacts. In some countries exposed to tropical cyclone impacts (notably the United States and China), it is mainly coastal and immediate inland locations that bear the brunt of the storm-related surge and wind losses, while the economic activity of the entire nation spans a much larger area [27]. However, the lack of attention in arid and semi-arid areas and its damage cannot be underestimated.

It is evident that the frequency and intensity of flood disasters in Northern Xinjiang present prominent regional characteristics in terms of spatial distribution. In Bortala Mongolian Autonomous Prefecture, flood disaster events were most frequent, while the severity of these events was highest in the Ili River Valley. For counties (cities), flood disasters in Wenquan County occurred the most, and Yining County had the most vigorous disaster intensity. Grade 1 and 2 flood disasters appear most frequently in Bortala Mongolian Autonomous Prefecture, and Grades 3 and 4 appear most frequently in Ili Valley. And the counties (cities) with the highest frequency of Grade 1 to Grade 4 disasters are respectively Wenquan County, Wenquan County, Altay City, and Gongliu County.

In the seasonal and monthly distribution, disaster occurrence frequency and intensity showed diversity. The occurrence frequency and intensity of Grade 1 to 3 disasters are in unimodal distribution, with peak events occurring in July and the most frequently occurring in July; the frequency and intensity of Grade 4 disasters are in a bimodal distribution, with the most frequent month in July, the second most frequent month in April. And the month with the most vigorous disaster intensity is April, and the second strongest month is June.

Regarding inter-annual changes, the annual occurrence of flood disasters in Northern Xinjiang from 1981 to 2019 showed a significant linear increase, increasing by 7.7 times every 10 years. The annual occurrence of disasters is closely related to the number of heavy rain and rainstorm days from March to October. The annual increase in heavy rain and rainstorm days may result in the annual increase in flood disasters. The annual occurrence number of flood disasters of different grades showed differences. Both Grade 1 and Grade 2 showed a significant linear increase trend, increasing by 5.2 and 2.0 times per 10 years, respectively. However, the occurrence number of Grade 3 and Grade 4 annual occurrences fluctuated around the climate average, showing no linear trend of increase or decrease.

The Chinese government has actively responded to global disaster risk reduction strategies and made active explorations in the discovery of flood risk management. Over the past 30 years, China has recognized that disaster risk reduction measures are investing in a more secure society. Moreover, Chinese authorities are increasingly emphasizing local disaster prevention measures. Local governments can better prepare for disasters and participate more in disaster reduction operations, and disaster prevention and relief have saved many lives [28]. The impact of this shift was particularly evident in the 2020 flood disasters, and China is also ramping up investment in critical infrastructure to enable it to mitigate flooding better. It is still essential to continue efforts in some areas, including developing an in-depth understanding of the laws of disasters, gradually shifting from disaster response and preparation to risk prevention and mitigation, and enhancing emergency response capabilities, which can improve systems for disaster mitigation.

Acknowledgments

This work was supported by the Macao Polytechnic University Research Project Fund (Fund number: RP/FCHS-03/2022) and the Jin Feng Hua Yun Development Fund Project (Fund number: Hyj202201).

Conflict of interest

The authors declare no conflict of interest.

References

1. Tabari H. (2020). Climate change impact on flood and extreme precipitation increases with water availability Sci. Rep. 10 13768. https://doi.org/10.1038/s41598-020-70816-2.
2. Yang T., Li Q., Chen X., De Maeyer P., Yan X., Liu Y., Zhao T., Li L., (2020). Spatiotemporal variability of the precipitation concentration and diversity in Central Asia. Atmos. Res. 241, 104954. https://doi.org/10.1016/j.atmosres.2020.104954.
3. Carmichael M. J., Pancost R. D., & Lunt D. J. (2018). Changes in the occurrence of extreme precipitation events at the Paleocene–Eocene thermal maximum. Earth and Planetary Science Letters, 501, 24-36. https://doi.org/10.1016/j.epsl.2018.08.005.
4. Loptson, C. A., Lunt, D. J., & Francis, J. E. (2014). Investigating vegetation–climate feedbacks during the early Eocene. Climate of the Past, 10(2), 419-436. https://doi.org/10.5194/cp-10-419-2014.
5. Myhre G., Alterskjær K., Stjern C. W., Hodnebrog Ø., Marelle L., Samset B. H., Sillmann J., Schaller N., Fischer E., Schulz M. & Stohl A. (2019). Frequency of extreme precipitation increases extensively with event rareness under global warming. Scientific reports, 9(1), 1-10. https://doi.org/10.1038/s41598-019-52277-4.
6. Wei L., Hu K. H., & Hu X. D. (2018). Rainfall occurrence and its relation to flood damage in China from 2000 to 2015. Journal of Mountain Science, 15(11), 2492-2504. https://doi.org/10.1007/s11629-018-4931-4.
7. Wu X., Jiang Y., Yu X., Zhang. L., & Li Y., (2019). Study on Risk Zoning of Mountain Flood Disaster in the Yigeziya River of Xinjiang Based on Flood Area Inundation Model. Journal of Arid Meteorology, 37(4), 663.
8. Ministry of Water Resources of the People's Republic of China. China Flood and Drought Disaster Prevention Bulletin 2020 [EB/OL]. (2021-12-08) [2022-04-15].
   http://www.mwr.gov.cn/sj/tjgb/zgshzhgb/202112/t20211208_1554245.html
9. Duan W.L, He B., Nover D., et al., Floods and associated socioeconomic damages in China over the last century, Nat. Hazards 82 (1) (2016) 401–413. (in Chinese)
10. Zhai P., Yu R., Guo Y., Li Q., Ren X., Wang Y., Xu W., Liu Y., & Ding Y. (2016). The strong El Niño of 2015/16 and its dominant impacts on global and China's climate. Journal of Meteorological Research, 30(3), 283-297. https://doi.org/10.1007/s13351-016-6101-3.
11. Zhang L., Liu Y., Zhan H., Jin M., & Liang X. (2021). Influence of solar activity and El Niño-Southern Oscillation on precipitation extremes, streamflow variability and flooding events in an arid-semiarid region of China. Journal of Hydrology, 601, 126630. https://doi.org/10.1016/j.jhydrol.2021.126630.
12. Chen Y. & Ma Y. (2021). Temporal and spatial variation characteristics of different grades of rainstorm and flood disasters in Xinjiang. Geography of Arid Regions (06), 1515-1524. (in Chinese)
13. Zhang Q., Singh V. P., Li J., Jiang F., & Bai Y. (2012). Spatio-temporal variations of precipitation extremes in Xinjiang, China. Journal of Hydrology, 434, 7-18. https://doi.org/10.1016/j.jhydrol.2012.02.038.
14. Sun G., Chen Y., Li W., Pan C., Li J., & Yang Y. (2013). Spatial distribution of the extreme hydrological events in Xinjiang, northwest of China. Natural hazards, 67(2), 483-495. https://doi.org/10.1007/s11069-013-0574-5.
15. Wei Y., GONG. M., & LI H. (2021). Spatial-temporal distribution and influence of disastrous rainstorm in Southern Xinjiang during 2010-2019. Journal of Arid Meteorology, 39(06), 930. (in Chinese)
16. Wang H., Chen Y., & Chen Z. (2013). Spatial distribution and temporal trends of mean precipitation and extremes in the arid region, northwest of China, during 1960–2010. Hydrological Processes, 27(12), 1807-1818.
   https://doi.org/10.1002/hyp.9339.
17. Zweifel L., Meusburger K., & Alewell C. (2019). Spatio-temporal pattern of soil degradation in a Swiss Alpine grassland catchment. Remote sensing of environment, 235, 111441. https://doi.org/10.1016/j.rse.2019.111441.
18. Chandrashekar V. D., & Shetty A. (2018). Trends in extreme rainfall over ecologically sensitive Western Ghats and coastal regions of Karnataka: an observational assessment. Arabian Journal of Geosciences, 11(12), 1-13.
   https://doi.org/10.1007/s12517-018-3700-6.

19. Li S., Tang Q., Lei J., Xu X., Jiang J., & Wang Y. (2015). An overview of non-conventional water resource utilization technologies for biological sand control in Xinjiang, northwest China. Environmental Earth Sciences, 73(2), 873-885. https://doi.org/10.1007/s12665-014-3443-y.
20. Li G., Yu Z., Wang W., Ju Q., & Chen X. (2021). Analysis of the spatial distribution of precipitation and topography with GPM data in the Tibetan Plateau. Atmospheric Research, 247, 105259. https://doi.org/10.1016/j.atmosres.2020.105259.
21. Shi P. (2019). Disaster risk science. Springer.
22. Yang L., Liu J. Some advances of water vapor research in Xinjiang. Journal of Natural Disasters, 2018, 27(02): 1-13. (in Chinese)
23. Wu M., Chen Y., & Xu C. (2015). Assessment of meteorological disasters based on information diffusion theory in Xinjiang, Northwest China. Journal of Geographical Sciences, 25(1), 69-84. https://doi.org/10.1007/s11442-015-1154-2.
24. Ling H., Deng X., Long A., & Gao H. (2016). The multi-time-scale correlations for drought–flood index to runoff and North Atlantic Oscillation in the headstreams of Tarim River, Xinjiang, China. Hydrology Research, 48(1), 253-264. https://doi.org/10.2166/nh.2016.166.
25. Chen B., Chen K., Wang X., Wang X. Spatial and temporal distribution characteristics of rainstorm and flood disasters around Tarim Basin. Pol. J. Environ. Stud. Vol. 31, No. 3 (2022), 2029-2037. https://doi.org/10.15244/pjoes/143579.
26. Jia J.P. Statistics. Tsinghua University Press. 68, 2004.
27. Pielke R. (2021). Economic 'normalisation' of disaster losses 1998 – 2020: A literature review and assessment. Environmental Hazards, 20(2), 93-111. https://doi.org/10.1080/17477891.2020.1800440.
28. Jia H., Chen F., Pan D., Du E., Wang L., Wang N., & Yang A. (2022). Flood risk management in the Yangtze River basin—Comparison of 1998 and 2020 events. International Journal of Disaster Risk Reduction, 68, 102724.
https://doi.org/10.1016/j.ijdrr.2021.102724.

Development process and trend of medium and low yield field reconstruction in Shaanxi Province

Na Lei[1,2,3,4,*], Yanan Li[1,2,3,4], Jing Zhang[1,2,3,4]

[1]Shaanxi Provincial Land Engineering Construction Group Co., Ltd., Xi'an, China
[2]Institute of Land Engineering and Technology, Shaanxi Provincial Land Engineering Construction Group Co., Ltd., Xi'an, China
[3]Key Laboratory of Degraded and Unused Land Consolidation Engineering, the Ministry of Land and Resources, Xi'an, China
[4]Shaanxi Provincial Land Consolidation Engineering Technology Research Center Xi'an, China

Abstract. The transformation of medium and low yield grain fields is an important part of cultivated land regulation, which is basically consistent with the requirements of high standard farmland construction. The cultivated land consolidation has experienced from overcoming the single limiting factor to exploring the integrated factors combining the adjustment of ownership and then to the ecological landscape construction target, which has put forward different requirements on the engineering construction, land capacity improvement and management organization innovation integration of cultivated land consolidation. At present, the concrete measures of land leveling, irrigation and drainage, field road and forest protection network have been studied in the field engineering construction of medium and low yield grain fields in Shaanxi province. In the aspect of soil fertility improvement, effective measures such as ditch treatment, soil fertility improvement, soil layer recombination and crop optimization were studied. In the aspect of management organization innovation, the research of property right model combining different yield reform mode and different cultivated land consolidation mode has promoted the optimization and transformation of medium and low yield grain fields in different degrees.

Key words: Medium and low yield fields; engineering construction; land capacity enhancement; landscape construction; Shaanxi Province

1. Introduction

By the end of 2018, there were 60.7555 million mu of cultivated land in Shaanxi Province, and the area of medium and low yield land reached 42 million mu, accounting for 70% of the total area of cultivated land. It is certain that the yield of existing cultivated land will be significantly increased by comprehensive management, improvement of production conditions and increase of yield per unit area and multiple cropping index. In the cultivated land of Shaanxi province, the proportion of medium and low yield grain fields is more serious, especially distributed in the wind-sand beach area along the Great Wall of northern Shaanxi, loess plateau area of northern Shaanxi, Weibei tableland area, Guanzhong plain area and mountainous and hilly area of southern Shaanxi, which runs through the topographic characteristics of various regions of the province and different ecological zones, forming serious medium and low yield grain field barriers. At present, government departments and enterprises have carried out some effective exploration and treatment, but there are still the following deficiencies: first, there is a lack of engineering construction standards for the corresponding ecological area, and no effective zoning classification; Secondly, it fails to combine engineering construction with soil fertility improvement under different ecological conditions and lacks key technologies and integrated applications in soil fertility improvement. Third, it fails to effectively combine engineering construction, land capacity improvement and management organization integration and innovation, leading to frequent management and protection problems after engineering construction.

Therefore, how to effectively carry out the reconstruction project of medium and low yield grain fields in Shaanxi Province, identify the key limiting factors for the reconstruction and construction of medium and low yield grain fields in different ecological zones, develop the sequence combination of grain fields projects, soil capacity improvement and operation management and protection technology standards based on the characteristics of different ecological zones, and form the integrated system of the reconstruction and construction project of medium and low yield grain fields in different ecological zones in Shaanxi Province. In addition, key pilot projects should be established for effective promotion and demonstration, so as to increase the

* Corresponding author: 619648133@qq.com

quantity, quality and output of grain fields in Shaanxi Province, realize the transformation goal of "high yield, high efficiency and green" medium and low yield grain fields, and guarantee the food security of Shaanxi Province. It is a problem that needs to be focused on and solved at the present stage.

2. Development process and trend of transformation of medium and low yield grain fields

The transformation of medium and low yield grain fields at home and abroad is different from the land consolidation and land consolidation that started first a large number of scholars have done related research on this aspect. Chinese land comprehensive improvement related research self Since the 1980s, it has gone through four stages: initiation, development, evolution and extension replanning (1981-1985), recoordination (1986-1996), reengineering (1997-2009) with the four processes of re-coordination (2010 to now), it has become a national top-level strategy-oriented coordination.

A comprehensive platform and an important lever for promoting modernization, ecological progress, rural revitalization and urban-rural integration [1-2].Land consolidation is to ensure rural development and improve the efficiency of land use. It can increase the area of cultivated land, reduce the fragmentation of cultivated land, and improve agricultural production capacity [3].Overall land consolidation, including the comprehensive treatment of fields, water, roads, forests and villages, agricultural land consolidation, industrial and mining abandoned land reclamation benefits Use, urban and rural inefficient use of land redevelopment. Rural areas of land consolidation activities are mainly cultivated land consolidation, the implementation of cultivated land consolidation, can coordinate nervous people Land relations to alleviate land use conflicts [4]. At present, land consolidation is more limited to engineering and technical attributes, and its core objectives are mainly focused on material aspects such as expanding farmland scale, improving cultivated land quality and optimizing village layout. In some areas, farmland consolidation is only regarded as a means to occupy land space provided by urban construction. Traditional cultivated land consolidation blindly pursues the increase of cultivated land quantity and ignores the service function of ecosystem. It is suggested that the quality of landscape unit area and ecosystem service value should be improved in the process of land consolidation [5]. Scholars argued that the traditional operation mode of relying solely on the government-led "top-down" lacks effective social response and communication mechanism, resulting in the dislocation of the value orientation of the participants, which leads to the contradiction between the original intention of land consolidation supply and the actual situation. Traditional cultivated land consolidation model has a single financial input mechanism and insufficient incentives for project implementation and management subjects, resulting in low overall efficiency. In the traditional cultivated land consolidation mode, the government plays the dominant role, and farmers have little enthusiasm. Moreover, the scale of agricultural land management shows the characteristics of "small, scattered and fragmented", and the degree of agricultural organization and marketization is low. The lack of public participation causes a series of problems, such as light management of reconstruction and separation of construction and management, which affects the long-term and effective play of the comprehensive benefits of cultivated land consolidation [6]. Therefore, we should start from the Angle of land science, pay attention to the adjustment of farmland ownership, let farmers actively participate in the cultivation of farmland, become the leader, so as to improve the efficiency of land use.

After the reform of national institutions, the high standard farmland construction and high efficient water-saving projects have become the main carriers of farmland construction. At present, the "top-down" government-led construction mode before the institutional reform is still inherited, which mainly focuses on four major engineering measures, such as land leveling, field roads, farmland water conservancy and forest protection network, ignoring the ownership adjustment, the combination of engineering construction and fertility improvement, and the combination of fertility improvement and management organization innovation. The result is that cultivated land consolidation especially in the transformation of low-and medium-yield grain fields can not play an effective role.

3. Development process and trend of farmland engineering construction

Overseas, the development of farmland capital engineering construction has been relatively perfect, especially in the field of ecological agriculture is relatively complete. Compared with foreign countries, the reconstruction and development of low-yield grain fields in our country mainly focuses on the high standard farmland construction and land space planning projects. It has experienced three stages. The second stage is combined with the ownership adjustment to adjust the comprehensive factors; In the third stage, ecological landscape construction was carried out on the basis of adjustment and improvement of comprehensive factors. But now our country needs to change from the past four farmland projects construction, land leveling, field road, farmland water conservancy and protected forest network to a more comprehensive farmland project construction combined with local conditions and ownership adjustment [7]. Because the land is too scattered, the ownership relationship is too complicated, and the corresponding labor force is insufficient.

Therefore, although high-standard farmland construction has made some achievements in recent years, some areas have basically entered the stage of perfecting agricultural supporting infrastructure, but there are still some problems in the construction process. For example, some local departments of farmland construction standards are different, the planning content is uneven, the effect is too ignored, the shape Type attention degree is too high, late supervision and management is not enough. It is urgent to develop grain field projects for different ecological areas Standards to accurately support the implementation of the transformation project of low - and medium-yield grain fields.

4. The development process and trend of grain field fertility enhancement

In recent years, due to the continuous high intensity utilization, use and cultivation imbalance, fertility imbalance, soil degradation serious, straw resources Due to rational utilization of source, agricultural non-point source pollution and water and soil loss, the soil point pollution rate of cultivated land exceeded the standard It reached 19.4%, and there were more agricultural film residues in northwest and other places. The decrease of soil organic matter content, especially the low quality grade of some supplementary cultivated land, has seriously affected the cultivated land output. Especially, the sand beach area along the Great Wall of northern Shaanxi, loess plateau area, Weibei tableland area, Guanzhong plain area and mountainous and hilly area of southern Shaanxi are faced with different problems such as fertility decline, soil organic matter content decline, biodiversity reduction and soil texture damage.

Liu Yansui et al. took the hilly and gully region of loess as an example to carry out experimental research on geographic engineering and countermeasures for high-quality agricultural development, including six soil-relationship fertility enhancement technologies, including climate-crop optimization, soil structure improvement, topographic-crop optimization, soil quality improvement, soil-crop optimization, and benefit-crop optimization [8]. In summary, domestic and foreign scholars have carried out researches on measures to improve the fertility of medium and low yield grain fields from the aspects of soil texture, crop optimization, returning materials, etc. However, they mainly explore the fertility improvement technology of medium and low yield grain fields from the micro perspective of soil science and so on. They fail to develop targeted fertility improvement technology from different ecological regions in Shaanxi, combine exploration with management organization innovation, and identify management and protection mechanism.

5. Integration model and trend of medium and low yield grain field renovation project and management organization innovation

As a key part of cultivated land improvement, the transformation of medium and low yield grain fields is to timely supplement cultivated land and revitalize stock An important approach to land, to optimize urban and rural land use, to strengthen intensive land use, and to improve land productivity is essentially for people In the re-debugging of ground relations, the error of any link will affect regional stability and development. Among them, land property right is the core and main line of cultivated land renovation, land property right space Replacement is the way of cultivated land consolidation, and property right manifestation is the important direction of cultivated land consolidation. How to integrate the three Joint innovation is the key to promote smooth progress of land consolidation. Currently our country is not Few pilot in the process of reform for land consolidation accumulated valuable experience, including: Qingyuan City in Guangdong Province to villagersThe self-government approach promotes land integration, adjusts scattered land, liberalizes management rights, and promotes appropriate scale agricultural operations Integrated development of rural industries [9]; Zhao Jiamao as village collective property rights system reform In accordance with the principle of "clear ownership, clear rights and responsibilities, strict protection, flow of elements", the property rights are defined Clear, through the land transfer and pricing of shares will land stock conversion, incremental rights issues, land shares and other kinds of The method was transformed into villagers' equity, which promoted the demutualization of rural assets and land share ownership, and revitalized various resources Su [10]; Nongnong Tun, Longzhou County, Guangxi, has adopted the "small and large" soil of farmer autonomy Land consolidation mode, through farmers independent "small and large" to achieve centralized contiguous, and land transfer To expand the scale of operation and realize the organizational innovation of production relations is the "bottom-up" and "top-down" of farmers and the government a typical representative of the combination of "lower" [11]; In Yangchun County, southern Anhui Province, organized land transfer is implemented on the basis of virtual right confirmation. Virtual ownership gives farmers quantified shares The right, and the organization of land circulation is relying on the "integration of parts" to achieve land integration, in the collective land ownership The basis of activation not only protects the interests of small farmers, but also connects modern agriculture.

In view of the exploration of different innovation modes in local areas, the exploration carried out by land consolidation as the carrier can be described as experienced and effective. In general, the successful implementation of land consolidation mode relies on the clear definition of property rights, but also needs to bring into play the integration function of land ownership to land contract rights and land management rights to achieve efficient and orderly land transfer, which is precisely the advantage of China's collective land ownership over private land ownership Of course, each mode has its scope of application and defects, and there are similarities and differences among various regulation modes. Therefore, how to further deepen the integration and innovation mode of medium and low yield grain field renovation project (engineering standards and land capacity improvement), property rights reform and land transfer, explore the common and characteristic experience of different successful cases, dig deeply into the factors and combinations that can be copied, promoted and used for reference, and follow the principles of adapting to local conditions, gradual progress, easy replication and promotion? This paper summarized and condensed the integration model and regulation path of medium and low yield grain field renovation project and management organization innovation for different ecological areas.

6. Conclusion

The lack of research and judgment of the core link of the reconstruction project of medium and low yield grain fields, the unknown diagnosis of key obstacle factors and the lack of construction standards of grain fields in different ecological areas lead to the different departments in the same region have inconsistent understanding of the reconstruction of medium and low yield grain fields, lack of coordination and other problems, and the lack of targeted farmland construction standards in non-ecological areas. The lack of support of key technologies and integrated application of technologies leads to the lack of expected effects of construction funds. As the land ownership relationship is too complicated, the cultivated land area is too scattered, and the plots are fragmented, the integration degree of efficient utilization of engineering technology and management organization innovation of medium and low yield grain fields is low, which is an obstacle to the construction of large-scale high-standard farmland with guaranteed harvest in drought and flood in most areas of Shaanxi Province. In Shaanxi Province, there is a lack of corresponding typical ecological zone division of medium and low yield grain field transformation, engineering construction standards corresponding to different ecological zones, models, promotion and demonstration of integration of land fertility improvement technology and management organization innovation.

Therefore, remote sensing image and field survey will be used to identify the spatial distribution of medium and low yield grain fields in Shaanxi Province at this stage, and then divide different ecological zones into different types, such as sand beach area along the Great Wall of northern Shaanxi, loess plateau area of northern Shaanxi, Weibei tableland area, Guanzhong plain area and mountain hilly area of southern Shaanxi Province. The core links of medium and low yield grain field construction in different ecological zones of Shaanxi Province will be studied and identified and key obstacle factors will be identified. The standard and combination model of grain field construction in different ecological areas were studied, such as land leveling project, irrigation and drainage project, field road project and forest protection network project. Research and develop the key technologies for improving the fertility of medium and low yield grain fields in different ecological areas, such as soil structure improvement technology, soil fertility improvement technology, green biological efficient utilization technology, etc. To explore the collaborative innovation and efficient utilization mode of grain field construction project and management organization innovation (land transfer, property right reform, etc.), study the construction of responsible subjects for efficient utilization and management of grain field project, study the multiple collaborative management and protection mechanisms of government, village groups and management subjects, and then establish the demonstration zone of integration innovation mode of grain field construction project and management organization in different ecological areas. Play a role in promoting and leading.

Acknowledgements

This work was financially supported by Key Research and Development Program of Shaanxi, China (Program No. 2022ZDLNY02-01), Xi'an Science and Technology Plan Project, China (22NYGG0001).

References

1. F.Z. Xia, Y.M. Yang, Ji.M. Yan, A review on the Connotation of China's Comprehensive land Management in Recent 40 years: Evolution and development change , China Land Science, Commun. 32 (2018) 78-85.
2. Z. Chao, Y.H. LV, W.J. Yun, L.L. Gao, D.H. Zhu, J.Y. Yang, Analysis on research progress of Remote Sensing Monitoring of land consolidation, Agricultural Machinery Journal of Mechanical Engineering. Commun. 50 (2019) 1-22.
3. Z. Zhang, W. Zhao, X. Gu. Changes resulting from a land consolidation project (LCP) and its resource–environment effects: A case study in Tianmen City of Hubei Province, China, Land Use Policy, Commun. 40 (2014) 74-82.
4. J.Q. Wu, X.L. Wang, G.L. Xiong. Study on the impact of land consolidation on farmers' livelihood strategies based on sustainable livelihood framework, Bulletin of Soil and Water Conservation. Commun. 40 (2020) 269-277+284.
5. J. Wang, S. Yan, Y. Guo, The effects of land consolidation on the ecological connectivity based on ecosystem service value: A case study of Da'an land consolidation project in Jilin province, Journal of Geographical Sciences. Commun. 25(2015) 603-616.
6. B. Zhang, W. Niu, L. Ma, X. Zou, X. Kong, H. Chen, Y. Zhang, W. Chen, M. Zhao, X. Xia, A company-dominated pattern of land consolidation to solve land fragmentation problem and its effectiveness evaluation: A case study in a hilly region of Guangxi Autonomous Region, Southwest China. Land Use Policy. Commun. 88 (2019)104-115.
7. Y.S. Liu, H.L. Long, J.L. Li, Analysis of coupling model and guarantee mechanism of land consolidation and transfer under the background of agricultural transformation, Economy Geography. Commun. 40(2020) 50-57.
8. Y.S. Liu, L. Zhu, Y.H. Li, Analysis on basic theory and model of rural land consolidation in transition period, Advance of Geographical Science Exhibition. Commun. 31(2012)777-782.
9. X.W. Liu, T.L. Zhao, Exploration of Land integration and Reflection on Rural Land Consolidation: A case

study of qingyuan city, Guangdong Province, Northwest agriculture Journal of Lin University of Science and Technology (Social Science Edition). Commun. 18(2018)18-26.

10. Y.Z. Zang, Y.S. Liu, Y.Y. Yang, Y.S. Wang. Typical model of land consolidation for targeted poverty alleviation in China, Geography Research, Commun. 38(2019)856-868.
11. B.B. Zhang, W.H. Niu, X.Y. Zuo, The Merging and consolidation Model and effect evaluation of farmers' autonomous fine cultivated land in Guangxi, Agriculture Chinese Journal of Engineering. 35(2019) 265-274.

Practice of Water Ecological Restoration of Large Urban Eutrophication Lake——A case of study of Donghu Lake, Wuhan

Shenghua Hu [1, 2, *], Feichao Shu [1], Junwei Deng [2], Zhongkui Wu [2], Changwen Zhou [2], Xiaoming Li [1]

[1] Wuhan Bridge Engineering Co., Ltd, Wuhan 430071, China
[2] Wuhan Hangfa Ruihua Ecological Technology Co., Ltd., Wuhan 430063, China

Abstract: Submerged macrophytes are the main components of lake ecosystems, and restoration of submerged macrophyte community is an important part of restoring lake ecosystems. Taking Water Ecological Restoration Projects in Wuhan Donghu Lake as an example, Combined with the current situation of the Donghu Lake environment, the submerged phytoremediation area, the restoration species of each sub-lake and the submerged plant configuration plan were determined. The planting of submerged macrophyte adopts the process of "selecting the area - removing debris and repelling fish ecologically - improving the substrate - setting the line - planting". Since the project was implemented , the submerged macrophyte grow well, the coverage of plants reaches more than 70%, and the transparency can reach more than 120cm, which meets the design requirements basically.

Key words: Hydroecology restoration; Submerged macrophyte; Urban lake

1. Introduction

Donghu Lake is the second largest lake in Wuhan city (with a water area of about 33. 63km^2), which has played an important role in urban development. However, with the development of the city, the ecosystem function of Donghu Lake has been seriously degraded, which needs to be systematically repaired. There are some problems in aquatic ecology, such as single community structure of aquatic plants, small distribution range and serious degradation of aquatic vegetation[1]. The submerged plant restoration project is a process of increasing species diversity of submerged plants[2] and an effective means to improve the water quality and ecological environment of urban eutrophic lakes[3, 4]. At present, the application and successful implementation of aquatic plant community restoration technology in China are mostly concentrated in small and medium-sized lakes, such as Nanjing Yueya Lake[5], Ma'anshan East Lake[6], Guangzhou Xiaocui Lake[7] and urban park landscape water bodies[8]. How to effectively restore submerged plant communities in large freshwater lakes is still a blank in China. With the country's emphasis on water environment, it advocates the concept of ecological civilization of respecting nature, conforming to nature and protecting nature. Therefore, the water ecological project of Donghu Lake in Wuhan has a strong demonstration significance and function, which can provide reference and engineering application demonstration for the treatment of eutrophic shallow lakes in large cities.

2. Project overview

Wuhan Donghu Lake Water Ecological Restoration Project is one of the sub-projects of Donghu Lake Water Environment Improvement Project, which covers the whole Donghu Lake. The submerged plant restoration area includes 8 sub-lakes, including Shaojihu Lake, Tanglinghu Lake, Xiaotanhu Lake, Guozhenghu Lake, Miaohu Lake, Lingjiaohu Lake, Houhu and Tuanhu Lake. The restoration of aquatic vegetation in Donghu Lake follows the principle of planting submerged plants, partially dotted with floating plants and emergent plants. According to the present situation of lake water environment, the submerged vegetation restoration area is determined to be about 3. 2km^2.

3. Goal of water ecological restoration

By constructing a perfect ecological community, a submerged plant community with complex spatial structure, rich species and good water purification effect will be formed, and finally the goal of repairing the water environment of Donghu Lake will be achieved.

* Corresponding author: otto2003@126. com

https://doi.org/10.1051/e3sconf/202339401011

4. Restoration design of submerged plants

4.1 Selection principle of submerged plants

The first step in the implementation of aquatic vegetation restoration is the selection of species for aquatic vegetation restoration. According to the field situation of the restoration area, the following principles are followed:selecting the species or species combination that once appeared as dominant species in this area; choosing species with strong colonization ability, wide ecological range and strong tolerance as pioneer species; selecting species with fast amplification speed, large biomass and easy management as constructive species;selecting species with long life cycle and stable biomass as redundant species to improve the level of community diversity;trying to use native species, strictly control or put an end to the use of exotic species, especially invasive species;choosing to purify the water quality and stabilize the sediment as the main starting point, taking into account the landscape effect.

4.2 Submerged plant species

According to the historical investigation results[1] and previous research data[9], submerged plants suitable for planting in Donghu Lake were screened. The Donghu Lake in Wuhan was once widely distributed with submerged plants such as Vallisneria natans, Ceratophyllum demersum, Vallisneria denseserrulata, Myriophyllum spicatum, Potamogeton malaianus and Potamogeton maackianus. Based on the existing research results[10-13], the environmental status of each lake, and the above selection principles, it is determined that six submerged plants, namely, Sophora alopecuroides, Myriophyllum spicatum, Haematococcus verticillata, Potamogeton microphylla and Potamogeton malayi, are the submerged plants planted in the water ecological restoration project of Donghu Lake.

Compared with a single species, the combination of submerged plants has a better purification effect on pollutants such as organic matter, nitrogen and phosphorus[14], and the algae inhibition efficiency is significantly higher than that of single planting[15]. Submerged plant combinations such as "Sophora alopecuroides and Ceratophyllum", "Haematococcus verticillata and Ceratophyllum verticillatum" and "Sophora alopecuroides and Ceratophyllum verticillatum" all have good purification effects on water pollutants. Therefore, the submerged plant planting combination of each sub-lake in Wuhan Donghu Lake is finally determined. The results are shown in Table 1, and the submerged plant planting area of each sub-lake is shown in Figure 1.

Tab. 1 The kinds of submerged macrophyte in every sub-lake of Donghu Lake in Wuhan

Lake name	Major repair species	Collocation species
Guozhenghu Lake	*Vallisneria natans*、*Myriophyllum spicatum*	*Hydrilla verticillata*、*Ceratophyllum demersum*、*Potamogeton maackianus*、*Potamogeton malaianus*
Tanglinghu Lake	*Vallisneria natans*、*Myriophyllum spicatum*、*Hydrilla verticillata*	*Potamogeton maackianus*、*Potamogeton malaianus*
Lingjiaohu Lake	*Vallisneria natans*、*Myriophyllum spicatum*、*Potamogeton maackianus*	*Hydrilla verticillata*、*Ceratophyllum demersum*
Xiaotanhu Lake	*Vallisneria natans*、*Hydrilla verticillata*、*Ceratophyllum demersum*	*Potamogeton maackianus*、*Potamogeton malaianus*
Shaojihu Lake	*Vallisneria natans*、*Ceratophyllum demersum*、*Potamogeton malaianus*	*Potamogeton maackianus*
Tuanhu Lake	*Vallisneria natans*、*Hydrilla verticillata*	*Myriophyllum spicatum*、*Potamogeton malaianus*
Houhu Lake	*Vallisneria natans*、*Hydrilla verticillata*	*Ceratophyllum demersum*、*Potamogeton malaianus*
Miaohu Lake	*Vallisneria natans*、*Myriophyllum spicatum*	*Hydrilla verticillata*、*Ceratophyllum demersum*、*Potamogeton maackianus*

4.3 Planting submerged plant

(1) Planting process

Submerged plants:selecting the area - removing debris and repelling fish ecologically - improving the substrate - setting the line - planting.

The submerged plant planting area is separated from the main lake area of Donghu Lake by impermeable enclosure to reduce the influence of external environment on the planting area. The watertight enclosure is laid with steel pipe (DN80 galvanized steel pipe, 6m single) piles. Then remove the sundries and floating objects on the water surface in the watertight enclosure; Ecological measures such as trawling, artificial flapping on the water surface and ground cages are adopted to drive herbivorous fish and large-scale omnivorous bottom fish (carp, crucian carp, etc.) to designated lakes. Then, before the aquatic plants are planted, the improver is sown to improve the bottom material in the purse seine, and the corresponding seed-throwing seedlings, cutting seedlings and bed-sinking seedlings are prepared. Finally, the planting line is calibrated with bamboo poles and ropes in the planting area, and the submerged plants are planted according to the planting line. In the later stage, the seedlings of submerged plants should be managed and maintained, and the floating and dead seedlings should be salvaged in time and replanted.

(2) Planting area and planting density

https://doi.org/10.1051/e3sconf/202339401011

Submerged plant recovery areas are mainly distributed in areas with a water depth of about 2m, high water transparency, suitable nearshore sediment, less wind and waves and far from the waterway. The planting density of submerged plants in Donghu Lake is 16 clumps/m^2 and 5 plants/clump.

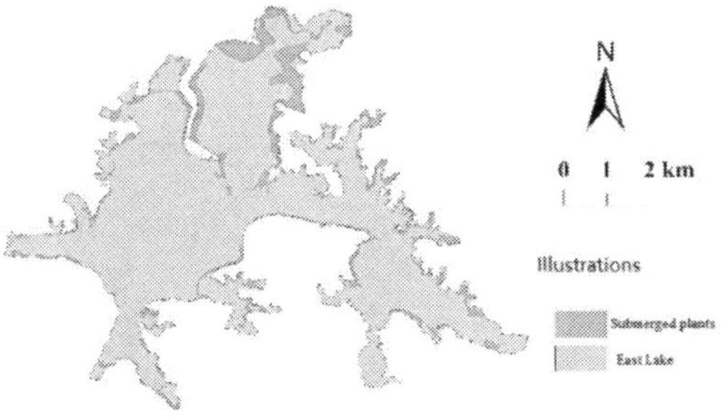

Fig. 1 The distribution of submerged macrophyte restoration areas of Donghu Lake in Wuhan

(3) Planting method

The sediment types of Donghu Lake can be roughly divided into riprap-gravel hard sediment, natural silt sediment and the combination of them. The roots of submerged plants in riprap gravel base can not be planted, so it is necessary to use degradable non-woven fabric to wrap and throw seeds. The natural basement has a lot of subsoil and soft soil, which is suitable for cutting. Therefore, the restoration of submerged plants in Donghu Lake mainly adopts seed throwing method and cutting method, and the submerged bed planting method is adopted in the local deep water area. According to the underwater topography and sediment characteristics of different sub-lake areas, seed throwing method and submerged bed planting method are adopted in areas with deep water depth;Seed throwing method and cutting method can be used in other areas. Three construction methods of submerged plant are shown in Figure 2.

(a) Cutting method-artificial cutting

(b) Cutting method-auxiliary tool cutting

(c) Seed throwing method

(d) Immersed bed planting method

Fig. 2 Three planting methods of submerged macrophyte of Donghu Lake in Wuhan

5. Effect evaluation

Through the construction of perfect ecological community, scientific planting and maintenance technology, the sub-lakes of Donghu Lake are targeted and improved, and the water environment of Donghu Lake is obviously improved. Submerged plants grow well, the plant coverage is obviously improved, and the transparency of water body is obviously improved.

Taking the demonstration area of Donghu Lake Ezui area as an example, in the submerged plant restoration area, the plants grow well, the underwater forest appears, the water quality is clearly distinguished from the enclosure, the transparency of the water in the enclosure is significantly improved, and the overall sensory effect is better. The coverage of submerged plants in the impermeable enclosure can reach more than 70% and the transparency can reach 120cm. The growth of submerged plants in the remediation area is shown in Figure 3 and Figure 4.

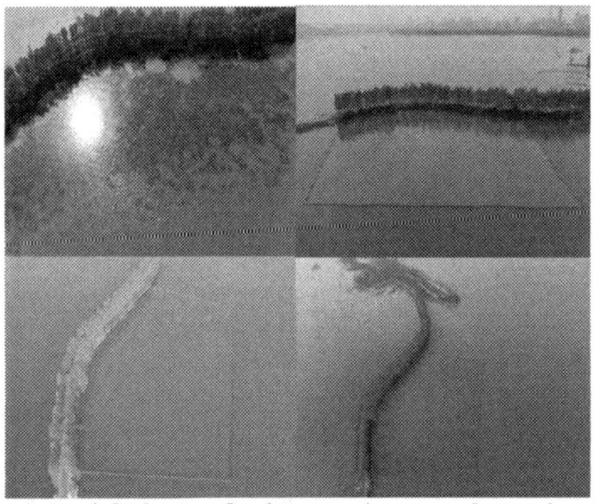

Fig. 3 Aerial photo of submerged macrophyte planting area of Donghu Lake in Wuhan

Fig. 4 Underwater photography of submerged macrophyte planting area of Donghu Lake in Wuhan

References

1. ZHONG Aiwen, SONG Xin, ZHANG Jing, et al . Diversity and distribution of aquatic plants in Lake Donghu in Wuhan in 2014 [J] . Research of Environmental Sciences, 2017, 30(3) : 398-405.
2. HU Shenghua, LIN Qingwei, DAI Zhigang, et al. Change of species diversity during the restoration process of submerged macrophyte in Xihu Lake, Hangzhou [J]. Ecology and Environmental Sciences, 2018, 27(8): 1440-1445.
3. HUANG Xiaolong, GUO Yanmin, WAN Bin, et al. Effect of submerged macrophytes restoration on the ecological environment of urban eutrophic lake[J]. Environmental Engineering, 2018, 36(07):17-21.
4. CHENG Xiaoying, LI Shijie, PU Peimin. Ecological restoration of urban eutrophic lakes-a case study on the physical and ecological engineering in Lake Mochou, Nanjing[J]. Journal of Lake Sciences, 2006, 18(3):218-224.
5. YUAN Peng, XU Liankui, KE Baoling , et al . Treatment and ecological restoration of black and odorous water body in Yueya Lake in Nanjing City[J]. Journal of Environmental Engineering Technology, 2020, 10(5) : 696-701.
6. WU Shuyuan, FENG Zhifei, ZHU Hongsheng. Engineering design of water purification and water ecological remediation for urban inner lake——A case of Ma'anshan East Lake[J]. Water Purification Technology, 2020, 39(S2): 149-154.
7. XIE Danping, JIANG Dong, LIU Aiping, et al. Study on Ecological Restoration of Xiaocui Lake[J]. CHINA WATER & WASTEWATER, 2009, 25(07):17-21.
8. ZHAN Meida, LIN Peibin, ZHANG yang, et al. Practice of ecosystem restoration and water quality improvement project on urban landscape water —— Taking landscape lake of Shenzhen Longtan Park as an example[J]. Environmental Ecology, 2020, 2(11):57-62.
9. YU Dan, TU Manghui, LIU Lihua, et al. Changes of floristic composition of aquatic plants in a shallow, eutrophic Chinese lake(Donghu Lake) from 1954 to 1994[J]. ACTA HYDROBIOLOGICA SINICA, 1998(03):219-228.
10. ZHANG Zhihao, WU Xiaofu, LI Wei. Functions of submerged macrophytes in in-situ ecological restoration of eutrophic waters[J]. Journal of Central South University of Forestry & Technology, 2018, 38(03):115-121.
11. WANG Shoubing, XU Ziran, ZHANG Jie. Dynamic changes of higher submerged macrophytes in Dianchi Lake in recent 50 years and implication for ecological restoration[J]. Water Resources Protection, 2016, 32(06):1-5.
12. ZHANG Miao, LI Mingmeng, LIU Lv, et al. Effects of in-situ ecological restoration in nanhu lake, a hypertrophic urban lake, China[J]. Oceanologia et Limnologia Sinica, 2021, 52(06):1399-1407.
13. ZHANG Meng, CAO Te, GUO Longgen et al. Restoration of Constructed Hydrophytes Community in East Lake of Wuhan and Experilnental Study on Water Quality Improvement[J]. ENVIRONMENTAL SCIENCE & TECHNOLOGY, 2010, 33(06):154-159.
14. QIAN Zhenyu, CHEN Mei. Purification Effect of Different Submerged Macrophytes and Their Combinations on Water Quality[J]. Journal of Anhui Agricultural Sciences, 2019, 47(01):64-67.
15. LI Wei. The Research on purification effect of 6 common submerged plants on eutrophic water in Hunan Province[D]. Central South University of Forestry&Technology, 2018.

https://doi.org/10.1051/e3sconf/202339401012

Heavy Metals Pollution in Soil Around the Lead-zinc Smelting Plant in guanzhong Plain

Xiaohui Cao [1, 2, 3, 4, 5, *], Meng Yuan [1, 2, 3, 4, 5], Yan Zhang [1, 2, 3, 4, 5]

[1] Shaanxi Provincial Land Engineering Construction Group Co., Ltd. China
[2] Institute of Land Engineering and Technology, Shaanxi Provincial Land Engineering Construction Group Co., Ltd. China
[3] Key Laboratory of Degraded and Unused Land Consolidation Engineering, the Ministry of Natural Resources, China
[4] Shaanxi Provincial Land Consolidation Engineering Technology Research Center, China
[5] Land Engineering Quality Testing of Shaanxi Land Engineering Construction Group Co., Ltd. China

Abstract: [Objective] T To analyze and determine the content of eight heavy metals Cd, Cr, Cu, Pb, Hg, Zn, As, and Ni in 40 soil samples collected around a lead-zinc smelting plant in Guanzhong through experimental analysis, and to preliminarily grasp the content of heavy metals in the soil around the smelting plant, and to conduct pollution research and evaluation.. [Method] In mid to late April 2020, 40 soil samples (with a sampling depth of 0-30 cm) were collected from the vicinity of a lead zinc smelter in Guanzhong. The heavy metal content was determined through transcoding, natural air drying, and grinding processes.. [Results] The average contents of Cd, Cr, Cu, Pb, Hg, Zn, As, and Ni in the soil around a lead zinc smelter in Guanzhong were 0.40, 85, 26.0, 48, 0.078, 108, 12.5, and 26 mg · kg-1, respectively. Except for Ni, the average contents of Cd, Cr, Cu, Pb, Hg, Zn, and As were higher than the background values of soil elements in China and Shaanxi Province, especially Cd, Hg, and Pb, The content is 4.26, 2.60, and 2.28 times the background values of soil elements in Shaanxi Province, and 4.12, 1.2, and 1.85 times the background values of soil elements in China. By calculating the standard deviation and coefficient of variation, it was found that the standard deviation and coefficient of variation of Cd, Pb, and Zn in the soil around the smelter were relatively large, indicating that these three heavy metal elements accumulated significantly in the soil.. [Conclusion] The soil around a lead-zinc smelter in Guanzhong is mainly contaminated by multiple heavy metal complexes, mainly Cd pollution. The atmospheric sedimentation around the smelter and dust from mining sites may be the main sources of pollution..

Keywords: lead-zinc smelting plant area; Soil; Heavy metals; Pollution assessment.

1. Introduction

Soil not only provides permanent habitat for human beings, animals and plants, but also provides a beautiful home for all things in the world. Because of this, the soil almost bears 90% of all kinds of pollutants from nature. The "three wastes" generated by various production activities, namely waste water, waste gas, waste residue, etc., will pollute the soil, and the pollution caused by heavy metals is the most serious [1]. Heavy metal pollution in soil refers to the phenomenon that the content of heavy metal elements in soil increases due to human activities, including trace pollution exceeding the background value and excessive pollution caused by long-term accumulation and migration [2]. Heavy metals can interact strongly with proteins and other enzymes in the human body to inactivate them, and they are easy to remain and accumulate, and cannot be biodegraded, causing irreversible damage to the human body and soil. Therefore, heavy metal pollution has become an important factor threatening human health and ecosystem . Therefore, it is urgent to solve this problem [3-4].

Research shows that the sources of soil heavy metal pollution are mainly divided into two aspects. The first is natural sources, such as weathering of soil forming parent material, chemical migration caused by hydraulic and wind transport, and the second is human interference, such as metallurgy, mining, pesticide application, transportation, etc. [5]. Compared with these two aspects, the latter has a greater impact, especially metallurgy and mining. The "three wastes" generated in the process of metallurgical and mining production, as well as a large number of heavy metal elements in dust and tailings, will cause serious pollution to the surrounding soil [6].

In recent years, heavy metal pollution incidents in soil at home and abroad have occurred frequently, which has attracted great attention. Heavy metal pollution in soil of smelting plant has also become a current research hotspot [7-11]. Therefore, in the middle and late April of 2020, this study collected, analyzed and tested the content of

* Corresponding author: 871306815@qq.com

heavy metals in the soil around a lead-zinc smelter in Guanzhong area, and evaluated the pollution status, in order to find the source of heavy metal pollution in the soil in this area, and provide a strong basis for the protection and treatment of heavy metals in the future..

2. Materials and methods

2.1 Overview of the study area

Changqing Town, Baoji City, is low in the middle and high in the east and west, belonging to a semi-arid and semi-humid monsoon climate, with an annual average temperature of 11.5℃and annual rainfall of 600-700 mm. In August 2009, the social event of "children's blood lead exceeding the standard" in Changqing Town caused heated discussion among people from all walks of life. This study took the soil around a lead-zinc smelter in the town as the object to analyze and determine the content of heavy metal elements in the soil.

2.2 Sample collection

In late April 2020, according to the actual investigation on the site, the topography and land use types of the smelter area, the sampling points are mainly distributed in the southeast and northwest directions of the smelter. Because the soil characteristics around the smelter are similar and the land use functions are the same, the "systematic random distribution method" is used to arrange the soil sampling points in this study. 20 soil sampling points are arranged in these two directions, and 40 soil sampling points are arranged in total. Four soil samples (sampling depth of 0-30 cm) were collected at each sampling point. After being fully mixed, about 1 kg of soil samples were retained by the "quartering method" and placed in self sealing bags to prevent the collected samples from being polluted by the outside world. Finally, 40 soil samples were collected in the southeast and northwest directions of the smelter.

The collected soil samples shall be brought back to the laboratory as soon as possible for unified transcoding, and then placed in a ventilated and dark environment at room temperature to dry naturally. The dried soil samples were grinded and passed through 2 mm, 1 mm and 0.149 mm nylon sieves and divided into three parts. One is used to determine the basic physical and chemical properties of soil, the other is used to determine the content of heavy metal elements Cd, Cr, Cu, Pb, Hg, Zn, as and Ni, and the other is reserved for preservation.

2.3 Sample handling and testing

The soil sample treatment and test methods are shown in Table 1 below.

In the process of testing the content of heavy metals in soil samples, Gss-34 (national standard soil sample) was inserted for quality control. All the reagents used in this study were analytically pure, and all samples were blank with corresponding reagents. The determination of parallel samples reached 20%.

The repeatability of the analysis results was evaluated by calculating the relative deviation (RD) of the test results of parallel samples. The relative deviation (RD) is calculated by the following formula:

$$RD = \frac{R_1 - R_2}{R_1 + R_2}$$

Where R_1 represents the result of R sample and R_2 represents the result of parallel samples of R. According to the statistical results, the relative deviation of all detected parallel samples is within the scope of quality control.

Table 1 Items and methods of soil nutrient determination

Items	Methods	Standard
pH	Acidity meter method	NY/T 1377-2007
organic matter	Potassium dichromate external heating method	LY/T 1237-1999
Hg	Atomic fluorescence spectrophotometry	GB/T 22105.1-2008
As		GB/T 22105.2-2008
Cd、Cr、Cu、Pb、Zn、Ni	ICP-MS	HJ 803-2016

2.4 Evaluation method of heavy metal content in soil

The geoaccumulation pollution index method can comprehensively reflect the pollution degree of heavy metal elements in soil. Other pollution assessment methods only consider human interference factors and background values of heavy metal elements in soil, while natural sources, such as changes in background values caused by diagenesis, are often ignored. The geoaccumulation pollution index law fully takes this factor into account and makes up for the shortcomings of other evaluation methods. Therefore, the method of geoaccumulation pollution index is used for evaluation in this experiment.

The geoaccumulation pollution index method was proposed by the German scientist m ü ller in the late 1960s to evaluate the pollution degree of heavy metals in solid sediments and other substances. The geoaccumulation pollution index method is an index for quantitative study of heavy metal pollution in water sediments [12], which has been widely used in the study of soil heavy metal pollution. The specific calculation formula is as follows:

$$I_{geo}=\log_2[C_i/(k\times B_i)]$$

Where: I_{geo} —geoaccumulation pollution index

C_i—actual measured concentration of heavy metal element I in soil

B_i—background value of soil heavy metal element i

k—constant, correction for the change of background value caused by diagenesis (generally 1.5)

3. Results and discussion

3.1 Content of heavy metal elements

Through analysis and measurement, it was found that the soil around a lead zinc smelting plant in Guanzhong is slightly alkaline, with a pH range of 7.65-8.68 and an average value of 8.24. The soil organic matter content is 12.5-58.0 $g \cdot kg^{-1}$, with an average value of 27.9 $g \cdot kg^{-1}$. The content range of heavy metal elements Cd, Cr, Cu, Pb, Hg, Zn, As and Ni in the soil around the smelter is 0.11-0.61 $mg \cdot kg^{-1}$, 50-135 $mg \cdot kg^{-1}$, 15.4-34.9 $mg \cdot kg^{-1}$, 22-77 $mg \cdot kg^{-1}$, 0.042-0.128 $mg \cdot kg^{-1}$, 6.6-15.8 $mg \cdot kg^{-1}$, 70-152 $mg \cdot kg^{-1}$, and 20-34 $mg \cdot kg^{-1}$, respectively, with an average content of 0.41, 86, 25.9, 47, 0.077, 106, 12.7, and 24 $mg \cdot kg^{-1}$. Among the eight heavy metal elements, except Ni, The average content of other heavy metals is higher than the background value of soil elements in China and Shaanxi Province, especially Cd and Pb, which are 4.26 and 2.27 times of the background value of soil elements in Shaanxi Province, 4.10 and 1.84 times of the background value of soil elements in China. According to the above table, the standard deviation and coefficient of variation of heavy metal elements Cd, Pb and Zn in the soil around the smelter are large, indicating that their accumulation in the soil is obvious. To sum up, the eight heavy metal elements measured in this study have been polluted and accumulated in different degrees in the soil of the region.

Table 2 Heavy Metal Content in the Soil Around a Lead Zinc Smelter in Guanzhong

heavy metal	Maximum value	Minimum value	Average value
Cd	0.61	0.11	0.41
Cr	135	50	86
Cu	34.9	15.4	25.9
Pb	77	22	47
Hg	0.128	0.042	0.077
Zn	152	70	106
As	15.8	6.6	12.7
Ni	34	20	24

3.2 Enrichment characteristics of heavy metal elements

The results of heavy metal pollution in the soil around the lead-zinc smelter based on the cumulative pollution index method are shown in Table 3 below. From the average value, the soil in the study area is only polluted by CD, Pb and Hg, and the pollution of other heavy metals is not reflected. However, from the maximum value, CD has reached a moderate to strong pollution level in some areas, Pb and Hg have reached a moderate pollution level, and Zn, Cu and CR have reached a mild to moderate pollution level. From the sampling area, the soil in the northwest is more polluted by heavy metals than that in the southeast. Among them, CD and Pb pollution is the most serious, and has reached the moderate to strong pollution level. According to the analysis, this phenomenon is most likely related to the local wind direction. The local wind direction is mainly northwest wind, which leads to the dust, dust and Aerosol and other pollutants have chemical migration in the process of wind transport. Therefore, the soil in the northwest is seriously polluted by heavy metals. In addition, the soil beside the highway is polluted by the tail gas emitted by various vehicles, resulting in the high content of Pb and Hg elements. In general, the pollution of Pb, CD and Hg in the soil around a lead-zinc smelter in Guanzhong is serious.

3.3 Variation characteristics of heavy metal elements with distance

There are a lot of heavy metal elements in the smoke and dust discharged from lead-zinc smelters. With the settlement and diffusion of these smoke and dust, heavy metal elements will gradually be enriched in the surrounding soil environment. After smoke, dust and other pollutants enter the atmosphere, the degree of soil pollution varies with the distance from the pollution source and the location. The most likely reason for this phenomenon is that natural environment, geography, monsoon and other conditions can affect the formation of microclimate in different regions, and then affect the migration direction of pollutants. It is understood that in regions with different longitudes and latitudes, altitudes, landforms and other conditions, the climatic conditions are different, and the pollution degree and mode of pollutants such as soot and dust in the atmosphere to the surrounding environment are also different. In this experiment, the topsoil 0-1500 m away from the northwest of a lead-zinc smelter in Guanzhong was collected, and the contents of CD, Hg and Pb were determined and analyzed. Finally, the horizontal distribution of the contents of these three heavy metals was obtained.

The specific method of collecting soil samples in this experiment is as follows: take a sample every 50 m at a distance of 0-400 m from the northwest of a lead-zinc smelter in Guanzhong, and take a sample every 100 m at a distance of 400-1500 m, a total of 20 soil samples are collected. Finally, it was found that the content of heavy metal elements Cd, Hg and Pb in the soil in the northwest of a lead-zinc smelter in Guanzhong decreased with the increase of the distance from the smelter. The closer the distance from the lead-zinc smelter, the higher the content of heavy metal elements in the soil, that is, the more serious the heavy metal pollution. It can be seen from the figure that within the range of 0-400 m from the smelter, the slope of the change curve of CD, Hg and Pb content is large, indicating that the content of heavy metal elements changes greatly and decreases rapidly, that is, the attenuation is sharp, while within the range of 400-1500 m from the smelter, the change curve of CD, Hg and Pb content is close to the level, that is, the change of heavy metal element content is significantly slowed down, and the attenuation trend is significantly weakened, so it can be inferred that, The smelter is an important pollution source that leads to high content of heavy metal elements in the soil in this area. To sum up, the contents of heavy metal elements Cd, Pb and Cu in the soil around a lead-zinc smelter in Guanzhong decreased with the increase of the distance from the pollution source, and the attenuation of the content of heavy metal elements had obvious stages.

4. Conclusion

(1) The average contents of CD, Cr, Cu, Pb, Hg, Zn, as and Ni in the soil around a lead-zinc smelter in Guanzhong were0.41, 86, 25.9, 47, 0.077, 106, 12.7, and 24 mg·kg^{-1}, respectively. Except for Ni, the average contents of CD, Cr, Cu, Pb, Hg, Zn and as were higher than the background values of soil elements in China[13] and Shaanxi Province, especially Pb and CD, which were 2.27 and 4.26 times higher than the background values of soil elements in Shaanxi Province, respectively, They are 1.84 and 4.10 times of the background values of soil elements in China, respectively. Compared with the background value, the order of the contents of eight heavy metals from high to low was cd>hg>pb>zn>cr>cu>as>ni.
(2) The evaluation results of geoaccumulation pollution index show that, from the average value, the soil around the smelter is only polluted by CD, Pb and Hg, and the pollution of other heavy metals is not reflected. However, from the maximum value, CD has reached a moderate to strong pollution level in some areas, Pb and Hg have reached a moderate pollution level, and Zn, Cu and CR have reached a mild to moderate pollution level.
(3) The analysis of the variation characteristics of heavy metal elements with distance shows that the contents of heavy metal elements Cd, Pb and Cu in the soil around the smelter decrease with the increase of distance from the pollution source, and the attenuation of heavy metal elements has obvious stages.
To sum up, the soil around a lead-zinc smelter in Guanzhong is mainly polluted by a variety of heavy metals, mainly CD. The atmospheric deposition and diffusion around the smelter, as well as the smoke and aerosol emissions, may be the main pollution sources.

Acknowledgements

Innovation Capability Support Program of Shaanxi " Construction of Shaanxi soil mass quality detection and evaluation sharing platform" (2021PT-053)

References

1. Zhou Jianli, Chen Tongbin. Research status and prospect of heavy metal pollution in vegetable soil and vegetables in suburbs of China [J]. Journal of Hubei Agricultural College, 2002, 22(5):476-479.
2. Tian Tian, Yang Ting, et al.Pollution characteristics and risk assessment of heavy metals in soil around the iron and steel plant -- a case study of farmland around Sanming Iron and steel plant in Western Fujian [J] .Acta agronomica Sinica, 2021, 11 (6): 42-46+95.
3. Yu Fazhan, Qi Fangyan, et al. Heavy metal pollution and its evaluation in the soil of urban park greenbelt in xuzhou city [J]. Urban Environment & Urban Ecology, 2009, 22(3):20-23.
4. Huang Yizong, Hao Xiaowei, et al. Remediation technology and practice of heavy metal contaminated soil [J]. Journal of Agro-Environment Science, 2013, 32(03):409-417.
5. Du Ping. Study on spatial distribution and speciation of heavy metal pollution in soil around lead zinc smelter [D]. Beijing:China Academy of Environmental Sciences, 2007.
6. Fan Shuanxi. Distribution characteristics and pollution assessment of heavy metals in the soil around the smelter in Changqing Town, Baoji City [J] . Environmental chemistry, 2014, 33 (5): 861-862.
7. Wang Lijun, Lu Xinwei, et al. Study on Heavy Metal Pollution in the Soil around the Lead and Zinc Smelter in Changqing Town, Baoji, Luo Dacheng [J]. Journal of Agro-Environment Science, 2012, 31(2):325-330.
8. Deng Chaobing, Li Lihe, et al. Pollution characteristics of heavy metals in paddy soil of typical lead zinc mining area [J]. Journal of Agricultural Environmental Sciences, 2009, 28 (11): 2297-2301.
9. Lei Dongmei, Duan Changqun, Wang Ming. Evaluation of soil fertility and heavy metal pollution in abandoned land of different mining areas in yunnan [J]. Journal of Agro-Environment Science, 2007, 26(2):612-616.
10. Ji Yanfang, Li Yonghua, et al. Behavior characteristics of heavy metals in soil rice system of Fenghuang lead zinc mining area [J]. Journal of Agricultural Environmental Sciences, 2008, 27 (6): 2143-2150.
11. Qu Jiao, Ma Zhenyu, et al. Analysis and evaluation of heavy metal pollution in vegetable soil around transportation trunk line in molybdenum mine area [J]. Journal of Agro-Environment Science, 2008, 27(1):178-181.
12. Müller G. Index of geoaccumulation in sediments of the Rhine River[J]. Geojournal, 1969, 2:108-118.
13. National Environmental Protection Agency, Environmental Monitoring of China. Background value of soil elements in China [M]. Beijing: China Environmental Science Press, 1990:87-90, 330-496.

Simulation of distributed photovoltaic power generation system

Gaoyuan Cai [1,*], Bingbing Li [1], Liqun Liu [2], Zhen Zhao [2], Lulu Han [2], Youfu Lai [2]

[1] Central South Electric Power Test Research Institute, China Datang Corporation Science and Technology Research Institute Co., Ltd, Zhengzhou, Henan, 450000, China
[2] Datang Huaxian Wind Power Generation Co., Ltd, Huaxian, Henan, 456400, China

Abstract: Combined with the operation characteristics of photovoltaic power generation system, the mathematical model and simulation model are established, and the control strategy of bipolar inverter with wide output characteristics and the maximum power point tracking (MPPT) control method of conductance increment method are designed. The maximum power output and stable operation of distributed photovoltaic power generation system are realized. The simulation results show that the photovoltaic power generation system model established in this paper can effectively reflect the characteristics of distributed photovoltaic power generation and MPPT control. With the change of lighting conditions, the distributed photovoltaic power generation system can accurately adjust the system output, which is of great significance for the research and operation of distributed photovoltaic power generation system.

Key words: Distributed photovoltaic; maximum power point tracking (MPPT); inverter control; modeling and simulation.

1. Introduction

With the continuous progress of distributed photovoltaic technology and the gradual decline of equipment cost, distributed photovoltaic has more prominent technical advantages in power generation performance, operation control and environmental protection. At the same time, the proportion of installed power generation capacity in our country is increasing year by year. According to the statistics of the National Energy Administration, in 2021, the newly added photovoltaic grid-connected capacity reached 54.88 million kW, a cumulative total of 3.06 trillion kW, an increase of 13.9% year on year, among which the newly added distributed photovoltaic grid-connected capacity was 29 million kW in 2021, accounting for 55%. In terms of newly installed capacity connected to the grid, distributed PV power generation has surpassed centralized PV power generation.

With the rapid growth of photovoltaic power generation, especially distributed photovoltaic power generation, in order to master the characteristics and operation characteristics of distributed photovoltaic power generation, strengthen the management and control of distributed photovoltaic power generation, and improve the operation capability and control level of distributed photovoltaic power generation, it is necessary to carry out research on the operation characteristics of distributed photovoltaic through modeling and simulation of distributed photovoltaic system [1-3].

In this paper, based on the operation characteristics of distributed photovoltaic, the mathematical model and simulation model of distributed photovoltaic are established, and the control strategy of bipolar inverter and the maximum power point tracking (MPPT) control method of incremental conductance method are designed. In addition, The simulation system of distributed photovoltaic power generation system is also built, and the change of lighting conditions, MPPT control and operation control of the simulation system are designed. Finally, the simulation results of distributed photovoltaic power generation system are analyzed[4-6].

2. PV array modeling

When the sunshine intensity and temperature are constant, the V-I curve of the output voltage and current of the photovoltaic array is called the volt-ampere characteristic of the photovoltaic array. Typical PV array V-I characteristic curve is shown in Figure 1.

As shown in Figure 1, V_{oc} and I_{sc} represent the open circuit voltage and short circuit current of the photovoltaic array. As can be seen from the V-I characteristic curve in Figure 1, the output current I of the photovoltaic array varies with the output voltage V. The output power of photovoltaic array is the product of output voltage and output current, that is, output power P=V*I, which is the area S of the shaded part in Figure 1.

* Corresponding author: hfcaigaoyuan@163.com

The key of photovoltaic power generation system MPPT is to find an output voltage V_m and output current I_m, so that the shadow area P_m is the maximum, and at this time, is the voltage V_m, current I_m and maximum power value s corresponding to the maximum power point.

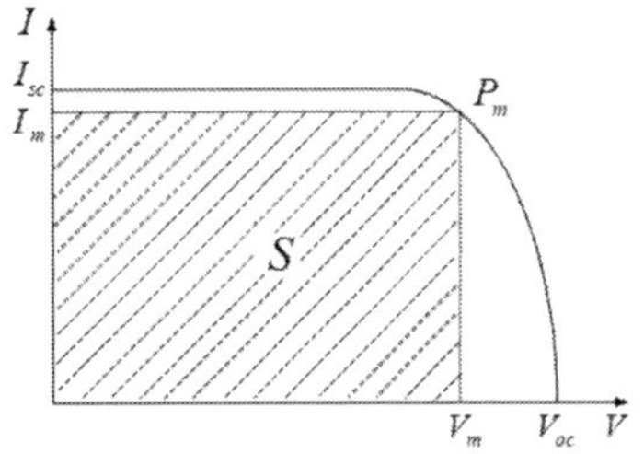

Fig.1 V-I characteristic curve

For reference to sunshine intensity and ambient temperature, the short-circuit current and open-circuit voltage of the photovoltaic array are respectively I_{scref} and V_{ocref}, and the working current and working voltage of the maximum power point are respectively I_{mref} and V_{mref}. Under any sunshine intensity and ambient temperature, the output voltage V and output current I of the photovoltaic array can be calculated by formula (1) ~ (7):

$$I = I_{scref}[1 - C_1(\exp(\frac{V - \Delta V}{C_2 * V_{oc}}) - 1] + \Delta I \quad (1)$$

$$T_c = T + t_c * R \quad (2)$$

$$\Delta T = T_c - T_{cref} \quad (3)$$

$$\Delta I = a * \frac{T}{R_{ref}} * \Delta T + \left(\frac{R}{R_{ref} - 1}\right) * I_{scref} \quad (4)$$

$$\Delta V = -b * \Delta T - R_S * \Delta I \quad (5)$$

$$C_2 = \frac{\left(\frac{V_{mref}}{V_{ocref}} - 1\right)}{\left(\ln(1 - \frac{I_{mref}}{I_{scref}})\right)} \quad (6)$$

$$C_1 = (1 - \frac{I_{mref}}{I_{scref}}) * \exp(-\frac{V_{mref}}{V_{ocref} * C_2}) \quad (7)$$

In Formula (1) ~ (7), R is the total sunshine intensity of photovoltaic array surface; T_c is the surface temperature of the photovoltaic array; T is the ambient temperature; t_c is the temperature coefficient of photovoltaic array; R_{ref} is the reference sunshine intensity of photovoltaic array; T_{ref} is the reference temperature of photovoltaic array; a is the current-temperature coefficient of photovoltaic array under the reference sunshine; b is the voltage-temperature coefficient of photovoltaic array under the reference sunshine; R_S is the series resistance of photovoltaic array; C_1 and C_2 are correction coefficients of photovoltaic array; ΔT is the temperature compensation value; ΔI is the current compensation value.

3. MPPT control

Under any sunshine intensity, temperature and other environmental conditions, the output power of the photovoltaic array changes with the change of working voltage, and the output power is maximum only at the maximum power operating point. Therefore, it is necessary to track the power control of photovoltaic power generation system, and the incremental conductance method is a stable and effective MPPT control method.

When the photovoltaic array works at the maximum power point, the derivative of its output power to the working voltage is zero, that is, $dP/dV = 0$. According to the current formula, Formula (8):

$$\frac{dP}{dV} = \frac{dVI}{dV} = V\frac{dI}{dV} + I = 0 \quad (8)$$

Therefore, by judging the $V * \frac{dI}{dV} + I$ value at different working voltages V, we can judge whether the voltage is at the maximum power point voltage, so as to realize the MPPT of photovoltaic array.

The principle of the incremental conductance method MPPT is shown in FIG. 2, $P_n(V_n, I_n)$ is the output power of the PV module in the NTH iteration, $P_m(V_m, I_m)$ is the output power of the PV module in the m iteration, ΔV is the voltage step, $P(PU)$ and $V(PU)$ are respectively the output power of the PV module and the unit value of the voltage. When the environmental conditions change quickly, the incremental conductance method has the advantages of realizing MPPT voltage tracking quickly, and its voltage fluctuation is small.

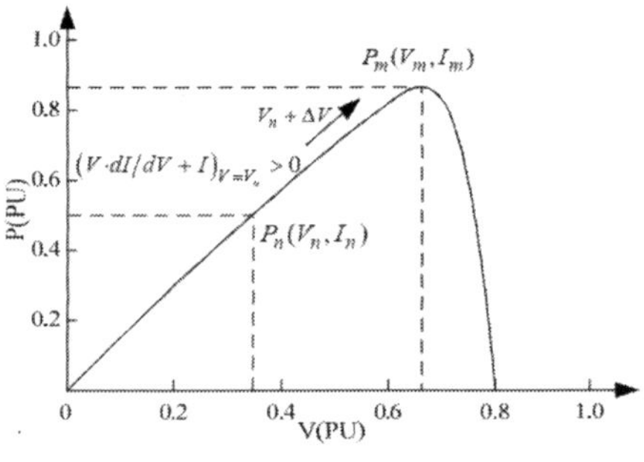

Fig.2 Incremental conductance method MPPT principle

4. Control strategy of photovoltaic power generation system

The two-stage distributed photovoltaic power generation system is composed of photovoltaic array, DC/DC conversion circuit, DC/AC inverter circuit, filter circuit and large power grid, as shown in Figure 3.

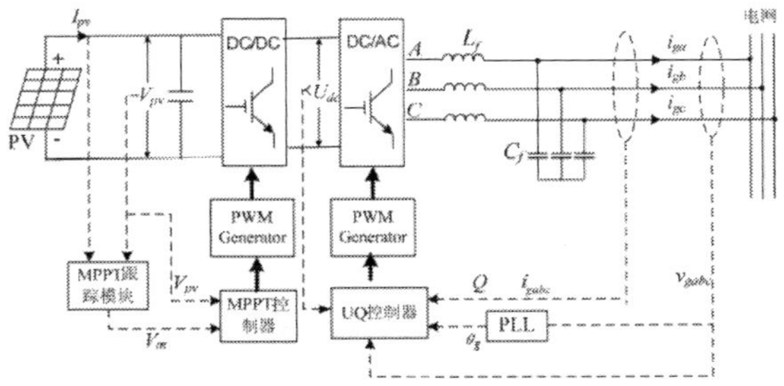

Fig.3 Control structure of distributed photovoltaic power generation system

The Boost circuit with more stable output and continuous output current is used for DC conversion in MPPT control circuit of distributed photovoltaic power generation system.

In application, considering that the DC/AC part can maintain the constant DC side voltage U_{dc} of the inverter, that is, the output side voltage in the DC/DC circuit is constant, the duty ratio D should be adjusted at this time to ensure that the DC voltage U_{pv} of the photovoltaic array can always track the output voltage value U_m at the maximum power point, so as to realize the MPPT of the photovoltaic array.

Inverter control adopts voltage-reactive power control, control inverter DC side voltage constant, that is, inverter output reactive power constant, adopt double closed-loop control, the outer ring is the voltage ring, the inner ring is the current ring, and the inner ring adopts decoupling control. The specific control strategy is shown in Figure 4.

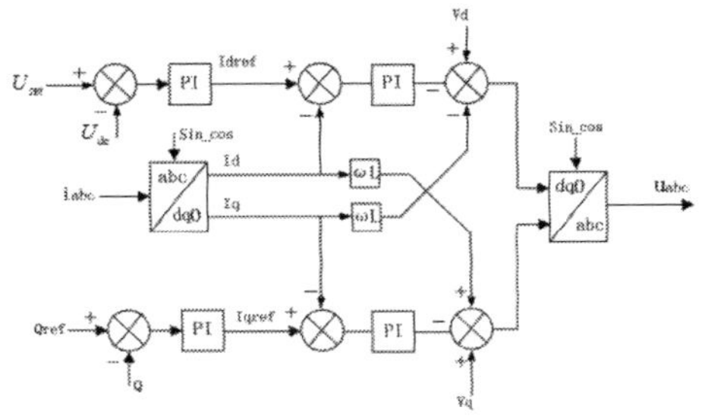

Fig.4 DC/AC control policy

5. Example simulation

The simulation model of photovoltaic power generation system is built. The distributed photovoltaic power generation system is simulated and analyzed. Set the initial moment, light intensity 1000W/m2, temperature 25°C; At t=3s, the illumination intensity decreases from 1000 W/m2 to 700W/m2. When t=6s, the light intensity increases from 700 W/m2 to 1000W/m2. At t=8s, due to scheduling requirements, the distributed photovoltaic power generation system carries out limited power output, which is required to limit the output power to 32kW. The system switches from MPPT control to limited power control operation. When t=10s, the power limiting output requirement of the distributed photovoltaic power generation system is lifted and the power limiting control is switched to MPPT control. The simulation results are shown in Figure 5.

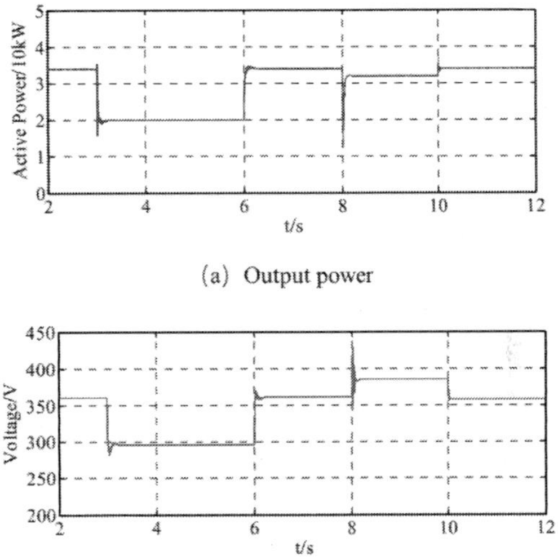

(a) Output power

(b) Photovoltaic array output voltage

Fig.5 Simulation results of distributed photovoltaic power generation system

As shown in Figure 5, at t=3s, the light intensity drops sharply, the output power of the distributed photovoltaic power generation system rapidly drops to 20kW, and the terminal voltage of the photovoltaic array drops to 300V. When t=6s, the illumination intensity returns to the initial state, and the output power of the distributed photovoltaic power generation system and the output voltage of the photovoltaic array also return to the initial output state. At the moment t=8s, the output power of the distributed photovoltaic power generation system quickly stabilizes at 32kW due to the requirement of power limiting. At the same time, the output voltage of the photovoltaic array also runs to the non-maximum power operating point voltage, indicating the effectiveness of the power limiting control strategy of the distributed photovoltaic power generation system. When t=10s, the power limit requirement is lifted, the system restores MPPT control, and the system running state is restored to the initial state, indicating that the distributed photovoltaic power generation system runs smoothly and the status switch is smooth.

6. Conclusion

The simulation model of distributed photovoltaic power generation system established in this paper accurately reflects the characteristics of distributed photovoltaic power generation and MPPT control. The output characteristics of photovoltaic array under different temperature and illumination conditions are simulated. The control strategy of bipolar inverter and the maximum power control method of conductivity increment method are used to establish a complete distributed photovoltaic power generation system. Distributed photovoltaic power generation system under the change of lighting intensity, power control and other environmental conditions, the simulation results show that the dynamic adjustment of the simulation model is fast and effective, and can prepare to reflect the operating characteristics of distributed photovoltaic power generation system under the change of external environment, to ensure the maximum power output and stability control of the photovoltaic power generation system. It is of great guiding significance to the research and operation of distributed photovoltaic power generation system.

References

1. Wu Fan. Research on real-time Simulation modeling of large-scale photovoltaic power generation system based on clustering algorithm [D]. Hefei: Hefei University of Technology,2019.
2. WANG Yu. Modeling and grid-connected Simulation of photovoltaic power generation [D]. Shijiazhuang: Hebei University of Science and Technology,2017.
3. Liu Chang, Chen Qizhuan, Chen Guiyue, et al. Modeling and simulation of photovoltaic fuel cell hybrid power generation System [J]. Acta Energiae Solaris Sinica,2018,39(11):3113-3119.
4. Lu Chang, Feng CPPCC, Zhi Yongjun. Dynamic Modeling and Simulation Analysis of grid-connected photovoltaic power generation system [J]. Sichuan Electric Power Technology,2016,39(2):15-19,58..
5. AI Yongle, Zhang Xin, Li Xing, et al. Research on frequency control method of high permeability photovoltaic hybrid power generation system [J]. Engineering Journal of Wuhan University,2022,55(3):269-274.
6. Zhang Lijuan, Shi Junliang, Hu Xiangling, et al. A flyback current type inverter with decoupling capability [J]. Renewable Energy,2022,40(3):415-420.

Preliminary application of grade equivalent theory in a local section of a mine and economic estimation of the deposit

Guan Wang [1], Huaqiang Song [1], Zongda Yang [1], Lize Wang [2], Pei Xue[1], Jingyuan Xu [1], Wei Jiang [1]

[1] Norin Mining Limited, Beijing 100053, China
[2] NORINCO International Cooperation Ltd, Beijing 100000, China

Abstract. By introducing the practical application of grade equivalent theory in a copper -cobalt mine, it can show the maximum economic value of the individual ore block and the mine, maximize the comprehensive utilization of limited resources and contains positive significance for extending the service life of the mine. Briefly introducing the steps of Micromine software in the process of reserve estimation and the feasibility of displaying the economic value of individual ore blocks. Estimation of reserve can be completed objectively and scientifically by Micromine software, grade and economic value of ore blocks can be displayed in three-dimensional image.

Keyword: Grade Equivalent; Economic Value of Ore Blocks; Micromine Software; Copper -Cobalt Mine in Drc.

1. Introduction

Head grade of a single metal element determined by the geological and reserve management departments was used for poly-metallic open pit mines during feasibility stage, mining and metallurgical circuit. However with variety of metal price, the improvement of mining and processing technology and other factors, it is necessary to consider the value of associated component elements which is the integrated economic value to estimate the value of mineral deposit, maximize the recovery of mineral resources, it is essential to determine the combined grade of polymetal which is so-called grade equivalent.

2. Methodology of Grade Equivalent Calculation

According to the definition of integrated grade of polymetallic ore, the integrated grade equivalent is calculated as follows:

$$g_z=g_0+\sum_{i=1}^{n} g_{0i}=g_0+\sum_{i=1}^{n}(f_i\cdot g_i)$$

gz: integrated grade equivalent; g0: grade of primary metal; g0i is the e grade quivalent of the ith associated component converted to the main component; gi is the grade of the ith associated component; fi is the equivalent factor of the ith component converted to the grade of the main component.

Grade information of components which are required to calculate integrated grade can be provided by geological data. Calculation of the equivalent factor requires consideration of the economic value of ore mined. Profitability method and price method are two general methods for calculating the equivalent factor.

2.1 Profitability Method

$$f_i=\frac{(k_i-r_i)\cdot\alpha_i\beta_i}{(k_0-r_0)\cdot\alpha_0\beta_0}$$

fi is the equivalent factor of conversion of the ith component into primary component; ki is price of the ith associated component; ri is smelting cost of the ith associated component; αi is processing recovery of the ith associated component; βi is the smelting recovery of the ith associated component; k0 is price of the primary component in the ore; r0 is smelting cost of primary component in the ore; α0 is processing recovery of primary component in the ore; β0 is the smelting recovery of the primary component in the ore.

2.2 Price Method

$$f_i=\frac{k_i\cdot g_i}{k_0\cdot g_0}$$

fi is the equivalent factor of conversion of the ith component into primary component; ki is price of the ith associated component; gi is the average grade of the ith associated component; k0 is price of primary component in the ore; g0 is the average grade of primary component in the ore.

3. Appilication

A copper-cobalt mine, situated in the northern parts of the Central Africa Copperbelt(CACB) which primary metal is copper with cobalt as associated element. It is necessary to improve utilization of cobalt metal as the price of cobalt surged in recent years and there is a downward tendency reflected in copper price. Grade equivalent theory is applied at this specific mine to enhance the economic efficiency of the mine while effectively extracting natural resources to maximize the comprehensive utilization of limited resources and avoid depletion of resources. The operation of optimization of the orebody and the display of the individual ore block values was made by mining software Micromine.

3.1 Collection of Drilling Data

Drilling data is collected from exploration line 23 to 47 of western main ore body. All drilling data in the area were electronically organized into four Micromine database format files which are Collar, Survey, Lithology and Assay in order to establish the drilling database.

Table 1 Drilling Database Format

Collar						
Hole ID	East	North	Elevation	Depth		
Survey						
Hole ID	Depth	Angle of dip	Azimuth			
Lithology						
Hole ID	From	To	Litho	Rock type		
Assay						
Hole ID	Sample num	From	To	Length	Cu	

3.2 Establishment of Drilling Database

The main steps to build the drilling database are as follows: ①Data import. Drilling data is imported into Micromine software to generate "dat" format file. ②Verification of data. Micromine software seeks errors automatically in the drilling data and modify them with verification function. ③ Database generation. Generating database through data which is verified with steps above.

3.3 Generation of Ore Body Wireframe

Display drill hole information including drill hole trajectory, lithology, grade distribution of sample within Micromine software. A string of sections is generated along exploration lines, ore body is interpreted according to geology and assay information, wireframe or three-dimensional solid model which is used to describe the outline of the ore body is created as follow criteria: ①cut-off: 0.5% cu; ②exploitable thickness: 2m; ③permissible maximum thickness of gangue parting: 2m.

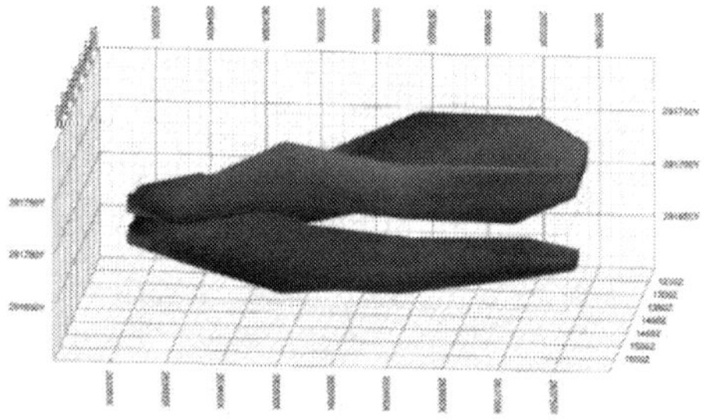

Figure1 wireframe of ore body

3.4 Block Model

After creating the block model which is used to estimate the resource by using the inverse distance weighted(IDW) interpolation. The interpolation should be restricted to a space where the ore grade are correlated and block within the ore body should be identified using a wireframe ore body, then the identified blcok should be used as the basis for statistical analysis and interpolation. Dividing whole block into numerous sub-blocks with same size which requires a comprehensive consideration of the morphology, complexity and drill hole spacing. Complete block model is comprises of each individual block that is interpolated by data adjacent. Drill hold spacing for resource estimation is 100m×50m, sample composite is 1m, morphological complexity of the ore body is medium and the block size is determined as 10m×2m×5m. Edge blocks are divided using sub-block factor to ensure the accuracy of the solid model boundaries. Radius of the search ellipsoid is set to 50m. Three-dimensional morphology of the ellipsoid is consistent with the morphology of the ore body basically, the long-axis azimuth is consistent with strike of the ore body and the sub-axis azimuth is consistent with angle of dip of ore body. To reduce the interpolation bias from the high-density sample point area, it is necessary to disperse the cluster samples and limit the minimum maximum number of sample points in each quadrant of the search ellipsoid to enable unbiased search for ellipsoidal valuation. In order to interpolate all empty blocks, interpolation process has been run three times with increased radius of search ellipsoid at a time. Block model is examined in each run until every block in the model has been interpolated a grade.

3.5 Report of Resource Estimation

Result of resource and average grade estimation is as follow

Table 2 Result of resource estimation

Ore body	Volume m^3	Tonnage t	Density m^3/t	Cu grade %	Contained Cu t
upper	1253962	3134905	2.5	1.67	52445.02
lower	1790627	4476567	2.5	2.30	102794.57
total	3044589	7611472	2.5	2.04	155239.59

3.6 Update of Block Model

Equivalent factor is determined combined by price and recovery rate of each component metal. Copper and cobalt prices are based on three-month average bid price of US$4,750/t and US$26,500/t on LME respectively. Average grade of copper and cobalt are 2.04% and 0.116% respectively. Copper to cobalt equivalent factor is 0.317. Partial integrated grade is calculated and shown as follow.

Table 3 Integrated grade information

Name or ore body	Sample number	from	to	litho	Cu %	Co %	EqCu %
ZKPW31-2	H26	86.6	87.6	RSC	0.752	0.481	0.904
ZKPW31-2	H31	91.6	92.6	RSC	1.180	0.378	1.300
ZKPW31-2	H34	92.6	93.6	RSC	3.890	0.553	4.065
ZKPW31-2	H39	97.6	98.6	SDB	5.200	0.440	5.339
ZKPW31-2	H100	156.7	157.7	CMN	4.410	0.358	4.523
ZKPW31-2	H102	157.7	158.7	CMN	7.130	0.465	7.277

Updated wireframe is created by applying integrated grade and cut-off is EqCu 0.5%. Optimized block mode and result of resource are shown as follow. Copper grade and metal contained are increased slightly mainly caused by low cobalt grade overall which has minor influence converted to integrated grade.

Table 4 Result of resource estimated by EqCu

Ore body	Volume m^3	Tonnage t	Density m^3/t	EqCu grade %	Contained Cu t
upper	1254522	3136306	2.5	1.71	53765
lower	1793313	4483284	2.5	2.32	104230
total	3046907	7619590	2.5	2.07	157995

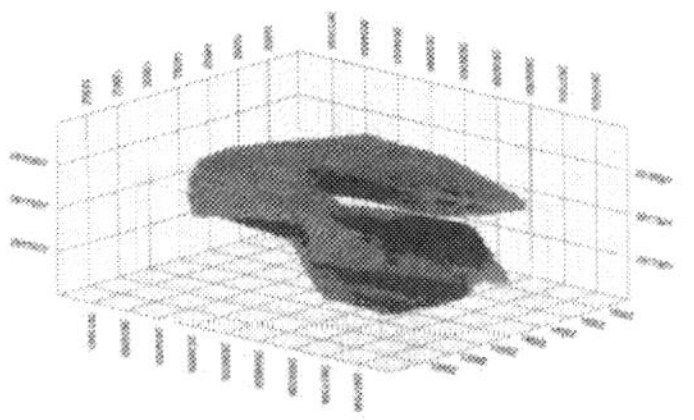

Figure 2 Block Model of EqCu

3.7 Economic Value Evaluation

Economic value or profit of the ore blocks is remaining value of revenue contained in each block minus mining cost, processing & smelting cost and sales cost value (Table 5). Construction infrastructure investment and cash flow during the operation period are not included.

Table 5 Economic Value of Block

Number of block	Sub-block factor	Orebody	Cu %	Cu metal t	Block value $	Mining cost $	P&S cost $	Sales cost $	Block profit $
1	0.03	lower	1.28	0.08	335.1	102.4	180.3	54.9	-2.5
2	1.00	lower	3.57	8.56	36594.3	4000.0	19688.2	5992.1	6914.1
3	0.99	lower	2.36	5.67	24231.8	3999.6	13037.0	3967.8	3227.4
4	1.00	upper	2.41	5.79	24747.8	4000.0	13314.6	4052.3	3380.9
5	1.00	upper	upper	2.96	12653.5	4000.0	6807.7	2071.9	-226.2
6	0.82	upper	upper	2.21	9462.0	3272.0	5090.7	1549.3	-450.0

Block value is the copper contained in individual block times copper price. Copper price is based on three-month average bid price of US$4,750/t on LME. Mining cost is unit cost of individual block which size is 10m x 2m x 5m and volume is 100m^3. Unit integrated mining cost in which contained mining, striping, drilling and blasting is 8$/$m^3$ referenced to the Comica mine and other mines located adjacent of Likasi. Strip radio is 4:1; Density of ore is 2.5t/m^3; Processing and Smelting(crush, mill, leach and SX-EW) cost 2300$/t copper which is referenced to average cost of mining company in DRC. Recovery rate is 90% from actual data. Sales cost 700$/t copper which is referenced to average cost of mining company in DRC. Profit Model and value can be displayed by Mircromine software as follow(partial). Profit value of upper and lower ore body is $15627787 and $55723125 respectively.

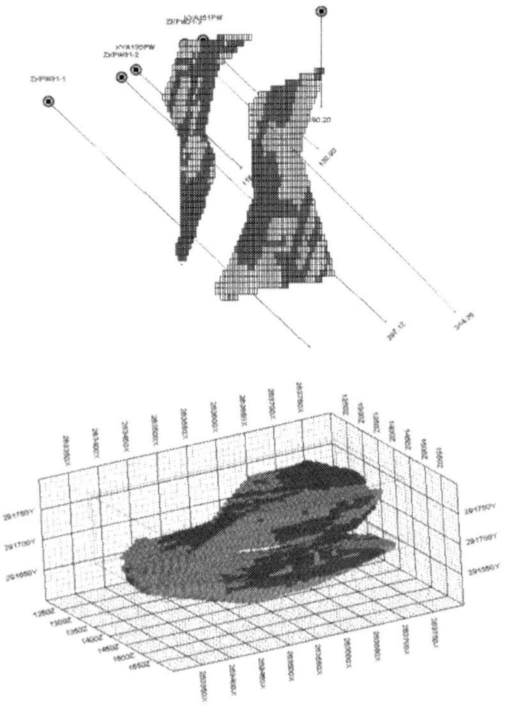

Figure 3 Profit Model of ore body

note： profit block is shown on red color; loss block is shown on white color

4. Conclusion

Comprehensive evaluation of the economic value of individual block by using grade equivalent is beneficial to maximize utilization of resources and contribute to postpone the depletion of resources as well as extend life of mine. Given mining, processing, smelting and sales costs are basically stable while metal prices keep volatile, updated integrated grade and profit value can be generated according to metal prices volatility. Visibility and rapid interpretation of Micromine software enables continuous update and optimization of block model as well as estimation of economic profit value of resource. It provides actual direction for mine operation. Comprehensive evaluation of resouce with grade equivalent enable rational exploitation of limited mineral resources.

References

1. Pang Chun-yong,Zhang yong-lin,Wu Jian-sheng. Rapid evaluation of the development value of polymetallic deposits with a single metal equivalent grade[J]. Minerals and Geology, 2001,15 (3) : 1-2.
2. Chen Dao-gui, Hu Nai-lian, Yin Sheng-hua. Determination and application of open-pit cut-off grade of polymetallic deposits under the new classification standard[J]. metal mine, 2007 (5) : 1-2.
3. Hu Bao-gen,Xi Mu-sheng. Application of Equivalent Grade in Dexing Copper Mine[J]. Copper project, 2006 (4) : 2-3.
4. Qin Tao. Research on cut-off grade optimization of underground mining based on dynamic programming method[D]. Shenyang: Northeastern University,2008.
5. Xie Ying-liang,Gao Yao. Discussion on Mineral Product Price Fluctuation and Mining Cutoff Grade Adjustment[J]. China Mining, 2003,12 (6) : 1-2.
6. Chu Dao-Zhong,Wang Qing. Dynamic programming optimization method of cut-off grade and its application[J]. China Mining, 2008,17 (9) : 1-2.
7. Chen Jian-hong,Xiao Cheng,Zhou Zhi-yong. New progress in grade economics research and dynamic intelligent calculation method of cut-off grade[J]. Metal Materials and Metallurgical Engineering, 2012 (40) : 2-4.
8. Liu Feng-gui. Application of DIMINE software in geological modeling of Qibaoshan lead-zinc mine[J]. Jiangxi Metallurgy, 2012,32 (5) : 2-4.

Study on the Development Path of Green Low-Carbon Environmental Protection Industry under the New Situation

Yin Wang*, Yao Ding

Southwest Municipal Engineering Design & Research Institute of China, Chengdu 610084, China;

Abstract: This paper aims to study the development path of green, low-carbon and environmental protection industries under the new situation. Currently, the green, low-carbon and environmental protection industry has become an important field for economic and social development. This paper proposes two key development paths, namely strengthening policy guidance and promoting technological innovation, by outlining the current status of the green, low-carbon and environmental protection industry. In terms of policy guidance, it is recommended that the government develop supportive policies and regulatory measures to guide and promote industry development. In terms of technological innovation, it is recommended that enterprises strengthen independent innovation and technological cooperation to promote the innovation and application of green, low-carbon and environmental protection technologies. Through comprehensive analysis and empirical research, new ideas and references can be provided for the development of the green, low-carbon and environmental protection industry.

Keywords: green, low-carbon, environmental protection industry, development path, policy guidance, technological innovation

1. Introduction

With the increasingly severe global climate change and environmental pollution, the green, low-carbon and environmental protection industry has become an important field of global economic development[1,2]. Under the new historical conditions, exploring the development path of the green, low-carbon and environmental protection industry and promoting its healthy and rapid development are urgent issues to be solved. This paper aims to study the development path of the green, low-carbon and environmental protection industry under the new situation and propose actionable recommendations to promote its development.

2. Opportunities for the development of green, low-carbon, and environmentally friendly industries

2.1 Global action on climate change

Global climate change has seriously threatened ecosystem security and economic and social development. According to the World Meteorological Organization's "Global Climate Status Report (2022)", the annual average land temperature in the world in 2022 was 1.67°C higher than that of 1850-1900. China even experienced a large-scale heat wave for 79 consecutive days[3]. To actively fulfill international agreements such as the United Nations Framework Convention on Climate Change and the Paris Agreement, China has strengthened its national contributions, continued to promote carbon peaking and carbon neutrality actions, integrated them into the overall economic and social development, key areas of transformation, technological innovation, comprehensive improvement of carbon sequestration capacity, and governance system reform, and taken multiple measures to strive to achieve the strategic goal of carbon peaking and carbon neutrality, and make due contributions to global climate change.

2.2 Global biodiversity conservation cooperation

Currently, the earth is undergoing the sixth mass extinction of species. According to the World Wildlife Fund's "Living Planet Report 2020 - Bending the Curve of Biodiversity Loss," we are destroying biodiversity at an unprecedented rate. In the future, only climate change will still cause about 1/5 of wild species to face extinction risks in the 21st century, and by 2070, 1/3 of plants and animals may face extinction[4]. Biodiversity plays a crucial role in maintaining ecosystem stability. As a species permanently disappears, nature's ability to sustain us will also be weakened. China practices multilateralism, actively carries out international cooperation in biodiversity conservation, actively fulfills the Convention

* Corresponding author: wangyin@ccccltd.cn

on Biological Diversity, and promotes the synergistic effectiveness of biodiversity-related conventions.

2.3 Concept of the Fourth Industry of Ecological Products Proposed

Chinese and foreign experts jointly proposed the concept of the Fourth Industry of ecological products. Compared with the traditional three-industry classification, the emergence of the Fourth Industry of ecological products can help us understand the relationship between green development, green economy, green industry, ecological economy, ecological industry, and environmental protection industry, understand the development status of China's ecological and green industries, and bring about an unprecedented wave of industrial structural reform and upgrading[5].

From the current situation in China, thermal power, steel, construction, etc. are the main sources of carbon emissions. Some studies have shown that the carbon emissions from the sewage treatment industry account for only 1%-2% of the total carbon emissions in the country[6]. Many people wonder whether the sewage treatment industry still needs to vigorously transform towards green and low-carbon directions? From the perspective of industrial layout, the essence of the traditional three industries is a dissipative structure, reflected in the "entropy increase" of a large amount of resource consumption and pollution emissions, which can only be reduced as much as possible through green and low-carbon circulation technologies but cannot be reduced to zero[7]. However, the Fourth Industry of ecological products, due to the involvement of the ecological production process, most of the industrial activities are based on natural solutions, so it is "negative entropy" or "zero entropy" for the biosphere and the ecological environment, with a very small or even positive impact. This means that some high-energy-consuming and high-polluting industries such as industry, can be transformed into ecological industries, realizing the transformation of industrial structure and promoting sustainable development.

3. Challenges Facing the Development of Green, Low-Carbon, and Environmental Protection Industries

3.1 Insufficient Policy Support

The development of green, low-carbon, and environmental protection industries faces the challenge of insufficient policy support. On the one hand, some regions and departments have not established a complete environmental policy system, which leads to a lack of attention and support for these industries. On the other hand, some regions and enterprises have lagged behind in implementing policies, resulting in the insufficient effectiveness of policies. In terms of resource utilization, the government's determination to promote and support is not strong enough[8]. Resource utilization involves multiple technological fields, and there are certain bottlenecks in technology. The government's support for technological research and development and application promotion of resource utilization is insufficient, which is also one of the factors that restrict the development of resource utilization. However, even after solving technological bottlenecks and economic issues, resource utilization technology still faces the difficult challenge of marketization[9]. The government's comprehensive management ability, including supervision, means, and other aspects, urgently needs improvement. Resource utilization involves waste disposal, energy utilization, and other aspects. Insufficient supervision can easily lead to resource waste, environmental pollution, and other problems. In addition, resource utilization requires a significant amount of investment, but the current government's financial support for resource utilization is insufficient, making it challenging to support large-scale resource utilization projects.

3.2 Low Rate of Technology Commercialization

A low rate of technology commercialization is a significant challenge facing the field of technological innovation in China. A low rate of technology commercialization means that technological innovation results are difficult to implement and cannot effectively contribute to economic and social development. The main reasons for the low rate of technology commercialization are as follows: (1) insufficient investment in innovation. Although China's investment in technological innovation has increased year by year, there is still a gap compared to developed countries, especially in the field of basic research. This results in inadequate support and protection for researchers during the research process, making it difficult to produce transformable technological innovations. (2) insufficient technological innovation capability. China's technological innovation capability and level are still relatively backward, and there is still a certain gap compared to developed countries. This has led to relatively fewer breakthroughs and achievements in innovation, resulting in a low rate of technology commercialization. (3) inadequate mechanism for technology commercialization. China's mechanism for technology commercialization still needs improvement, including technology transfer, intellectual property protection, and support from the capital market. These deficiencies make the commercialization of technological innovation difficult and also undermine the confidence of companies and investors in technology commercialization[10].

3.3 Low initiative on the part of the enterpris

Currently, the imperfect carbon trading mechanism in China is characterized by varying carbon emission standards among different countries or regions, making it difficult to achieve unified measurement and comparison of carbon emissions, thereby hindering the realization of carbon trading. Furthermore, the unreasonable allocation of carbon trading quotas may lead to market supply-demand imbalance and price volatility, making it difficult for companies to formulate long-term development strategies. Inadequate supervision may result in fraudulent or fake practices, affecting the fairness and transparency of the market and lowering market credibility. In addition, the small scale of the carbon trading market, the lack of trading entities, and insufficient market liquidity, make it difficult to establish an effective price discovery mechanism, thereby impacting market development[11]. These factors have contributed to low market activity in China's carbon trading market, with an average price of CCERs of only 55 yuan, far below the price of 187 euros for European CERs, making it difficult to stimulate companies to actively invest in carbon sequestration projects, leading to a more wait-and-see attitude.

4. Green, low-carbon, and environmentally friendly industry development pathway

4.1 Transformation of water and environmental protection processes and technologies

The transformation of water and environmental protection processes and technologies mainly involves the following aspects: firstly, promoting the application of clean energy. Clean energy is one of the keys to achieving the dual-carbon goals. The water and environmental protection industry can reduce carbon emissions by adopting clean energy such as solar and wind power. Secondly, promoting green, low-carbon processes. Green, low-carbon processes are another key to achieving the dual-carbon goals[12]. The water and environmental protection industry can reduce carbon emissions and chemical oxygen demand by promoting green, low-carbon processes such as membrane separation technology and biotreatment technology. Thirdly, achieving the recycling and utilization of wastewater resources. The recycling and utilization of wastewater resources is an effective way to achieve the dual-carbon goals. The water and environmental protection industry can reduce energy consumption and carbon emissions by achieving the recycling and utilization of wastewater resources such as water reuse and energy recovery. Fourthly, promoting digital transformation. Digital transformation is another key to achieving the dual-carbon goals. The water and environmental protection industry can improve resource utilization efficiency and reduce carbon emissions by promoting digital transformation such as intelligent monitoring and data analysis.

Overall, the transformation of water and environmental protection processes and technologies is one of the keys to achieving the dual-carbon goals. It needs to be comprehensively promoted, continuously innovated, and strengthened in terms of technology research and development and transformation application, so as to promote the sustainable development of the industry.

4.2 Development of Ecological Protection and Restoration Projects

The development of ecological protection and restoration projects includes the following aspects. Firstly, forest ecology protection and restoration: forests are one of the most important carbon sinks on Earth, and protecting and restoring forest ecosystems can help reduce carbon emissions. Ecological protection and restoration projects can include forest protection, restoration, afforestation, and other measures. Secondly, wetland ecology protection and restoration: wetlands are important ecosystems that can help reduce greenhouse gas emissions, absorb harmful substances, and prevent soil erosion. Ecological protection and restoration projects can include wetland protection, restoration, and other measures. Thirdly, ocean ecology protection and restoration: the ocean is one of the largest carbon sinks on Earth, and protecting and restoring ocean ecosystems can help reduce carbon emissions. Ecological protection and restoration projects can include ocean protection, restoration, marine ecological aquaculture, and other measures. Fourthly, urban ecological restoration: cities are one of the main sources of carbon emissions, and urban ecological restoration can help reduce carbon emissions and improve urban environmental quality. Ecological protection and restoration projects can include urban greening, urban park construction, and other measures.

In general, the development of ecological protection and restoration projects is one of the important ways to achieve the dual-carbon goals. It requires strengthening technological research and development, promoting technology transfer and application, enhancing policy support and funding, and promoting the implementation of ecological protection and restoration projects to achieve sustainable development of ecology, economy, and society.

The value realization paths for diversified ecological products can include the following aspects: firstly, promoting green and low-carbon products, which are an important way to achieve the dual-carbon target. This can be achieved by reducing product carbon emissions, chemical oxygen demand and other indicators to achieve low-carbon production and consumption. Green and low-carbon products include ecological agricultural products, low-carbon energy products, sustainable development products, etc. Secondly, promoting the development of a circular economy, which is another important way to achieve the dual-carbon target. This can be achieved by recycling resources, reducing waste, and processing waste to achieve a low-carbon economy. Diversified ecological products can be produced and consumed through promoting the development of a circular economy, such as waste recycling and utilization, ecological agriculture, etc. Thirdly, promoting digital transformation, which is a key factor in achieving the dual-carbon target. This can be achieved by intelligent production and management to improve resource utilization efficiency and reduce carbon emissions. Diversified ecological products can improve product quality and reduce carbon emissions by implementing digital transformation, such as intelligent production and digital marketing. Fourthly, strengthening brand value and cultural connotation, which can be achieved by enhancing brand image, brand story, cultural heritage, etc. This can increase product added value and market competitiveness, leading to sustainable development. In general, diversified ecological products are an important way to achieve the dual-carbon target, requiring strengthening of technological research and development and transformation, as well as policy support and investment, to promote their implementation and achieve sustainable development in ecology, economy, and society.

5. Conclusion

Against the backdrop of global climate change and biodiversity degradation, to promote the deep development of the ecological and environmental protection industry, several actions should be taken. Firstly, a comprehensive policy system for resource utilization should be established, and policies should be promoted and disseminated to encourage enterprises and society to participate in resource utilization. Secondly, financial support for green and low-carbon projects should be increased, special funds should be established, and social capital should be guided to invest in resource utilization. Thirdly, research and development of ecological and environmental protection industry technology should be strengthened, a technology innovation mechanism should be established, and ecological and environmental protection technology should be continuously improved. Fourthly, the intelligent upgrading of the ecological and environmental protection industry should be promoted, and technologies such as the Internet of Things, big data, and cloud computing should be utilized to improve pollution control, environmental monitoring, and service efficiency, guiding environmental management and decision-making. Finally, efforts should be accelerated to improve and perfect the carbon trading market mechanism, encourage market participants to innovate and develop carbon financial instruments, and provide support for the subsequent development of the ecological and environmental protection industry.

References

1. He, J. K. (2016). Global low-carbon transition and China's response strategies. Advances in Climate Change Research, 7(4), 204-212. https://doi.org/10.1016/j.accre.2016.06.007.
2. Li, Y., & Li, Y. (2013). Low-carbon City in China. Sustainable Cities and Society, 9, 62-66. https://doi.org/10.1016/j.scs.2013.03.001.
3. Yang, X., Wang, X. C., & Zhou, Z. Y. (2018). Development path of Chinese low-carbon cities based on index evaluation. Advances in Climate Change Research, 9(2), 144-153. https://doi.org/10.1016/j.accre.2018.05.004.
4. Yinan Li, Keren Chen, & Nan Zheng. et al. (2021). Strategy Research on Accelerating Green and Low-Carbon Development under the Guidance of Carbon Peak and Carbon Neutral Targets. IOP Conference Series: Earth and Environmental Science, 793 012009. DOI 10.1088/1755-1315/793/1/012009
5. Guo, Q.; Zhou, M.; Liu, N.; Wang, Y. (2019). Spatial Effects of Environmental Regulation and Green Credits on Green Technology Innovation under Low-Carbon Economy Background Conditions. Int. J. Environ. Res. Public Health, 16, 3027. https://doi.org/10.3390/ijerph16173027
6. Wang, Y., & Zhi, Q. (2016). The Role of Green Finance in Environmental Protection: Two Aspects of Market Mechanism and Policies. Energy Procedia, 104, 311-316. ISSN 1876-6102. https://doi.org/10.1016/j.egypro.2016.12.053.
7. Wang, H., Chen, Z., Wu, X., & Nie, X. (2019). Can a carbon trading system promote the transformation of a low-carbon economy under the framework of the porter hypothesis?—Empirical analysis based on the PSM-DID method. Energy Policy, 129, 930-938. doi: 10.1016/j.enpol.2019.03.007.
8. Jiang, B., Sun, Z., & Liu, M. (2010). China's energy development strategy under the low-carbon economy. Energy, 35(11), 4257-4264. https://doi.org/10.1016/j.energy.2009.12.040.
9. Yu, L. (2014). Low carbon eco-city: New approach for Chinese urbanisation. Habitat International, 44, 102-110. https://doi.org/10.1016/j.habitatint.2014.05.004.
10. Wu, Z.; Tang, J.; Wang, D. (2016). Low Carbon Urban Transitioning in Shenzhen: A Multi-Level Environmental Governance Perspective. Sustainability, 8, 720. https://doi.org/10.3390/su8080720
11. Chen Liu, Gill Valentine, Robert M. Vanderbeck, Katie McQuaid, Kristina Diprose. (2019) Placing 'sustainability' in context: narratives of sustainable consumption in Nanjing, China. Social & Cultural Geography 20:9, pages 1307-1324.
12. Xiao, J., Zhen, Z., Tian, L., Su, B., Chen, H., & Zhu, A. X. (2021). Green behavior towards low-carbon society: Theory, measurement and action. Journal of Cleaner Production, 278, 123765. https://doi.org/10.1016/j.jclepro.2020.123765.

https://doi.org/10.1051/e3sconf/202339401016

Effects of water bath digestion and DEENA pretreatment on mercury content detection in soil

Yabo Zhang [1, 2, 3, 4, *]

[1] Shaanxi Land Engineering Construction Group Co., Ltd., Xi'an 710075, Shaanxi, China
[2] Shaanxi Institute of Land Engineering Technology Co., Ltd., Xi'an, 710075, Shaanxi, China
[3] Key Laboratory of Degraded and Unused Land Remediation Engineering, Ministry of Land and Resources, Xi'an, Shaanxi, 710075, China
[4] Shaanxi Land Consolidation Engineering Technology Research Center, Xi'an 710075, Shaanxi, China

Abstract. There are many sample preparation methods for heavy metal detection of soil samples, but each has its advantages and disadvantages. Based on the above two pretreatment methods, the detection limit, precision, accuracy and actual soil sample measurement value of mercury detection in soil by atomic fluorescence were compared with atomic fluorescence, and the results showed that the automatic digestion and pretreatment system (DEENA) could digest soil samples to meet the requirements of GB/T 22105.1-2008 standard. Moreover, there was no significant difference between the two digestion methods on the detection results of mercury in soil, and the automatic digestion and preparation system (DEENA) was convenient, operable and highly automated, which was more advantageous for sample digestion.

1. Introduction

In recent years, with the continuous attention to soil quality issues, the detection of heavy metal pollution in soil has become one of the key contents of environmental monitoring and agricultural environmental protection. Mercury is a highly toxic element that is biotoxic and reserved, and measuring mercury levels in soil is an integral part of soil quality assessment.

At present, the detection methods of mercury in soil mainly include graphite furnace atomic absorption method, atomic fluorescence method and inductively coupled plasma mass spectrometry [1], among which atomic fluorescence method has the advantages of simple spectral line, high sensitivity, low detection limit and wide linear range, and this method is widely used in the detection of trace mercury. Soil samples need to be digested before monitoring, and the digestion methods mainly include water bath digestion, graphite digestion, electric hot plate digestion and microwave digestion[2-5]. In this study, the soil samples were pretreated by water bath digestion method and graphite digestion method, and the mercury content in soil samples was determined by hydride generation atomic fluorescence method, and the advantages and disadvantages of the two digestion methods were compared and analyzed, which provided reference for subsequent soil heavy metal detection.

2. Experimental part

2.1 Main instruments and reagents

2.1.1 Instrument

AFS-9760 atomic fluorescence photometer; DEENA 60 sample automatic graphite digester; Milli-Q Integral 15 Ultrapure Water Meter; BSA224S-CW electronic balance; DZKW-S-8 type water bath.

* Corresponding author: 987474549@qq.com

2.1.2 Test

Nitric acid (excellent pure), hydrochloric acid (excellent pure), potassium hydroxide, potassium hydrogen boride, mercury standard stock solution (1000 mg/L), ultrapure water (conductivity 1 μS/cm, resistivity 0.1 MΩ·cm).
Add 300 mL of hydrochloric acid, 400 mL of distilled water and 100 mL of nitric acid to the beaker to prepare aqua regia; Weigh 5 g of sodium hydroxide in a beaker, add a small amount of water to dissolve, and then weigh 5 g of potassium borohydride in sodium hydroxide solution, dilute it with water to 1 L to obtain sodium hydroxide-potassium borohydride mixed solution; The cleaning solution is a 5% (volume fraction) nitric acid solution.

2.1.3 Samples

After the soil dries, it is ground through a 0.149 mm sieve and stored in a brown jar. Prepare soil reference materials GBW07454 (GSS-25) and GBW07454 (GSS-32) simultaneously.

2.2 Resolution

(1) Water bath digestion
Weigh 0.500 0~0.509 9 g of the sample to be measured in a 50.0 mL plug colorimetric tube, wet the sample with a little water, add 10 mL of aqua regia, add stopper and shake well, boiling water bath for 2 h, shake once during the period, take out and cool, use deionized water to set the volume to the scale line, mix well, stand for clarification, and make blank samples in the same batch.
(2) Graphite digestion method
Weigh 0.500 0~0.509 9 g of the sample to be measured in a 50.0 mL digestion tube, put the digestion tube into the graphite digester, and the digestion steps are: add 1mL of ultrapure water, add 10mL of aqua regia, and vibrate for 30 s at 50% intensity and 30% height; Heat at 100 °C for 135 min; Shake for 30 s at 50% strength and 30% height, cool for 20 min, set the volume to 50 mL with pure water, and make blank samples in the same batch.

3. Results and discussion

The samples treated by two different pretreatment methods were determined using the same atomic fluorescence spectrometer and setting the same working parameters, and the specific working parameters of the instrument are shown in Table 1.

Table 1 Working parameters of atomic fluorescence photometer

Parameter	data
Negative high voltage /V	290
Total current /mA	30
Auxiliary cathode current /mA	0
Carrier gas flow /(mL·min^{-1})	500
Shielding gas flow /(mL·min^{-1})	900
Atomic gasifier height /mm	10
Reading time /s	16
Delay time /s	4
Measure the number of repetitions	2

3.1 Atomic fluorescence assay results

3.1.1 Drawing of standard curves.

The mercury standard solution was diluted three times to obtain 20 μg/L mercury standard solution, and 0, 0.5, 1.0, 2.0, 3.0, 4.0, and 6.0 mL of mercury standard solution were piped into a 50 mL volumetric flask, and the volume was fixed to the scale line, so as to obtain a gradient solution with mercury mass concentration of 0, 0.2, 0.4, 0.6, 0.8, 1.2 μg/L, and the first curve of fluorescence intensity and mercury mass concentration was obtained by atomic fluorescence determination, and the curve correlation coefficient was 0.999 2. It can be seen that mercury has a good linear relationship within 0.0~1.2 μg/L.

3.1.2 Determination results of detection limits

The blank sample was measured 15 times consecutively, and the ratio of the standard deviation to the slope of the standard curvature was calculated 3 times the 11 measured values, that is, the detection limit was obtained [6]. The results show that the detection limit of the atomic fluorescence method is below 0.002 μg/L, which meets the standard requirements.

3.1.3 Determination results of precision and accuracy

Six replicates were performed on GSS-25 (mercury identification value of 0.043±0.003 mg/kg) and GSS-32 (mercury identification value of 0.026±0.003 mg/kg), and the mean values of mercury in GSS-25 and GSS-32 were 0.044 and 0.026 mg/kg, respectively, the relative standard deviation was 3.74% and 3.44%, and the absolute values of relative error were 0~4.65% and 0~3.85%, respectively. Explain that the detection method meets the precision and accuracy requirements.

3.2 Actual soil sample measurement results

Thirteen actual soil samples (numbered SJ-1~SJ-13) were subjected to water bath digestion and graphite digestion, and detected with atomic fluorescence photometer, and the results are shown in Table 2.

Table 2 Actual sample measurement results

Sample number	Measured value /(mg·kg^{-1})		Relative error /%
	Graphite digestion	Water bath digestion	
SJ-1	0.086	0.084	1.19
SJ-2	0.080	0.085	3.26
SJ-3	0.155	0.169	4.40
SJ-4	0.061	0.057	3.22
SJ-5	0.146	0.132	4.98
SJ-6	0.065	0.071	4.18
SJ-7	0.054	0.051	2.82
SJ-8	0.030	0.031	1.64
SJ-9	0.107	0.098	4.56
SJ-10	0.123	0.134	4.28
SJ-11	0.188	0.208	5.00
SJ-12	0.080	0.082	1.23
SJ-13	0.166	0.171	1.35

3.3 Comparison of two pretreatment methods

3.3.1 Comparison of digestion procedures

The graphite digestion method uses a fully automatic digestion and pretreatment system for digestion, the main advantage is its rapid temperature rise, and it can complete the digestion of 60 samples at a time, which is suitable for batch testing samples. The method is simple to operate, high degree of automation, the instrument can automatically add reagents, temperature regulation, quasi-determination volume, and three-dimensional graphite heating module by establishing simple commands, uniform heating. The water bath digestion method uses a water bath for heat digestion, and its advantage is that it is cheaper than fully automatic graphite digestion, and the number of samples can be digested up to 120 at a time. The disadvantage is that the heating is uneven, and the addition of reagents and volume fixing need to be completed manually in turn, which is time-consuming and laborious, and the efficiency is low. Therefore, from the perspective of instruments, the automatic graphite digester is more convenient, operable and highly automated, which is more advantageous for sample digestion.

3.3.2 Comparison of test results

Two certified standard samples were detected by two methods, and their precision and accuracy met the standard requirements. After digestion of 13 actual soil samples, the relative deviation of the measured values of the two groups was within 5.0%, indicating that there was no significant difference in the influence of the two digestion methods on the detection results.

4. Conclusion

Atomic fluorescence method was used to determine mercury in standard solution, and the linear relationship between mercury within 0~1.2 μg/L was good, and the detection limit of the method was below 0.002 μg/L, which met the requirements. Water bath digestion and graphite digestion were used to digest soil reference materials and actual soil samples, and there was no significant difference in the impact of the two digestion methods on the detection results of mercury in soil, among which the automatic graphite digester was more convenient, operable and highly automated, and had more advantages for sample digestion.

Acknowledgements

Relying on the project: "Construction of Shaanxi soil mass quality detection and evaluation sharing platform （2021PT-053） "

References

1. LIU Qinglin. Comparison of the application of two sample preparation methods in atomic fluorescence mercury detection[J], Anhui Chemical Industry, 2017, 43(1): 83-84.
2. HUANG Yong, CHEN Bei. Determination of mercury in soil by water bath digestion[J], China Environmental Monitoring, 2007, 23(3):12-13.
3. Bo Weiping. KONG Xiaoxiao,CHEN Shixin,et al. Determination of arsenic mercury in soil by water bath digestion-atomic fluorescence[J],Environmental Monitoring and Early Warning,2016,8(3):34-36.
4. LU Kuang, HU Chenlu, LI Jingjing, et al. Application of atomic fluorescence for mercury and arsenic in environmental monitoring[J]. Resource Conservation and Environmental Protection,2019(6):4.)
5. ZHANG Huiqiang, ZHANG Qinming, MA Wenpeng, et al. Comparison of direct mercury measurement and atomic fluorescence method for the determination of total mercury in water[J], Guangzhou Chemical Industry,2019,47(10):108-110.)
6. HAN Bing, TAN Hui, et al. Evaluation of uncertainty of mercury detection limit by AF-7500 atomic fluorophotometer[J], Tropical Agricultural Engineering, 2013 (37):6:1-4.

Study on the standard of the four relationships of reservoir and the lower limit of effective reservoir in Nantun Formation of Oilfield A

Yalou Zhou [1], Guangjuan Fan [1,2], Ang Gao [1], Wentong Zhao [1], Ting Dong [1]

[1] School of Earth Science, Northeast Petroleum University, Daqing 163318, China;
[2] Key Laboratory of Reservoir Formation Mechanism and Resource Evaluation of Heilongjiang Province, Daqing 163318, China

Abstract. By using the data of logging, cuttings logging, thin section identification and analysis in the study area to study the four characteristics of the Nantun Formation reservoir and their relationship, the lower limit standard of effective reservoir physical property is established by the logging interpretation model of the reservoir mud content, porosity, permeability and saturation, and the mercury injection method, oil production method, oil test analysis method and cumulative frequency statistics method. The lower limit standard of effective reservoir electrical property is determined by using fluid identification chart. The results show that the lithology of the second member of Nantun Formation is mainly sandy conglomerate, and the first member of Nantun Formation is mainly sandstone, and the sandstone is more developed. The oil content is mainly full of oil, oil spots and oil stains. The reservoir porosity is between 4~20% and permeability is 0.01~100 mD, showing low porosity and low permeability. The lower limit of effective reservoir in Nantun Formation is porosity of 7%, permeability of 0.05mD, sonic time difference of 63μs/ft, density of 2.55 g/cm3 and water saturation of 20%.

Keywords: Wuerxun Sag; Nantun Formation; Four-property relationship of the reservoir; log interpretation model; standard for lower limits of effective reservoir.

1. Introduction

An oilfield is located in the Wuerxun depression in the central fault depression zone of the Hailaer Tamchage basin in China. The strata of the Tamulanhe Formation, Tongbomiao Formation, Nantun Formation, Damoguaihe Formation, Yimin Formation, and Qingyuangang Formation developed in the Lower Cretaceous. Three sets of reservoir cap assemblages are developed, with the Nantun Formation in the middle assemblage being the most abundant in oil and gas. The Damoguaihe Formation is a regional cap rock, while the Nantun Formation is the main source rock and reservoir. Mudstones in the study area are mostly grayish green, dark gray, and black in color, representing the underwater sedimentary environment. With little or no gravel, and high compositional and structural maturity, it represents a medium to low energy underwater sedimentary environment. Wedge shaped cross bedding, trough shaped cross bedding, deformed bedding, and ball pillow structures are developed, and unknown scouring surfaces are common. The sand body is relatively well sorted, with a fine particle size, and has traction current deposition characteristics with a certain transportation distance. The C-M diagram shows that progressive suspension transportation is the main body, followed by uniform suspension transportation and rolling transportation. Particles are characterized by uniform suspension and progressive suspension, and reservoir sedimentation is dominated by traction flow transportation mechanism. The provenance direction of the target layer mainly comes from the north of the research area, which is in the northeast or northwest direction within the development area. The target layer is braided river delta front subfacies deposition, mainly developing sedimentary microfacies such as underwater channel, mat bar, mat sand, and interdistribution, with sand bodies mainly concentrated in underwater channel microfacies. The relationship between the four characteristics of a reservoir (lithology, electrical property, physical property, and oil bearing property) and the determination of the effective reservoir lower limit are of practical significance for oil and gas exploration and development. Previous studies on the sedimentary characteristics of the study area have yielded sufficient results, but there has been no systematic study on the four characteristics of the reservoir in the study area and how to determine the lower limit criteria for effective reservoirs. This paper uses data from logging, cuttings logging, thin section identification, and analytical testing in the study area to study the four relationships of the

Nantun Formation, and establishes a reservoir logging interpretation model to determine the effective reservoir lower limit standard.

2. Study on the four characteristics and relationship of reservoir

2.1 Four characteristics of reservoir

(1) Lithological characteristics of the reservoir. Based on the analysis of core and logging data from 8 core wells in the study area, it is shown that the main lithology of the reservoir in the South Second Member is sandy conglomerate, while the main lithology of the reservoir in the South First Member is siltstone and fine sandstone. Compared to the two, sandstone is more developed. The particles of lithic sandstone are mainly sub angular to sub circular and of particle supported type. The proportion of rock debris in the rock particles is 29.2%, the proportion of quartz is 24.7%, the proportion of feldspar is 23.8%, and the proportion of filler is 22.3%, Among them, the muddy filling material is 18%, and the carbonate filling material is 4.3%.

(2) Oil bearing characteristics of the reservoir. Based on the analysis of oil bearing core sections and cuttings logging data from 8 core wells in the study area, it is shown that the oil bearing grades of the Nantun Formation are mainly divided into five categories: saturated oil, oil immersion, oil spots, oil traces, and fluorescence, with saturated oil accounting for 21% and oil immersion accounting for 10%; Oil spots accounted for 26%; Oil stains accounted for 22%, and fluorescence accounted for 21%. The main oil bearing grades of the Nantun Formation are saturated oil, oil spots, and oil traces.

(3) Reservoir physical characteristics. According to the histogram of the shale content of the reservoir in the Nantun Formation, the shale content of the reservoir in the South 2nd Member (Figure 1) mainly ranges from 10% to 30%; The main range of the South First Member (Figure 2) is between 10% and 20%. Compared to the two, the mud content of the South Second Member reservoir is higher. According to the distribution histogram of porosity and permeability of the Nantun Formation (Figure 3-6), it can be seen that the porosity of the reservoir in the Nantun Formation is mainly concentrated in 4% ~ 20%, and the permeability is mainly in 0.01mD~100mD; The porosity and permeability of the South 1st Member is better than that of the South 2nd Member, and the Nantun Formation is a low porosity and low permeability reservoir.

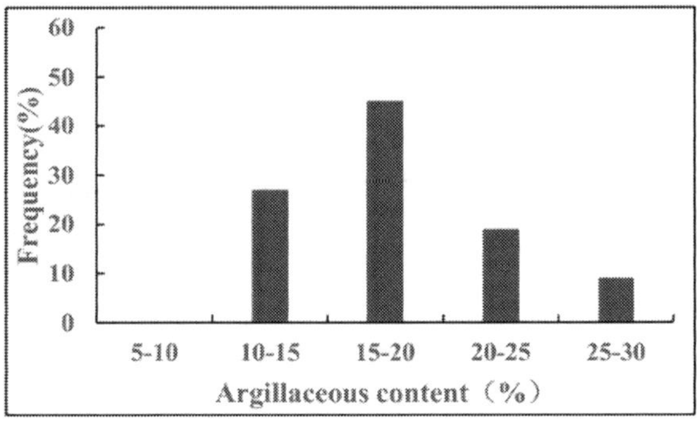

Fig.1 Distribution histogram of mud content in South Section II

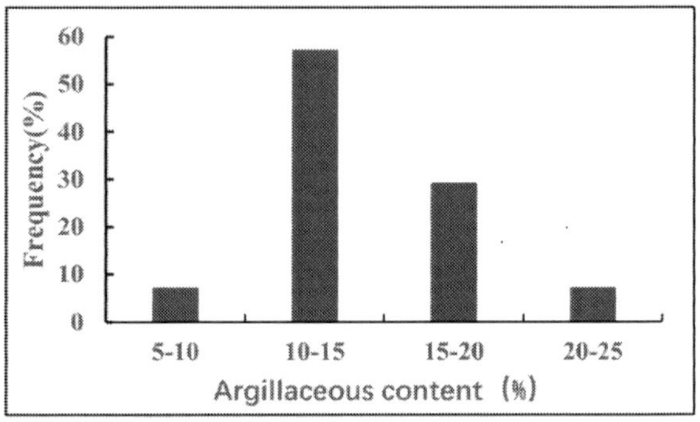

Fig.2 Distribution histogram of mud content in South Section I

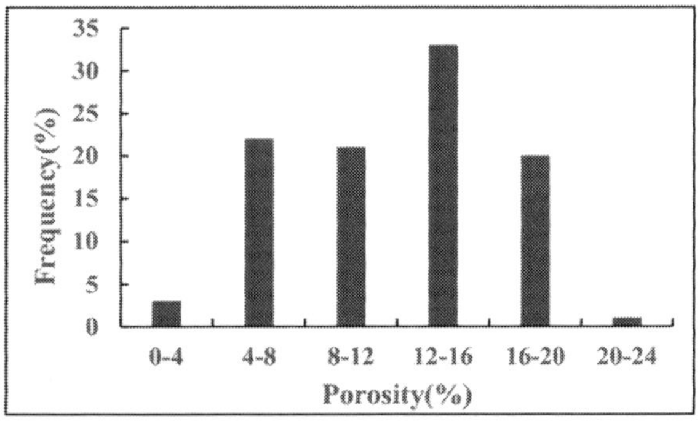

Fig.3 Porosity Distribution Histogram of South Section II

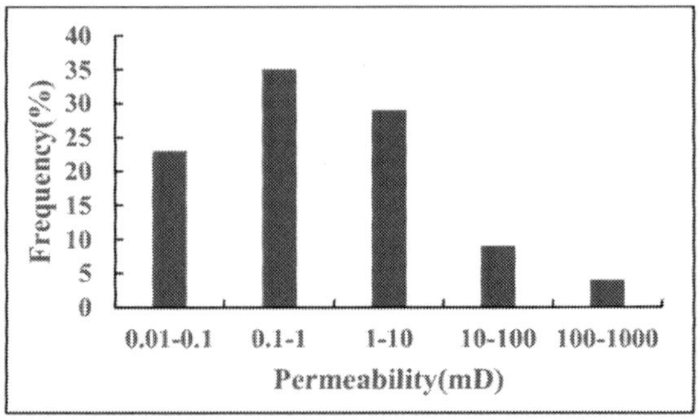

Fig.4 Permeability Distribution Histogram of South Section II

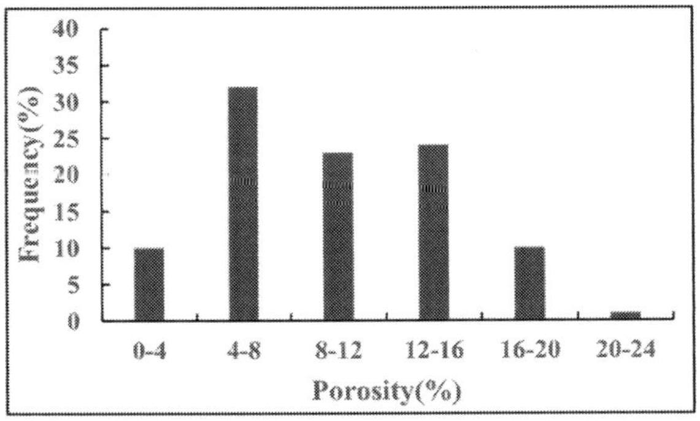

Fig.5 Porosity Distribution Histogram of South Section I

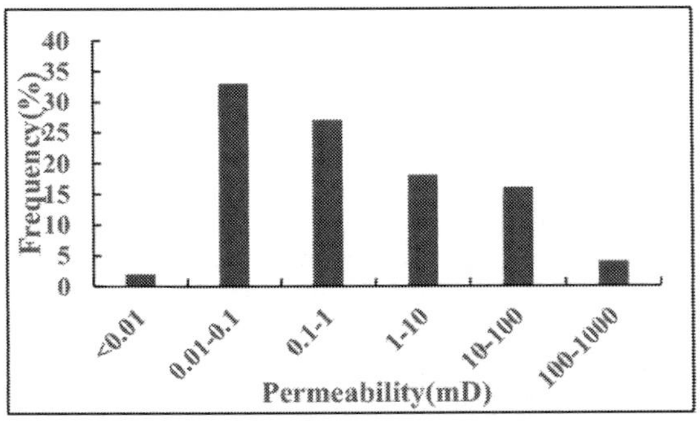

Fig.6 Permeability Distribution Histogram of South Section I

(4) Reservoir electrical characteristics. Reservoir lithology, oil bearing property, and physical property characteristics all have an impact on reservoir electrical properties. Based on the analysis of core well X27, it is found that both layer 113 and layer 115I of well X27 are oil layers.The resistivity amplitude value is large and the oil content is good; Low density, high neutron value, high porosity.

Layer 113:The natural gamma ray is 119.6 API, the shallow lateral resistivity is 32 Ω·m, the compensated neutron is 17.1%, the compensated density is 2.39 g/cm^3, and the acoustic transit time is 74.8μs/ft.

Layer 115:The natural gamma ray is 119.6 API, the shallow lateral resistivity is 32 Ω·m, the compensated neutron is 17.1%, the compensated density is 2.39 g/cm^3, and the acoustic transit time is 74.8μs/ft.

2.2 Relationship between four reservoir properties

(1) The relationship between lithology and oil bearing property. According to the distribution histogram of lithology and fluid types (Figure 7), it can be seen that the oil content of reservoirs is mainly concentrated in sandstone and glutenite reservoirs, with about 79% of the sandstone reservoirs containing oil, and only about 21% of the sandstone reservoirs containing water or dry layers. From the oil level and lithology distribution diagram (Figure 8), it can be seen that the main oil content levels of glutenite are oil immersion and a small amount of oil stains, and the oil content levels of sandstone are saturated oil, oil stains, and oil stains Oil bearing and fluorescence, from which it can be seen that the oil bearing grade of sandstone is higher than that of conglomerate.

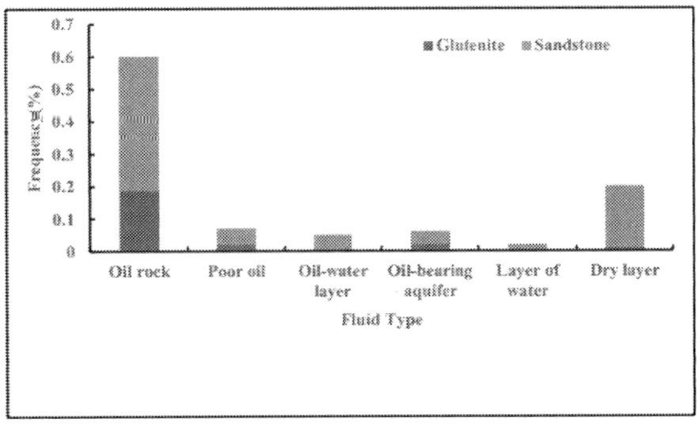

Fig.7 Relationship between lithology and fluid type

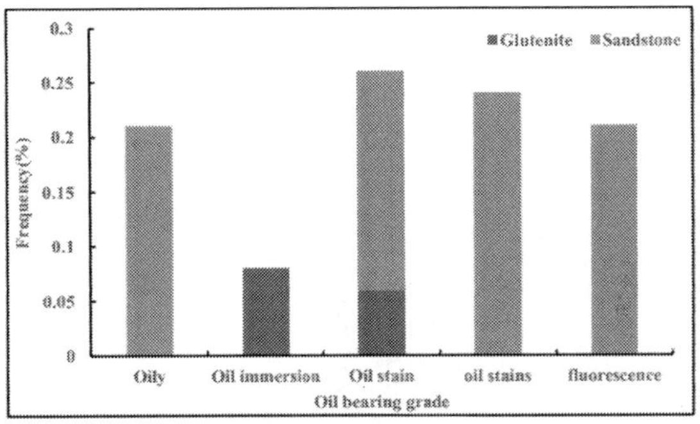

Fig.8 Relationship between lithology and oil grade

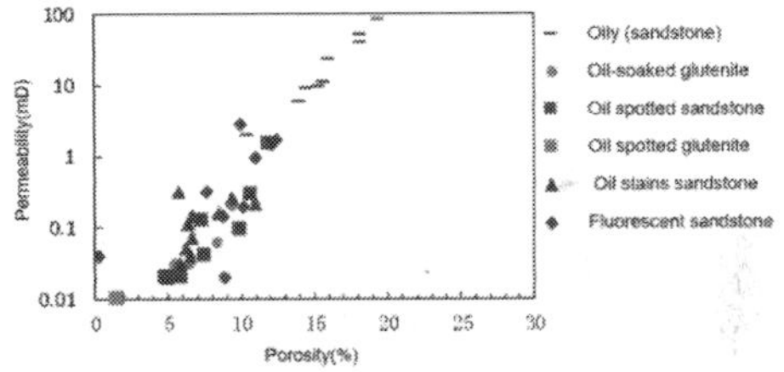

Fig.9 Cross plot of porosity and permeability of different oil-bearing occurrences

(2) The relationship between physical properties, lithology, and oil bearing properties. From the intersection diagram of porosity and permeability for different oil bearing occurrences (Figure 9), it can be seen that the better the physical properties of the reservoir (the higher the porosity and permeability value), the better the lithology and oil content of the reservoir. The physical properties of sandstone are better than those of sandy conglomerate, and lithology controls the physical properties; For sandstone, the porosity and permeability values of oil rich and oil stain are relatively high, while for oil stain and fluorescence, the overall porosity and permeability values are relatively low.

3. Logging interpretation model for reservoir parameters

3.1 Interpretation model of shale content

According to the research on the relationship between the four properties, the shale content of the reservoir is between 10% and 25% in both the South 1st and South 2nd Member. Overall, the shale content is relatively heavy, so it is necessary to consider the impact of the shale content on the calculation of reservoir parameters. Using natural gamma curve to calculate the shale content is not affected by fluid properties, which is the most common method. The specific formula is as follows (Equation 1~2):

$$V'_{sh} = \frac{GR-GR_{min}}{GR_{max}-GR_{min}} \quad (1)$$

$$V_{sh} = \frac{2^{Gcur*V'_{sh}}-1}{2^{Gcur}-1} \quad (2)$$

V_{sh} —Shale content,%;GR—Natural gamma, API; GR_{max} —Natural gamma value at mudstone, API; GR_{min} —Natural gamma value at sandstone, API;Gcur—Stratigraphic age coefficient, 2.0 for old formation.

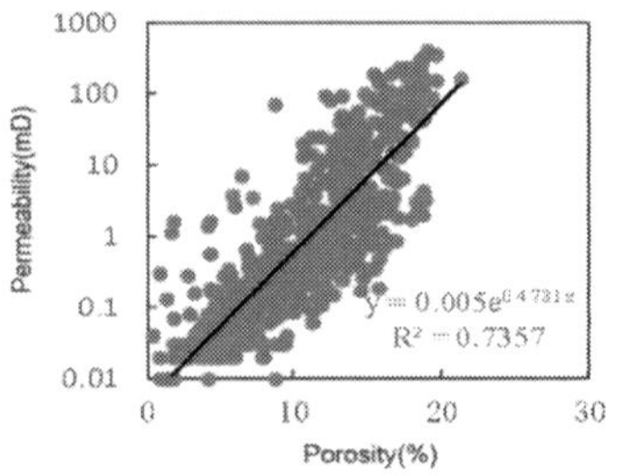

Fig.10 Porosity interpretation model

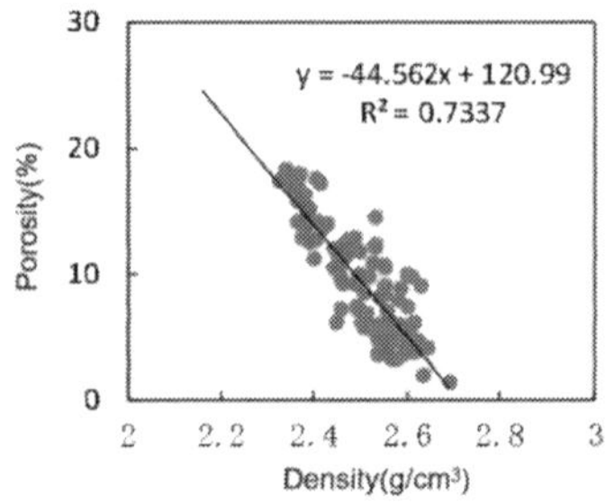

Fig.11 Permeability interpretation model

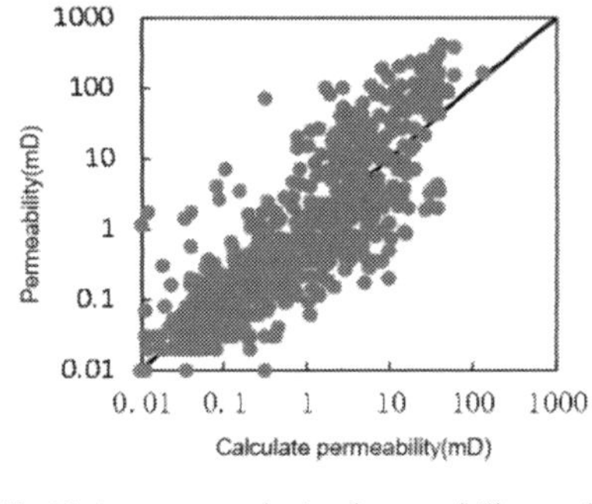

Fig.12 Accuracy analysis of permeability model

3.2 Porosity interpretation model

The reservoir lithology of South 1 and South 2 is mainly composed of sandstone and glutenite, while the core data is mainly from South 1. Therefore, a unified porosity model should be established to calculate the porosity of sandstone and glutenite reservoirs. Firstly, the core is deeply positioned. On this basis, a quantitative interpretation model for porosity in the study area is established using core analysis porosity and acoustic, neutron, and density regression methods, and the density method is preferred for porosity calculation. The regression formula is shown in Formula 3. From Figure 10, it can be seen that there is a good relationship between density and porosity, with a correlation coefficient of 0.86.

$$Y=44.562x+120.99; R^2 = 0.7337 \quad (3)$$

3.3 Permeability interpretation model

Permeability is an important parameter reflecting the permeability of a reservoir, which is closely related to the pore structure of rocks. For rocks with little change in pore structure, there is a good correlation between permeability and porosity, and this relationship can be used to estimate reservoir permeability through porosity. Using the method of core analysis porosity and permeability regression to establish a permeability interpretation model (Figure 11), the regression equation is shown in Formula 4; Figure 12 shows the intersection diagram of model calculated permeability and core analysis permeability. From the intersection diagram, it can be seen that the distribution of permeability is on the 45 degree line, indicating that the model can be used for the quantitative interpretation model of permeability in the Wuerxun area.

$$Y=0.005e^{0.4781x}; R^2 = 0.7337 \quad (4)$$

3.4 Saturation interpretation model

Aiming at the characteristics of low porosity and low permeability in the reservoir of the Nantun Formation, this paper selects the Archie formula to establish a reservoir saturation model. For the Archie formula, the first thing to determine is the value of the formation water resistivity.

(1) Determination of formation water resistivity

According to the analysis data of water samples from the Nantun Formation (Table 1), it can be seen that the equivalent sodium chloride mineralization of the South Second Member is mainly distributed between 12385~20172 ppm, with an average value of 16315 ppm; For the southern section, the equivalent sodium chloride mineralization is mainly distributed in 11218~23315 ppm, with an average value of 20203 ppm. Using the formation water salinity combined with Formula 5, it is determined that the formation water resistivity of the South Second Member is 0.12 Ω·m, and the formation water resistivity of the South First Member is 0.08 Ω·m.

Tab.1 Data Table of Mineralization Distribution of Formation Water in the Study Area

Layer	Range of mineralization	Equivalent Nacl mineralization range	Average value of equivalent nacl mineralization
South Section 2	22636.95~40099.27 ppm	12385.23~20171.67 ppm	16314.19 ppm
South section 1	11218.32~23123.80 ppm	11218.32~23314.58 ppm	20203.1 ppm

$$R_w = 82 * (10^{3.562-0.955*lgP_w+0.0123}) / (1.8*T+39) \quad (5)$$

Rw—Formation water resistivity,Ω·m;Pw—Mineralization degree, ppm;T—Temperature,℃.

(2) Determination of saturation parameters

Using the electrical experimental data of sandstone and glutenite in the study area, study the relationship diagram between formation factors, porosity, and resistivity increase coefficient, water saturation (Figure 13~14). Under dual logarithmic coordinates, they are linear characteristics. Therefore, use the regression formula as the value of the saturation parameter in the study area, namely, a=1.233, b=0.85, m=1.556, n=1.967.

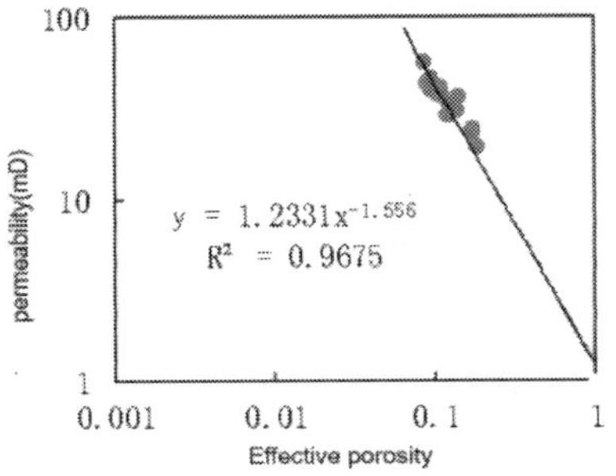

Fig.13 Cross plot of formation factors and porosity

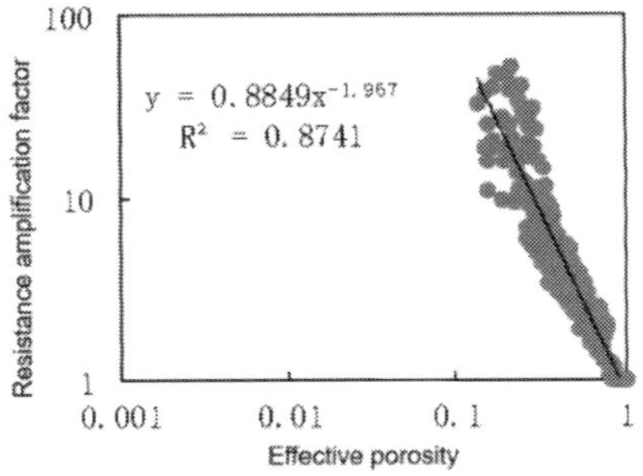

Fig.14 Cross plot of resistance increase coefficient and water saturation

(3) Resistivity forward modeling

The rock electrical parameter values established using rock electrical experimental data, combined with the formation water resistivity values, are substituted into the Archie formula to obtain a chart for resistivity forward simulation. The collected oil testing data are placed on the resistivity simulation line, and the saturation limits for each fluid type are determined. For oil reservoirs, the water saturation is less than 50%; The water saturation of the same oil water layer is greater than 50% and less than 70%; The water saturation of the oil bearing water layer is greater than 70% and less than 80%; The water saturation of the water layer is greater than 80%.

4. Determination of the lower limit of effective reservoir

The determination of the lower limit of effective oil and gas reservoirs is a key task for oil and water reservoir identification, reserve calculation, and development plan formulation. This study mainly uses data such as core testing and analysis, logging, and oil testing and production testing to determine the lower limit value of effective reservoirs through statistical analysis, construction of relationship diagrams, and other means. It mainly uses mercury intrusion method, oil occurrence method, oil testing analysis method, cumulative frequency statistics, and fluid identification charts to establish the lower limit standard of effective reservoirs.

4.1 Determination of reservoir lower limit by mercury intrusion method

Generally, capillary pressure curves and test data obtained from mercury intrusion experiments are used to analyze the microscopic pore structure characteristics of rocks to obtain the minimum flow pore throat radius. Based on the minimum flow pore throat radius, the corresponding porosity and permeability values, namely the lower limit value of reservoir physical properties, are further calculated.

The size of the minimum flow pore radius is mainly determined by the cumulative contribution rate of permeability. It is generally believed that the corresponding pore throat radius when the cumulative contribution rate of permeability reaches 99% is the minimum flow pore throat radius. Therefore, through statistical mercury intrusion data in the study area, the corresponding pore throat radius when the cumulative contribution rate of permeability of each rock core is 99% is obtained. Through statistics, it is determined that the minimum flow pore radius is 0.016 μm (Figure 15). By establishing the intersection diagram of pore radius and porosity for each rock sample (Figure 16) and the intersection diagram of pore radius and permeability (Figure 17), it is determined that the lower limit determined using mercury intrusion data is porosity equal to 7.4% and permeability equal to 0.05mD.

https://doi.org/10.1051/e3sconf/202339401017

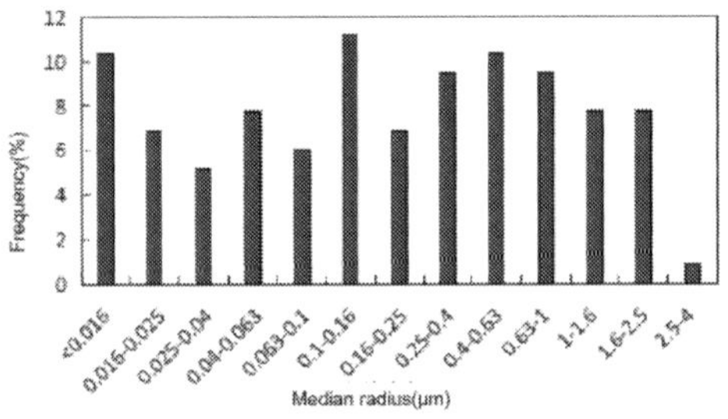

Fig.15 Median radius distribution map

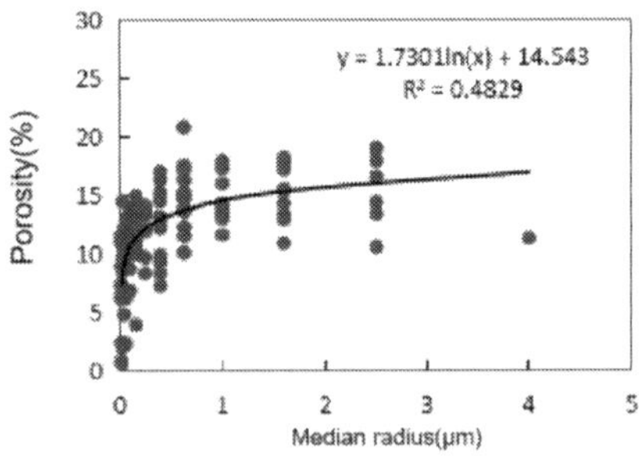

Fig.16 Cross plot of median radius and porosity

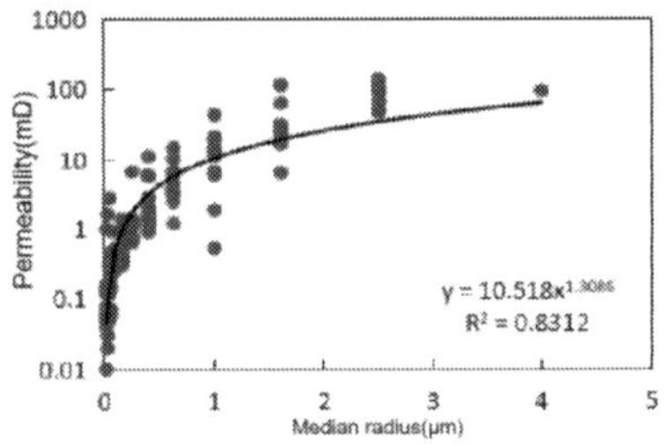

Fig.17 Cross plot of median radius and permeability

4.2 Determination of reservoir lower limit by oil bearing occurrence method

Based on the different levels of oil occurrence descriptions of logging cores and cuttings corresponding to different oil production capabilities, reservoirs described as saturated with oil, rich in oil, immersed in oil, or spotted with oil are classified as industrial oil layers, reservoirs with oil spots or traces are classified as low producing oil layers, and reservoirs with fluorescence and no display are classified as dry layers. The oil bearing occurrence method selected in this article mainly uses the porosity and permeability data from the physical property analysis of the core of the target layer to construct the porosity and permeability intersection diagram for different oil bearing grades (Figure 18), and determines the lower limit value of the reservoir physical property based on the boundary between the industrial oil layer and the low production oil layer. The limits determined using this method are porosity equal to 7.0% and permeability equal to 0.04mD.

4.3 Determination of reservoir lower limit by oil testing analysis

The oil testing method is to use the data of oil testing and core physical property analysis to determine the lower limit of physical property by constructing a cross plot of porosity and permeability for dry layers (non effective reservoirs) and producing layers (effective reservoirs), and using the porosity and permeability values corresponding to the boundary between the dry layers and producing layers. Due to the small amount of core data available at the oil testing horizon, the porosity of the reservoir is inversely calculated using the reservoir and dry layer logging response limits of sandstone and glutenite, and the permeability of the oil testing horizon is determined using a permeability model. The corresponding lower limit for the reservoir is that the porosity is equal to 5.7%, and the permeability is equal to 0.08mD.

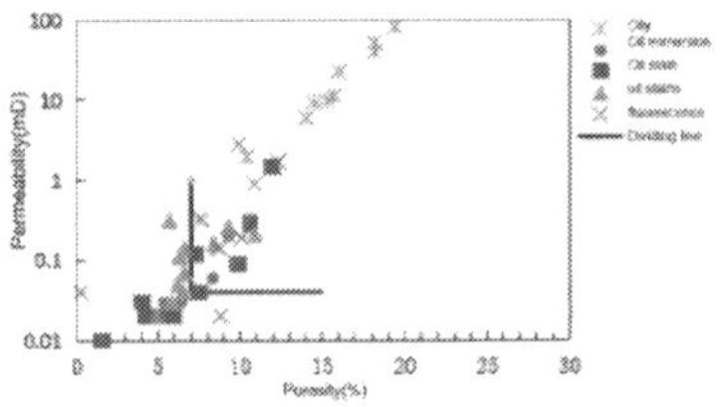

Fig.18 Cross plot of permeability and porosity by oil occurrence method

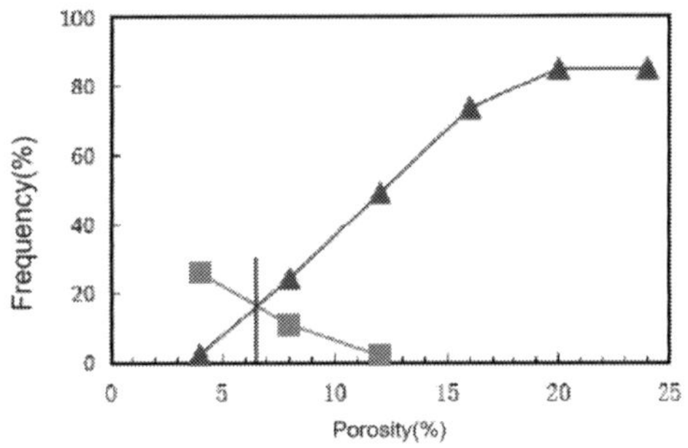

Fig.19 Schematic Diagram for Determining Lower Limit of Reservoir by Cumulative Frequency Method

4.4 Determination of reservoir lower limit using cumulative frequency statistics

This method utilizes the permeability limit determined above of 0.04mD, forward accumulation of coring data greater than 0.04mD, and reverse accumulation of coring data less than 0.04mD. The intersection of the forward and reverse accumulation curves is used as the boundary for determining whether the reservoir physical properties are effective. Figure 19 shows the sample distribution frequency greater than 0.04mD and the sample distribution frequency less than 0.04mD, with the intersection point of 6.5%. Therefore, the lower limit determined by this method is that the porosity is equal to 6.5% and the permeability is equal to 0.04mD.

4.5 Determining the lower limit of reservoir electrical property using fluid identification chart

 https://doi.org/10.1051/e3sconf/202339401017

The qualitative identification of fluid types mainly starts with oil testing data, and determines the differences in electrical characteristics of oil and water layers based on the study of the four characteristics. By summarizing the logging responses sensitive to fluids, the qualitative identification of fluid types is achieved using cross plot technology.

Based on real-time while drilling curves, compensated neutron, compensated density, acoustic transit time, and other logging curves that are sensitive to oil and water layers, a qualitative identification chart for oil and water layers has been established (Figure 20~21), thereby determining that the lower limit standard for acoustic transit time is 63 μs/ft, the lower limit standard for density is 2.55g/cm^3.

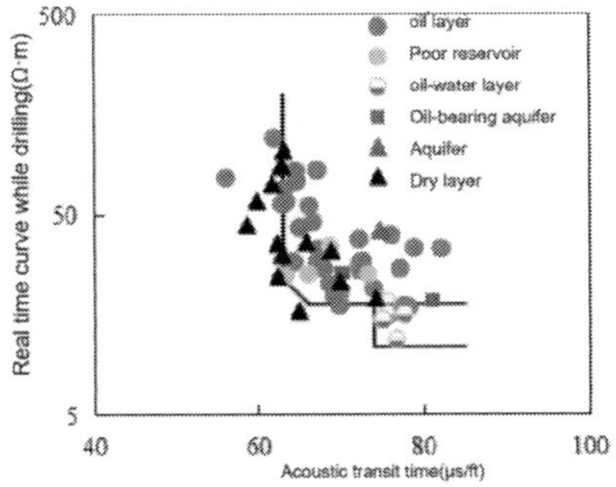

Fig.20 RT-AC Fluid Identification Plate

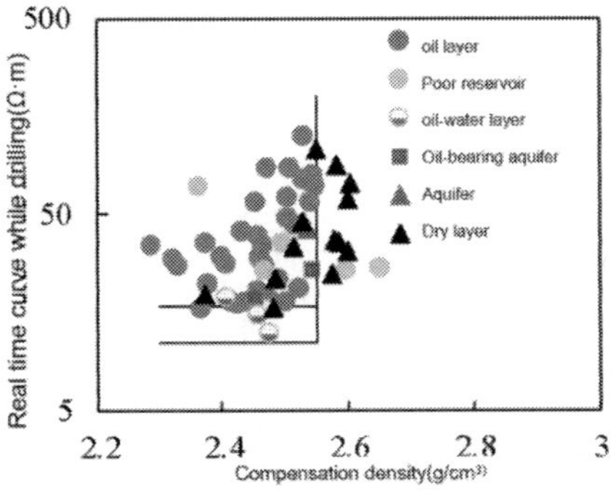

Fig.21 RT-DEN Fluid Identification Plate

Comprehensive mercury injection method, oil bearing occurrence method, oil testing analysis method, and cumulative frequency statistics method have established the lower limit standard for reservoir physical properties. Combined with the fluid identification chart, the lower limit standard for reservoir electrical properties has been determined, and the lower limit discrimination standard for the reservoir in the study area has been established (Table 2).

Tab.2 Standard Data Sheet for Lower Limit of Effective Reservoir

Reservoir logging response	Porosity/%	Permeability/mD	Acoustic transit time/(μs/ft)	Density/(g/cm^3)	Water saturation/%
Lower limit standard	7	0.05	63	2.55	20

5. Conclusion

(1) By using core logging data, cuttings logging data, thin section identification data, and analytical testing data in the study area to analyze the characteristics and relationships of the four reservoir properties (lithology, physical property, electrical property, and oil bearing property) of the Nantun Formation, the study indicates that the lithology of the second member of the Nantun Formation is mainly sandy conglomerate. The southern section is mainly composed of sandstone; The main oil-bearing properties are saturated oil, oil spots, and oil stains; The porosity of the reservoir is about 4~20%, and the permeability is about 0.01~100mD. The reservoir is a low porosity and low permeability reservoir; The electrical properties of oil bearing reservoirs are characterized by large resistivity amplitude values and good oil bearing properties; Low density, high neutron value, and high porosity. The lithology and physical properties of the reservoir are positively correlated with the oil content of the reservoir. Where the lithology of the reservoir is good, the porosity and permeability values of the reservoir are high, and the oil content level of the reservoir is high. In addition, the density value and neutron value of the reservoir with high oil content level are low.

(2) Research on logging interpretation models for reservoir parameters shows that there is a good relationship between density and porosity based on the porosity logging interpretation model, with a correlation coefficient of 0.86; The permeability logging interpretation model found that the permeability is mainly distributed around the regression line of 45°; By determining the formation resistivity and using Archie's formula to determine the limit of water saturation in the reservoir, the water saturation of the reservoir is less than 50%; The water saturation of the same oil water layer is greater than 50% and less than 70%; The water saturation of the oil bearing water layer is greater than 70% and less than 80%; The water saturation of the water layer is greater than 80%.

(3) Through comprehensive analysis of mercury intrusion method, oil bearing occurrence method, oil testing analysis method, and cumulative frequency statistics, the lower limit standard of effective reservoir physical properties in the study area is: porosity is 7%, permeability is 0.05mD; By using the fluid identification chart, the lower limit standard for effective reservoir electrical properties is determined as follows: acoustic transit time is 63 μs/ft, with a density of 2.25 g/cm^3.

Acknowledgements

This work was financially supported by the National Natural Science Foundation of China (42002154) and Open Fund Project of Heilongjiang Key Laboratory of Oil and Gas Reservoir Formation Mechanism and Resource Evaluation (KL20190101). The authors would like to thank the editor and the reviewers of this paper for their valuable comments that helped improve the quality of the paper.

References

1. Li Zandong,Bao Chuhui,Wang Dianju,et al. Controlling effect of structure palaeogeomorphology for sand bodies of Wuerxun—Beier Sag in Haila-er Basin[J]. Journal of Central South University Science and Technology,2016,47(7):2357-2365.
2. Wu Haibo, Li Junhui,Liu He. Formation condition and distribution rule of lithologic and stratigraphic reservoirs in Wurxun—Beir Sag[J]. Journal of Central South University Science and Technology, 2015,46(6):2178-2187.
3. Zhang Chen,Wei Kuisheng. Characteristics and reservoir formation condition in the southern Wuerxun Depression[J]. Acta Petrolei Sinica, 2005, 26(2):47-52.
4. Zhang Chen,Li Chunbai,Chu Meijuan,et al. Analysis on Characteristics and Controls of High-Frequency Sequences of Lower Cretaceous Series in Wuerxun Depression[J]. Acta Sedimentologica Sinica, 2005, 23(4):657-663.
5. Fang Wei. Reservoir characteristics of the first member of the south Wudong slope belt in Wuerxun depression[J]. Journal of Yangtze University Science and Technology, 2013,10(7):16-17.
6. Liu Yiming,Ye Jiaren,Zhang Fei,et al. Fluid Identification and Evaluation of Tight Sandstone Reservoir in Es2 of Qibei Slope Belt in Qikou Sag[J]. Earth Science,2022,47(5):1762-1776.
7. Yin Hongrong,Zhou Hu,He Weiqing,et al. Logging parameter interpretation model and the lower limit standard for reservoir evaluation of Chang 82 reservoir in Zhuang 9 Well Block of Heshui Oilfield[J]. Journal of Yangtze University Science and Technology, 2021,18(2):36-41.
8. Li Yingxu,Zhao Junxing,Wei QianSheng,et al. The reservoir"four properties"relation and the lower limit of effective reservoir in He 1 Member of Xiashihezi Formation in central area of Hangjinqi,Ordos Basin,China[J]. Journal of Chengdu University of Technology: Science & Technology Edition, 2021,48(6):675-682+731.
9. Chen Luoyuan,Qian Yugui,Liao Luyao,et al. COMPREHENSIVE EVALUATION OF HE7 EFFECTIVE RESERVOIR OF MIDDLE PERMIAN LOWER SHIHEZI FORMATION IN WELL BLOCK OF SU-14 OF SULIGE GAS FIELD[J]. Mineralogy and Petrology, 2021, 41(2):109-117.
10. Hui Zengbo,Sun Mengsi,Feng Congjun,et al. Study on four reservoir characteristics and oil-water layer identification of the third member of the east Cha71 fault block in Chaheji oilfield[J]. World Well Logging Technology, 2021,42(2):56-59.
11. Liu Chengchuan,Chen Jun,Li Huaji,et al. Logging evaluation of the tight sandstone reservoir in the gas reservoir of the Shaximiao formation within Zhongjiang gas field,China[J]. Geophysical Prospecting For Petroleum, 2020,59(1):131-140.
12. Chen Jia,Feng Congjun,Yu Tianjun,et al. The lower limit of physical properties of tight reservoir in He 8 Member of Y113-Y133 gas well area,Yanchang Gas Field,Ordos Basin[J]. Natural Gas Geoscience, 2022,33(6):955-966.
13. Zhang Shiming,Wang Jiangong,Zhang Yongshu,et al. Determination of petrophysical property cutoffs of lacustrine dolomite intercrystalline pore reservoir in the Xiaganchaigou Formation,western Qaidam Basin [J]. Acta Petrolei Sinica,2021,42(1):45-55+118.
14. Zhou Nengwu,Lu Shuangfan,Wang Min,et al. Limits and grading evaluation criteria of tight oil reservoirs in typical continental basins of China[J]. Petroleum Exploration and Development,2021,48(5):939-949.
15. [15]Meng Lei. Lower Limits of Physical Properties of Sandstone Reservoirs in the Second Member of Sangonghe Formation in Moxizhuang Oilfield[J]. Xinjiang Petroleum Geology,2020,41(5):557-564.
16. Lu Feng,Wang Jian,Zhang Juan,et al. Characteristics and lower limit of physical properties for reservoirs of 1st Member of Sangonghe Formation in Mosuowan area[J]. Fault-Block Oil & Gas Field, 2021,28(3):300-304.
17. Zhang Yiming,Fu Xiaodong,Guo Yongjun,et al. Petrophysical property limits of effective tight oil reservoirs in the lower part of the first member of Cretaceous Tengger Formation,A'nan Sag,Erlian Basin,North China[J]. Petroleum Geology & Experiment, 2016,0(4):551-558.
18. Cai Zhendong,Zhao Chenyang,Du Yu,et al.Study on the Effective Thickness of Yan 10 Reservoir in M Area of Ordos Basin Maling Oil Field [J]. Liaoning Chemical Industry,2015,0(4):451-455.
19. Bai Ze,Tan Maojin,Shi Yujiang,et al. Genesis of low-resistivity oil pays and a fluid identification method for tight sandstone reservoirs:A case study of the Chang 8 Formation in the Longdong West area,Ordos Basin[J]. Geophysical Prospecting For Petroleum, 2022,61(4):750-760.
20. Zhang Congxiu,Hao Jinmei,Liu Zhiheng,et al. A Study on the Logging-Based Identification Method for Reservoir Fluid Properties of the Yan'an Formation in the Huanxi–Pengyang[J]. Petroleum Drilling Techniques, 2020,48(5):111-119.

 https://doi.org/10.1051/e3sconf/202339401018

Typical case analysis of oilfield tracer testing technology

Shichao Zhang

Daqing Oilfield International Exploration and Development Company, Daqing, China

Abstract. The development history and principle of interwell tracer testing and single-well tracer testing were investigated, and the application conditions and application advantages of different tracer testing were analyzed through the analysis of two types of tracer testing technologies and test results in the production process of SMi oilfield with medium porosity and low permeability, so as to provide reference for the study of tracer development in the same type of oilfield. Tracer testing technology can master the connection between injection well and production well, effectively judge the connection relationship between Wells, and provide a basis for well pattern adjustment; The seepage velocity of injected fluid in the formation can be directly observed, and the dominant injection direction can be determined, which provides reference for solving the plane contradiction. It can judge whether there is fracture system or high permeability strip, which provides ideas for high precision numerical simulation to avoid risks. It can master the remaining oil situation of single well and judge the washing efficiency of chemical agents effectively on the basis of saving investment.

Key words: Interwell tracer test; Single-well tracer test; Plane contradiction; Hypertonic band; Oil saturation

1. Introduction

Tracer testing technology was introduced in China in the 1980s. The early oilfield tracer technology only qualitatively described the subsurface movement direction of the injected fluid and the nonhomogeneity of the reservoir. With the development of technology and the increasing demand of oilfield development, oilfield tracer technology has gradually become an important reservoir engineering means in the process of secondary and tertiary oil recovery, and has been rapidly popularized and applied. [1]

Due to different development concepts and national conditions, foreign oil fields tend to pursue extreme benefits, and pay more attention to where there is oil, which layer of oil, how to quickly recover oil. The SMi field is a large onshore sandstone reservoir with medium porosity and low permeability. Production peaked in 1989, then began to decline, and stabilized for some time after adjusting the well pattern. In 2006, the water content exceeded 90%, and the effect of extracting liquid decreased gradually. From 2009 to 2013, a large range of multi-well group interwell tracer test was carried out to determine the advantageous waterflood direction and solve the interzonal waterflood contradiction, which provided support for the follow-up flood control and stable production and suppressed the production decline. In 2017, the oilfield has entered the ultra-high water cut stage. In order to prolong the life cycle of the oilfield, three times of oil production has been attempted in the plot block. On the basis of the preparation of the reagent, a single well tracer test was carried out to verify the oil washing efficiency of the reagent and judge the economic benefits of the three times of oil recovery. The tracer test of this oilfield is comprehensive in the development process, which can provide reference for other domestic oilfields in the same development process.

2. History, principles and classification of tracer testing

In the 1950s, chemical tracers such as inorganic salts, dyes, halocarbons and low molecular alcohols were used to qualitatively understand the movement of underground fluids. In the 1970s, the radioisotope tracers represented by tritium and tritium compounds appeared. In the 1980s, the stable isotope tracers represented by deuterium and its compounds were studied and applied. Due to their shortcomings, the first three generations of oilfield tracers could not meet the requirements of oilfield exploitation, which promoted the production of trace substance tracers. Since the mid-1960s, oilfield tracer technology has made great progress in many countries in the world. [1]

Tracers are substances that are easily soluble and can be detected at very low concentrations to indicate the presence, flow direction and flow velocity of the dissolved fluid in the porous medium. Commonly used tracers can be divided into distributive tracers and non-distributive tracers. Non-distributive tracers are only soluble in water, while distributive tracers are soluble in both water and oil. [2] Field tracer testing can be divided into two categories: interwell tracer testing and single-well tracer testing. Interwell tracer testing is used to determine interwell connectivity and permeability anomalies, while single-well tracer testing is used to study remaining oil saturation near the well.

2.1 Interwell tracer testing

The interwell tracer test (FIG. 1) involves injecting a tracer slug into the injection well and continuously monitoring the tracer production in the surrounding production well. The tracer concentration curve over time is called the tracer production curve. The production curve of the tracer was analyzed to gain knowledge of the reservoir. [2] The early analytical methods of interwell tracer testing could only be used to qualitatively judge the reservoir ground connectivity and the existence of deep bands. In 1984, Briham et al. used analytical methods to quantitatively analyze the tracer production curve, and the parameters were the average formation parameters between oil and water Wells. In 1997, the numerical simulation method was used to solve the formation parameters in China with objective accuracy. In 2001, a semi-analytical method combining numerical and analytical methods appeared. In recent years, on the basis of semi-analytical method, combined with reservoir modeling, profile analysis, numerical simulation and other methods to take advantage of each other to form a comprehensive interpretation method. [3]

Cross-well tracer testing can determine and analyze the heterogeneity of the reservoir in both horizontal and longitudinal directions. Determine whether there is a high permeability layer in the formation and calculate its permeability and other formation parameters; It can also be used to determine the type and dosage of profile control agent; If a distributive tracer and a non-distributive tracer are injected at the same time during the interwell tracer test, the non-distributive tracer is only soluble in water, because the two tracers differ greatly in oil solubility. However, distributive tracers are soluble in both water and oil and produce late. According to the production time difference and distribution coefficient of the two tracers, the remaining oil saturation in the study area was obtained.

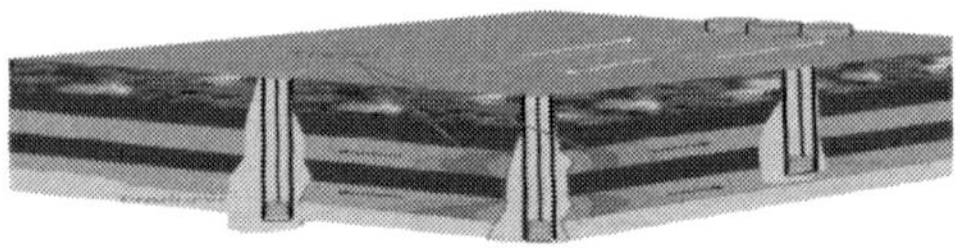

FIG. 1 Schematic diagram of tracer testing between Wells

2.2 Single-well tracer test

Single-well tracer tests (Figure 2), which measure remaining oil saturation with tracers injected and produced from the same well. Usually, a low molecular ester is injected as the first tracer, and then dissolved in water to form an alcohol as the second tracer. The distribution coefficients of the two tracers in oil and water are different. The first tracer is oil-philic and the second tracer is water-philic. The two tracers are separated during mining, and there is a time difference when their peaks reach the surface, according to which the remaining oil saturation can be obtained.

The determination of residual oil saturation by tracer method using the chromatographic effects of the formation was first proposed by Cooke. This method was successfully used by H.A. Daans to determine residual oil saturation in a model in the laboratory in 1967, applied to field tests in 1968, and tested three times in the same well in 1969 with success. Exxon Production Research patented the technology in 1971. From 1968 to 1993, more than 200 Wells were logged in the United States, Canada, Venezuela, North Sea and other oil fields. [4] Shengli Petroleum Administration Bureau began to study the single-well method to test residual oil saturation in 1981, and determined the technological process and interpretation method in 1987. Later, it was popularized and applied in Xinjiang Petroleum Administration Bureau and Yumen Petroleum Administration Bureau successively. [5] At present, the interwell tracer testing technology can not only store energy to determine the thickness and permeability of water-salt layer, the diameter of large pore, the channeling of injection well pipe and the perforation in the pipe, and the orientation of fractures, but also reflect the change of water injection quantity inside the waterflood layer, the contradiction between each waterflood layer and the distribution of remaining oil in water drive oilfield. [6]

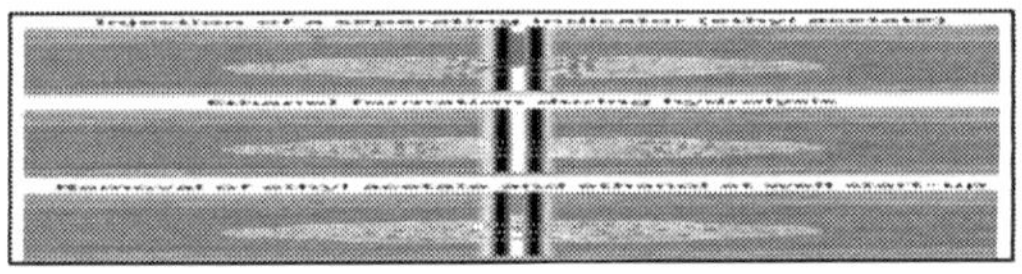

FIG. 2 Schematic diagram of single-well tracer testing

3. Case analysis

SMi Oilfield began to be put into production in 1982, and the reservoir saw water immediately after it was put into production. Most of the producing Wells had a short period of water-free oil production, and the water cut rapidly increased from 17.2% in the initial period to 80% (Figure 3). Production peaked in 1989, when water content remained below 40%, and then production began to decline. Water cut exceeded 80% in 2004 and 90% in 2010 (Figure 4). Comprehensive analysis shows that there may be hypertonic bands, which lead to this development phenomenon.

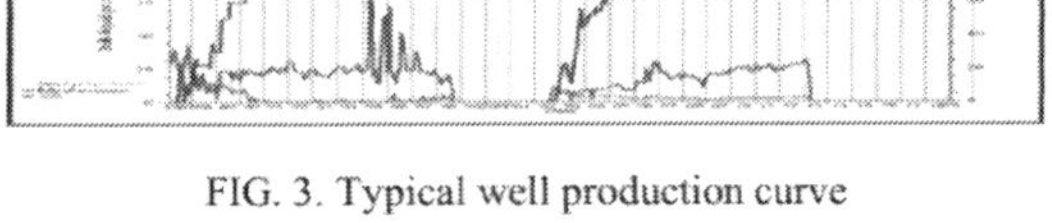

FIG. 3. Typical well production curve

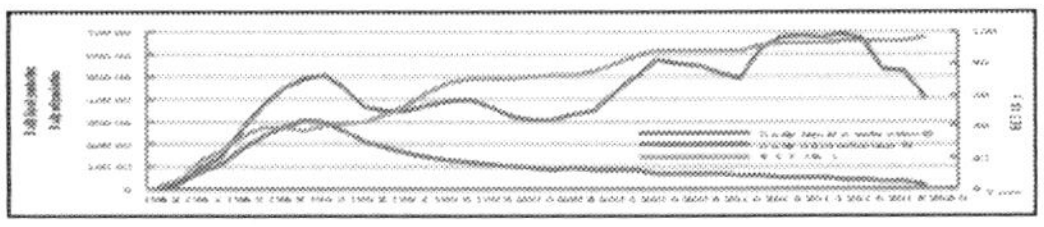

FIG. 4 Production curve of SMi oilfield

3.1 Interwell tracer testing

From 2009 to 2013, a total of 146 oil-well interwell tracer studies were conducted on 25 well groups in three target zones to investigate and study reservoir physical properties and assist in determining reservoir geological conditions. It plays a key role in the follow-up plugging and postponing production decline. Select injection Wells with injection rates greater than 100m³/d and inject tracers (e.g. Urea, guanidine thiocyanate, ammonium nitrate, sodium fluorescein) that are pre-dissolved in water into the Wells. 4-10 production Wells were selected around the injection well as monitoring Wells, and samples were taken within three months after the tracer injection. The reservoir physical property parameters were studied based on the tracer test. (Take layer 1# as an example)

Based on the tracer study results in well S, the following conclusions can be drawn: There is a good hydrodynamic connection between the S injection well and the surrounding Wells. The main flow line traced in the first 90 days percolated along the abnormally high conductivity interval. The average permeability of the seepage channel through which the tracer flowed was 6.96μm2, several hundred times higher than the designed level (0.187μm2). In the direction of Wells D, N and A, most of the tracers (58-66%) were percolated through channels with abnormally high conductivity (15-200μm2). Throughout the study period (90 days), tracer production accounted for approximately 71% of the injected volume (1000kg), and the total permeability of the tracer dominant channel was 137m3/d, accounting for 74% of the average daily injected water in well 5292. The main flow directions in the studied reservoir are as follows: northwest (13% of the injected water); Northeast direction: (44% of water injected); Southeast direction: (15% of the injected water); Southwest direction: (2% of water injected).

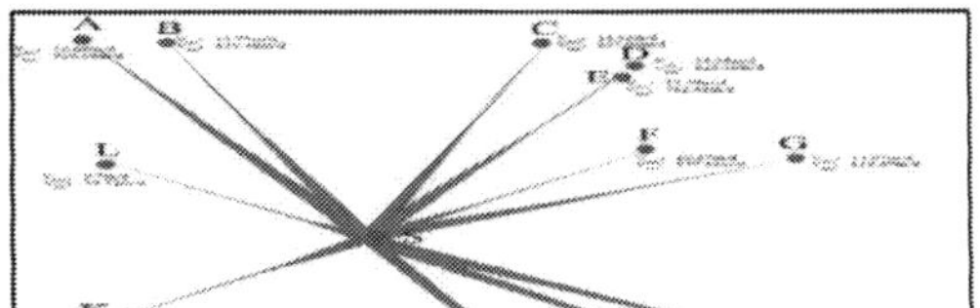

FIG. 5 Flow velocity and flow direction of the tracer in well S of Layer 1#

Table 1 Calculated permeability of small layer

Pound sign	Maximum permeability, μm^2	Minimum permeability, μm^2	Mean permeability, μm^2	Calculated permeability value, μm^2
A	187.2	3.3	11.4	
B	152.7	2.6	6.5	
C	146.7	2.5	5.7	
D	177.3	3.3	8.4	
E	135.4	2.3	5.0	
F	58.1	1.6	3.4	0.187
G	145.3	2.5	5.8	
K	274.6	4.7	11.4	
L	287.5	5.0	11.2	
N	202.7	3.6	9.7	
O	103.3	1.9	3.7	
Q	6.5	1.2	1.5	

The results of zone 1# tracer study showed that there was good connectivity between the injection well and the surrounding production well, and in most cases, the communication was through high conductivity channels. The distribution characteristics of seepage fluid and velocity indicate that the plane heterogeneity is serious. The average permeability of the pervious channels through which the tracer flowed varied from 4.6μm2 to 7.03μm2, which was several hundred times higher than the designed level (0.187μm2). The main distribution directions of low resistance seepage channels are northwest and southeast. The results of the interwell tracer test were consistent with permeability ranges from core, logging, and well testing studies, and the data were mutually verified.

3.2 Single-well tracer test

In 2018, the combined water content of the field exceeded 95%. In order to continue the field development cycle, the field is looking for a tertiary production partner. 2021 Partners successfully developed a polymer agent compatible with the formation on the basis of laboratory tests. To verify the efficacy of the agent, single-well tracer testing was conducted on selected Wells in late 2022.

Its work flow includes:

① Well selection: The study area should be located above the oil-water interface, without single-layer outburst, well casing seal condition, adequate formation energy, 2-6 meters near the well, 10 meters of effective thickness, and unfractured; (2) Select the appropriate tracer, verify its composition of five elements in laboratory, determine the hydrolysis rate and obtain the separation coefficient; ③ large dose injection of formation water; (4) Injection of tracer into the formation; (5) technical shut-in and interval sampling; ⑥ Interpretation of research results.

Site construction can be roughly divided into four stages: the first stage is the technical settlement of single well and the injection of a large amount of formation water into the formation; The second stage is to inject tracer into the formation. The third stage of chemical reaction technology settlement stage; Phase IV single well recovery and sampling phase. (Figure 6)

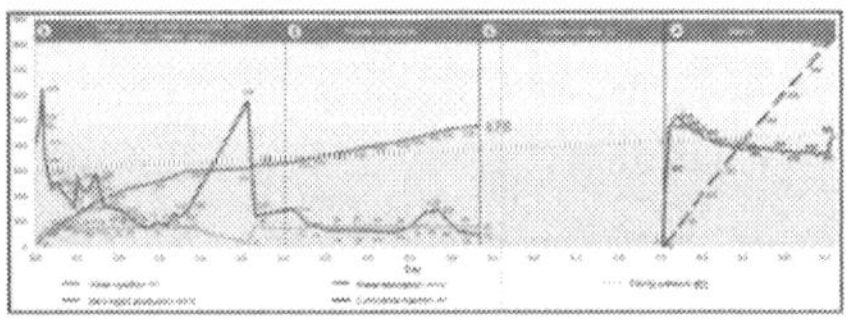

FIG. 6 Schematic diagram of single-well tracer test construction technology

The calculation formula is:

$$S_{oil} = \frac{((Q_{ethyl})/(Q_{ethanol} - \partial) - 1)}{((Q_{ethyl})/(Q_{ethanol} - \partial) - 1 + K_d)}$$

Among them:

Q_{ethyl} Is the extreme value of the ethyl acetate concentration curve, m^3;

$Q_{ethanol}$ Is the peak of ethanol concentration curve, m^3;

∂ Is wellbore volume, m^3;

K_d Is the separation coefficient。

Single-well tracer test results showed an 8.2% decrease in near-wellbore residual oil saturation from the initial stage of production, translating to 15.5% enhanced oil recovery. The increase in oil recovery by chemical flooding was 13%-22%. The two results are consistent, which proves the reliability of the single-well tracer test and verifies the feasibility of the agent.

4. Conclusion

1) Interwell tracer test can master the connection between injection well and production well, effectively judge the connection relationship between Wells, and guide well pattern adjustment and new well exploration and deployment;
2) The dominant injection direction can be determined according to the different seepage velocity and water absorption of the injected fluid in the formation, which plays an auxiliary role in adjusting the development strategy;
3) Cross-well tracer testing can determine whether fracture systems and high-permeability bands exist in the formation, which can provide strong support for numerical simulation;
4) Single-well tracer test can measure multiple parameters such as residual oil saturation, bound water saturation and formation water salinity in the near-well zone. The reserve and distribution of residual oil plays an important role in determining whether an enhanced oil recovery method is economically viable.
5) With the continuous progress of tracer testing technology, the testing methods are diversified and the application scenarios are more convenient, which provides a broader prospect for the field research of tertiary oil recovery technology.

References

1. [1] Yu Ruixiang, Zhang Taishan, Zhou Weisheng. Oilfield tracer technology [J]. Industrial Water Treatment, 2007, 27(8):12-15. (in Chinese)
2. [2] Ye Huimin. Residual oil saturation was measured with a distributive tracer. Fault-block Oil and Gas Field, Vol. 9, No. 4, 2002.7.
3. [3] Liu Tongjing, Zhang Xinhong, Jiang Hanqiao, Dai Xingbin, Zhang Peixin. New development of interwell tracer test Technology.2007.8 Vol.20 No.3.
4. [4] Wei Lichun, Han Rengong, Zhang Ping. Determination of residual oil saturation before and after combined flooding in Gudong small well spacing test area using single-well tracer method.
5. [5] Zhang Ping, Yuan Shigao. Single-well chemical tracer logging technique.1996.4. ISBN 7-5021-1636-2
6. [6] Zhao Guoyu. Application of interwell tracer technology in oilfield production. Petroleum Exploration and Development, Vol. 26, No. 4, 1998

https://doi.org/10.1051/e3sconf/202339401019

Study on Well Pattern System Optimization of Fuyang Reservoir in H Oilfield

Jing Zeng

Daqing Oilfield International Exploration and Development Company, Daqing 163000, China

Abstract. The selection of well pattern system is not only the key content in the process of oilfield development, but also the core problem in the design of oilfield development. It is also the basis for the effective utilization of tight reservoir technology. Through scientific analysis of the geological structure and dynamic characteristics of Yang reservoir in H oilfield, the well pattern system can be reasonably optimized and the development of Chinese oilfield can be vigorous. Starting from the selection of well pattern system for oil and gas field development, this paper analyzes the optimization of well pattern system parameters of Fuyang reservoir in H oilfield.

Key words: Fuyang reservoir, oil and gas field development, well pattern system optimization

1. Introduction

The development of oil and gas fields provides a solid guarantee for our energy supply and lays a good foundation for our economic development. The development and research of low-permeability petroleum resources plays an important role in promoting the healthy and sustainable development of our petroleum industry. The study of well pattern system has always been the key research subject in the process of oil and gas field development. The selection and design of well pattern system should fully respect the geological conditions and geographical characteristics of the exploitation area. The well pattern system chosen should be economically sound. To facilitate the development of oil and gas fields more smoothly.

2. Selection of well pattern system for oil and gas field development

The selection, deployment and adjustment of well pattern system is the key part of oil and gas field production and development, and also an important measure to promote the economic benefit of oil and gas field development enterprises. The form of well pattern is mainly affected by the geological characteristics of oil and gas fields. By observing whether the geometric shape of the well pattern is regular or not, the well pattern in the production site is usually divided into two types: regular well pattern and irregular well pattern. Different oil and gas reservoirs are exploited by different well patterns, and the nature of oil and natural gas is greatly different in nature. Different production methods should be adopted in the actual production process, and the well pattern of the oil field is much more complex than that of the gas field. In the production process of oil and gas fields, when the regular well pattern is used for mining, that is, the area injection well pattern. The common area injection well pattern is divided into linear well pattern, staggered well pattern, four-point well pattern, five-point well pattern, etc. The irregular well pattern is usually caused by the deformation of the regular well pattern. The specific well pattern adopted in actual production and exploitation should be analyzed according to the geological structure characteristics of oil and gas reservoirs, and then the mining method should be decided after detailed study [1].

In the initial stage of oil and gas field development, attention should be paid to the selection and deployment of well pattern. The selection and deployment of well pattern in the initial stage of development of oil and gas field has an important impact on the entire later production. In actual mining, the layout of basic well pattern is usually carried out first, and then the geological conditions and structural characteristics of the mining area are analyzed, and then the actual well layout is carried out. The application of the five-point well pattern in the regular well pattern is conducive to late adjustment and maximizes field production under certain conditions. According to the special geological condition of low permeability oilfield, well spacing should be shortened reasonably, and the density of well pattern should be increased to promote the best development effect. It is also necessary to carefully study formation fractures and rationally arrange injection-production well pattern [2].

The adjustment of well pattern is an essential link in the process of oil and gas field development. With the extension of development time, new problems will constantly appear, and the emergence of new problems requires constant adjustment of well pattern. The common well pattern adjustment methods include encryption and transfer. In the initial stage of oil and gas field development, relatively sparse well pattern is usually used, and the developed reserves are more concentrated. In the case of insufficient use of reserves and large amount of remaining oil and gas, encryption technology is often used to promote the stable and efficient operation of oil and gas fields. For fractured reservoirs, the current water injection method should be adopted to combine the sand body size and strike, and fully consider the uniformity of well pattern layout. For areas where fractures are not well developed, the relative uniformity of the well pattern should be considered. To maximize the development effect, appropriate adjustment methods should be selected according to the actual geological conditions [3].

According to the geological characteristics of Fuyang reservoir, the selection of well pattern system should take into account the well pattern system should adapt to the distribution characteristics of underground reservoir to the greatest extent, and promote the well pattern to have a higher degree of water flooding control, the well pattern system should also have a higher intensity of injection and production, and the well pattern should have a strong flexibility in the process of development adjustment.

3. Optimization of well pattern system parameters of Fuyang Reservoir in H Oilfield

3.1 Density of well pattern

The effect of waterflood development in oil fields is directly related to the density of well pattern. The cost of drilling also occupies a large part of the investment in the construction and development of oil and gas fields. Therefore, the density of well pattern plays a crucial role in the economic benefits of waterflood development technology. In order to achieve better development results, the density of well pattern should be strictly controlled to ensure the economic benefits of well pattern system effectively. The density of well pattern is a key content in the development design of oil and gas field, which involves the calculation of oil and gas field development indicators and the evaluation of economic benefits. The density of well pattern is related to the spacing and size of well pattern [4]. The density subdivision of well pattern can be divided into two kinds: reasonable well pattern density and limit well pattern density. Reasonable well pattern density means that under certain geological and development conditions, the total revenue of oil and gas fields minus the total input is the maximum, at which time the well pattern density will be obtained at the maximum profit. If the total input is equal to the total output and the profit is 0, the limit well pattern density will be obtained. As for the algorithm of limiting well pattern density, iteration method and curve intersection method are generally adopted. In general, the differential method is used to solve the consensus S in order to obtain the reasonable well pattern density. The calculation formula is different, and the calculation results are also different. In order to achieve accurate results, it is necessary to establish a mathematical model suitable for different types of oil and gas fields, considering the comprehensive factors, and select the appropriate well density through modeling analysis.

3.2 Well pattern density and recovery efficiency

With the expansion of the development degree of oil and gas fields, many domestic and foreign oil and gas fields have analyzed through practical studies that there is a great relationship between the well pattern density and the ultimate recovery factor. For homogeneous oil reservoirs, the influence of well pattern density on the ultimate recovery factor is not particularly obvious, but for heterogeneous oil reservoirs, especially in the case of complex geological structure, The density of well pattern has a great influence on the recovery efficiency, and this influence is most obvious in the middle and late period of oil and gas field development. In the middle and late period of oil and gas field development, the effect and quality of well pattern density development will directly affect the development effect and recovery efficiency of oil and gas field. If a thin well pattern is applied to an heterogeneous reservoir, the loss of reserves will increase. It can be seen from a large number of practices that the higher the well pattern density, the higher the recovery rate and the smaller the increase [5].

The density of well pattern should be adjusted according to the actual situation of the oilfield in the middle and late stage of oilfield production. Usually, the case is to carry out infill treatment. The practical research shows that the adjustment of the infill pattern can reduce the reserves loss, effectively improve the rate of oil production, and has a direct impact on the final recovery efficiency. Infill well pattern is an important way to solve the discontinuous oil field recovery efficiency. Improving the drilling rate of oil sand body is an important measure to improve the water drive recovery efficiency.

Waterflooding control degree refers to the percentage between the thickness of a producing well drilled on the sand body that can be effectively affected by water flooding and the total drilled thickness. This method mainly studies the influence of different well pattern densities and various development adjustment measures on the waterflooding control degree. In the actual development process of oil and gas fields, the density adjustment of well pattern is usually encrypted along with the development progress of the oil field. With each encryption, the control degree of the well pattern on water flooding will be continuously improved. The monkey method has obvious advantages, because it is relatively simple and the results are highly accurate. It has important reference value to compare the effect of different well pattern density on water flooding degree. The improvement principle of well pattern density is to take the total input of oil and gas field development as a function of well number, that is, well pattern density, convert the total output into a function of well pattern density through Serkachev formula, subtract the total input from the total output, and then adjust the well pattern density according to the profit value, which is also an economical and reasonable calculation method of well pattern density [6].

According to the basic principles of injection-production well pattern deployment in fractured reservoirs, diversified well patterns should be adopted on the basis of conventional injection-production well pattern deployment. For the square reverse five-point injection-production well pattern, the side-by-side direction is parallel to the fracture direction, and the well locations between rows are staggered. Well spacing should be controlled at twice the row spacing, and injection-production well rows should be alternated with each other, water injection should be parallel to the fracture direction, and the number of oil and water Wells should be 1:1. The well row direction of the linear injection-production well pattern is parallel to the fracture, and the well position between rows is staggered, and the row spacing ratio should be more than 2 times. The injection and expulsion Wells are alternated, and the linear water injection is conducted in parallel direction.

To rationally arrange vertical well pattern spacing, the rational spacing should be determined first. Fuyang reservoir belongs to low-ultra-low permeability reservoir, which has start-up pressure gradient. If the injection and production row spacing is too large, it will be difficult to establish an effective driving system. In the process of development, the production of a single well will decline, and the decline range will increase.

3.3 Horizontal well pattern

Horizontal well is an efficient oil and gas field development technology, which is widely used in the field of oil and gas field development. In view of this, horizontal well is introduced into the development well pattern, which is an expansion of vertical well pattern. With the continuous improvement of horizontal drilling technology and the gradual reduction of drilling cost, horizontal well pattern has received more attention from more people, and the academic circle has studied it more deeply. However, the domestic research on horizontal well pattern is more focused on theory, which provides a solid theoretical basis for the development of oil and gas fields, but the research on practice needs to be improved [7].

Reasonable well pattern has a serious impact on the exploitation and production of oil and gas fields, and can effectively promote the improvement of economic benefits of oil and gas fields. However, under the background of constantly updating oil and gas field development technology, the way of displacement is also constantly changing, and the development system of oil and gas fields is gradually complicated. From the perspective of system, a single well is an integral part of the well pattern system. Well pattern is a subsystem of the whole system of oil and gas field development, and the whole system of oil and gas field is a part of the complex system of society. Therefore, in order to solve the problem of well pattern system in oil and gas field development, we should start from the perspective of the whole. Firstly, we should optimize the single well, deal with the problem of each single well, and coordinate the relationship between the single well and the whole well pattern system. The optimization of well pattern should follow the principle of fewer Wells, large controlled drainage area and high recovery rate. The selection of well pattern density should start from the perspective of system, take diversified factors into consideration comprehensively, increase the study of irregular well pattern form, combine theoretical research with actual production, effectively solve the problem of well pattern system deployment and adjustment in the process of oil and gas field exploitation, and work out a better well pattern optimization strategy.

Conclusion: To sum up, the development of oil and gas fields is the main way to supply oil resources. In the actual development work, we should do a good job in the optimization design of well pattern system, to promote the fine system, and should be the best geological survey and inspection, development work according to the actual situation. Taking into account the density of well patterns and the economic benefits of reservoir controlled recovery, the establishment of a sound digital model, and strengthening the study of irregular well patterns, combining production experience with actual production, to promote a more reasonable and scientific well pattern study, explore more methods of well pattern system optimization for the development of our oil and gas field.

References

1. Yang Xinming, Gu Qinghong. Horizontal well design and well pattern system optimization in small fault block reservoir [J]. Journal of Yangtze University: Natural Science Edition, 2018, 15(23):4.
2. Yang Xinming, Gu Qinghong. Horizontal well design and well pattern system optimization in small fault block reservoir [J]. Journal of Yangtze University: Natural Science Edition, 2018, 15(23):4.
3. Yang Kai. Discussion on the design and optimization strategy of injection-production well pattern system for low permeability oilfield development [J]. Chemical Engineering Management, 2017(2):1.
4. ZHANG Qiyan. Study on adaptability of infill well pattern of Fuyang II Reservoir in Yushulin Oilfield [J]. Inner Mongolia Petrochemical Industry, 2017, 43(3):2.
5. XU Ying. Improving fracturing effect of Fuyang Reservoir in Daqing based on fine geological research results [J]. Chemical Engineering and Equipment, 2018(6):3.
6. He Ying, Li Min, Zhou Xisheng, et al. Optimization design of CO2 flooding well network in ultra-low permeability reservoir [J]. Journal of Daqing Petroleum Institute, 2011, 35(4):5.
7. SHAO Jinxiang. Study on optimization and adjustment of well pattern in late period of high water cut in Tainan Oilfield [J]. Inner Mongolia Petrochemical Industry, 2013, 39(16):3.

https://doi.org/10.1051/e3sconf/202339401020

Comprehensive treatment strategy for high water cut stage of medium and low permeability reservoir

Hui Xue

Institute of Geology, No.4 Oil Production Plant, Daqing Oilfield Co., LTD., Daqing Heilongjiang 163500, China

Abstract: Maintaining good conditions for petroleum exploitation is an important basis for achieving better exploitation results. At present, the main reservoir conditions of our main oil fields are relatively good, and have entered high water cut, middle and late development. Due to long-term exploitation, the reservoir has suffered a certain degree of destruction, resulting in the rapid rise of aquifers and sharp reduction of production. It has certain influence on the output of oil and well and the final recovery efficiency of oil field. In order to alleviate the contradiction of rapid reduction of oil production pressure and output in high water cut stage of medium and low permeability reservoir, the development efficiency of reservoir can be improved. By analyzing the high water cut stage of medium and low permeability reservoir, this paper puts forward the comprehensive treatment measures, which provides the idea for realizing the increase of production.

Key words: Medium and low permeability reservoir; High water content; Treatment strategy

1. Introduction

From a supply point of view, in addition to actively seeking new sources of oil and gas, both domestically and internationally, existing resources should be actively exploited to improve the efficiency of crude oil extraction. Increasing the crude oil production of existing oil and gas fields is of great significance for ensuring the safety of oil and gas resources and improving the exploitation efficiency and sustainable development of oil and gas resources. In the process of the continuous development of oilfield development in the high water cut stage, the calculation of geological reserves in the 1980s and off-balance sheet reserves in the 1990s can no longer meet the needs of current oilfield development [1]. Therefore, it is of great theoretical value as well as practical value to study the corresponding treatment countermeasures for the problems existing in the middle and low permeability reservoirs in the high water cut stage.

2. Main problems in high water cut stage of medium and low permeability reservoir

With the arrival of high water cut stage of medium and low permeability reservoir, the problem of rapid decline of recovery rate appears, which affects the in-depth study of oil stability and water control technology on reservoir. Taking the first North area of Daqing Oilfield as an example, the main problems are as follows:

2.1 Imbalance between reservoir and oil production reduces reservoir productivity

After primary and secondary encryption, and several large-scale water transfer, oil stabilization and water control measures, the remaining recoverable reserves of Daqing Sazhong Oilfield have increased from 7.77% to 9.20%. The reserve-production ratio of the field dropped from 12.87 to 10.8, and the possibility of increasing reserves became less and less, and the decline became larger and larger (Figure 1).

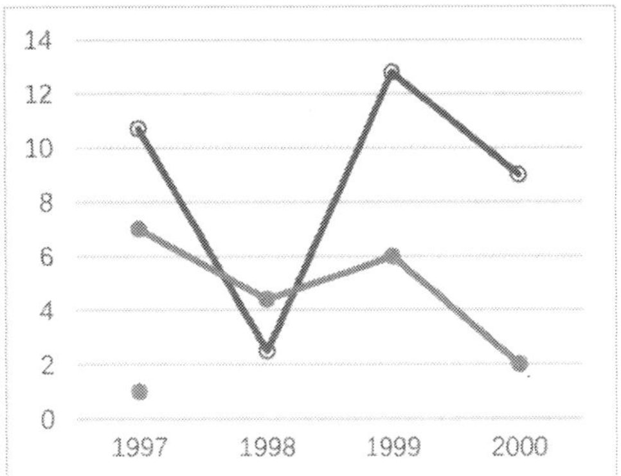

Figure 1. Oil field savings

 https://doi.org/10.1051/e3sconf/202339401020

2.2 The difference of water content in different strata becomes smaller, which increases the difficulty of water control

The difference of water content in the eastern part of the first member of the North fault decreased from 10.3% to 9.07% in the 1990s, and then to 2.48% in 2004. In the past 14 years, the comprehensive water cut in this region has only increased by 5.54%, with an average annual growth rate of 0.39%, which significantly inhibits the growth rate of water cut in this region. The possibility of further restraining the growth rate of water cut through structural adjustment in the future is very slim, and the growth rate of water cut will increase if no effective water control measures are taken.

2.3 The speed of casing damage increases, and it is difficult to control casing damage

After the second infill adjustment, the annual injection-production ratio of the first area in North China has been maintained at 1.52-1.24, and the cumulative injection-production ratio of 1.22 has been maintained at a high level for a long time. Before treatment, there were 103 overlying pressure injection Wells [2]. The intense waterflood development kept the formation pressure in the eastern block at the same level, and the total differential pressure in the block reached 0.21 Mpa before it was effectively controlled. Over time, the number of casing losses has increased. By 2000, the cumulative number of Wells had reached 238, and the rate had increased from 0.26% to 4.18%. With the increase of production, the control of casing damage becomes more and more difficult. (Figure 2).

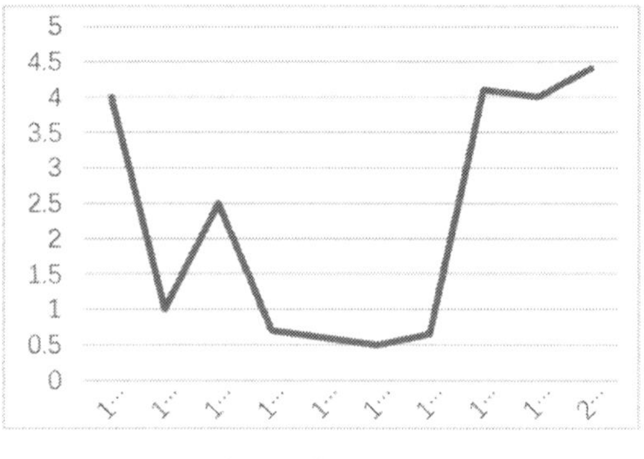

Figure 2 damage rate

2.4 High degree of transformation and low potential for increasing production

Today, the fracture rate of the old well is more than 71%, 90.8% of the Wells were re-fractured after the first confining fracturing, and 32.3% of the Wells were re-fractured after the second infill fracturing. To date, 54.3 percent of the Wells have been fractured, and nearly half of those Wells have been re-fractured. Recurring Wells accounted for 81.4 percent of all Wells. A set of tanks has a low potential for further development and adverse effects after eliminating factors such as compartments, flooded zones, cementing quality, and production techniques. The stimulation water yield per well was also increased from 68.9% to 81.2%.

2.5 Remaining oil with multiple reservoirs and high water cut that is difficult to excavate

Based on the data of three oil production Wells in the first North Area, it is concluded that the thickness of the eastern part of the fault is 83.5-87.7%. It was confirmed that the first and second oil reservoirs were flooded with water, and the degree of water erosion was large, accounting for 69.3%. In the thin differential layer, the number of scouring layers accounted for 78.8%, the thickness of sand layer accounted for 81.2%, the effective thickness accounted for 85.4%, and the scouring rate of surface layer accounted for 20.5%[3]. This shows that the residual reservoir in this area is very small, and it is very difficult to study and develop it.

2.6 Water injection and liquid production increase with the rise of water cut

In order to ensure the stable production of the oilfield under the condition of increasing water cut and accelerating production decline, measures to increase the water injection rate were continued in the east section of the first section of North Section. The annual water injection rate increased by 29.62.3 percentage points and the annual liquid output increased by 36.56 percentage points. Resulting in oil and gas field operating costs continue to increase, oil and gas field exploitation economic benefits decline. In order to explore the possibility of continuous improvement of oil recovery in high water cut stage of middle and low permeability reservoir, Sazhong Oilfield in Daqing has carried out regional comprehensive consolidation analysis on two fault blocks in the east and west of North No.1 Block. In order to realize the sustainable development of reservoir, it is discussed to improve the oil recovery in the high water cut stage of medium and low permeability reservoir.

3. Comprehensive treatment strategy

With the continuous improvement of the development degree of water drive reservoir, it is necessary to adjust the water drive reservoir after the high water cut period in order to control the water cut of water drive reservoir and make the production of water drive reservoir increase steadily. The well pattern deployment, injection-production relationship, and water flooding development effect in the development process are studied in detail (see Table 1), and the development layer is studied in detail, including porosity, permeability, oil saturation, etc. On this basis, the development prospect of the remaining oil in the oilfield is analyzed by using stratification, quantization, water flooding and other methods [4].

Table 1 Development layer parameters

No	Classification index	standard				
		A class	B class	Cclass	index	class
1	Control degree of water drive reserves /%	≥70	70-60	<60	71.02	B
2	Injection-production correspondence rate /%	≥65	65-45	≤45	42.34	C
3	Energy retention level and energy utilization degree	increase	steady	decrease	reduce	C
4	Water drive recovery factor /%	≥30	30-20	≤20	29.0	B
5	Rate of recovery of recoverable savings /%	≥7	6.9-6	<6	0.6	C
Comprehensive evaluation result						C

3.1 Reasonable well pattern layout and injection-production ratio

In the oil field of water drive development, the displacement of water in the water injection well is used to promote the smooth exploitation of oil flow in the oil well. It is necessary to adopt reasonable well pattern layout, improve the well pattern system, infuse it, and adjust the injection-production relationship, so that the well pattern density can be selected properly. Reasonable control should be carried out on the ratio of injection-production Wells, as long as water flooding can be achieved. If water injection is carried out too early, the oil reservoir will see water in advance or be flooded, thus losing the value of exploitation.

3.2 Cement sealing

Want to do a good job of technical support, make the production technology to better serve the oil and gas fields. For the technical measures supporting the development of the high water cut stage in the low permeability oilfield, the liquid flow in the well can be extracted to the surface by means of fluid extraction. In addition, cement plugging technology can be used to seal off discovered aquifers to unlock remaining geological reserves and increase field production.

3.3 Periodic water injection test

One of the important reasons for the influence on the oilfield development results in the high water cut stage of medium and low permeability reservoir is to study the heterogeneity of the reservoir through the periodic water injection test. In order to overcome the heterogeneity of the reservoir, the method of periodic water injection was tested in the water injection well. After injecting for a period of time, the water injection was stopped and observed, so that the high permeability area could see water in advance. Then, when the energy has dropped to a certain level, the next cycle is filled with water so that the effect is achieved and the oil is removed.

The development effect of periodic waterflooding is better than that of continuous waterflooding. The research shows that under the same water cut, the recovery degree of periodic waterflooding is higher than that of continuous waterflooding. The recovery factor of periodic waterflooding was 0.605% and that of continuous waterflooding was 0.573%. The improved recovery factor of periodic waterflooding was 3.2960% compared with that of continuous waterflooding. The results showed that there were many forces besides the displacement force, such as capillary force, elastic force and gravity force. In the process of periodic injection, the influence of capillary force is the first, followed by the influence of elastic force, and the influence of gravity on the process of periodic injection is not enhanced or weakened. The numerical results show that under traditional water flooding conditions, the pressure at the high and low permeability points is uniform due to the inhomogeneity in the formation. However, in the half cycle of periodic water injection stopping or reducing, the pressure at the high permeability decreases rapidly, while the pressure at the low permeability decreases slowly due to the difference in oil saturation and permeability. This causes the high osmotic pressure to be lower and the low osmotic pressure to be higher at the same time, creating an additional pressure difference between the two locations, so oily water moves from the low osmotic pressure with higher oil saturation to the high osmotic pressure. During the half-cycle of re-injection or increased water injection, the pressure at the high permeability site recovered quickly, while the pressure at the low permeability site recovered slowly, creating a reverse additional pressure difference. Meanwhile, in turn, the oil water flowed from the high permeability site and the oil saturation was lower than that at the low permeability site, which increased the formation pressure at the low permeability site. In the next cycle, More oil flows from low-permeability sites to high-permeability sites to conserve energy. Thus, in a complete cycle, a large amount of water is lost from the high permeability to the low permeability, and more oil flows from the low permeability to the high permeability, which expands the sweep volume of water flooding and improves the development effect of water flooding.

Similarly, under conventional waterflooding and multizone combined production conditions, the reservoir with high permeability is reduced faster than the reservoir with low permeability within half a cycle of stopping waterflooding. In the injected water recovery process, the pressure of the bad layer is higher than that of the good layer, the liquid production of the bad layer is improved, the water recovery rate of the good layer is better, the pressure drop of the bad layer is lower than that of the good layer, and the water absorption capacity of the bad layer is improved [5]. This changes the pressure distribution of the formation again and again, thus improving the development efficiency of the reservoir. Similar to intra heterogeneous reservoir, periodic injection can also cause cross penetration of high and low permeability intervals in plane, but its effect is less than that of intra heterogeneous reservoir. However, as long as the permeability level difference between high and low permeability zones is large and there is a large contact surface, the remaining oil in the low permeability zone will flow to the high permeability zone, so as to obtain higher recovery efficiency.

3.4 Polymer Injection

Through polymer injection, the viscosity of the injected water is increased, the sweep volume is expanded, and the displacement capacity is improved to increase the oil production. The technical measures to enhance oil recovery, polymerization agent, ASP flooding and CO2 flooding were adopted to increase production.

4. Conclusion

It is a common phenomenon for oil Wells at home and abroad to have high water cut in medium and low permeability reservoirs. In the high water cut stage, the oil field output is small and it is difficult to meet the actual production demand. Based on the actual situation of North No.1 area of Daqing Oilfield, this paper puts forward the comprehensive treatment measures for the oil development of middle and low permeability reservoirs in the period of high water content, partly solves the problems of development difficulty, and brings beneficial changes to ensure the normal production of oil fields and improve economic benefits.

References

1. Hong Yafei, Sun Dongsheng, Luo Yuhan, Tang Yun. Study on remaining oil distribution of low permeability thin interbedded reservoir in high water cut stage [J]. Complex Oil and Gas Reservoirs, 2022,15(02):72-77.
2. Zhang Lishuang. Optimization method of well pattern density in the late stage of high water cut development in low permeability reservoir: A case study of Area X [J]. Chemical Engineering and Equipment,2020(09):51-54.
3. Lin Rushan. Low permeability reservoir with high water cut period reservoir renovation technology optimization and parameter optimization research [J]. Journal of petrochemical technology, 2018, 25 (9) : 168.
4. Zhu Linlin, Fu Xianping. Study on reservoir reconstruction technology optimization and parameter optimization in middle and high water cut stage of low permeability reservoir [C]// Xi 'an Shiyou University, Shaanxi Petroleum Society.2018 International Conference on Oil and Gas Field Exploration and Development (IFEDC 2018), 2018:485-489.
5. Diao Guangzhi, Yao Xiang, Cui Wei, Dong Jie. Low permeability reservoirs of well pattern adjustment at the high water-cut stage study [J]. Journal of petrochemical industry applications, 2017, 4 (01) : 55 to 57.

Characteristics and adjustment strategies of reservoir agglomeration driving state in narrow channel sand body

Jinhua Wang

The Second Operation Area of No.6 Oil Production Plant, Daqing Oilfield Company Limited Liability Company, Daqing, Heilongjiang 163000, China

Abstract: With the continuous growth of economy, we are more and more aware of the importance of energy and its preciousness. In the production of crude oil, the narrow channel is the best location for oil storage, and the underlying sand body often has oil resources. In this paper, the effect of reservoir polymer flooding is taken as the purpose, the principle of polymer flooding technology as the theoretical basis, through the analysis of dynamic change characteristics, in order to improve the adjustment parameters, to achieve reasonable and efficient exploitation of oil Wells. The application of this technology, through the development of practice, has been recognized as an effective method of oil recovery.

Key words: Narrow channel sand body; Oil extraction; Poly flooding mining

1. Introduction

Without efficient use of energy, social development and construction cannot be realized. The application of the project of "West-east gas transmission and South-north water diversion" fully shows that energy economy will never be eliminated. In the context of high energy consumption and increasing demand in China, the improvement of polymer flooding technology, dynamic simulation and analysis of data changes for narrow channel sand reservoirs, and improvement of the entire mining process are bound to improve the technical practicability and improve the efficiency of crude oil exploitation [1].

2. Polymer flooding and present situation of oilfield

2.1 Principle of polymer flooding

Macroscopically, the purpose of polymer flooding is to increase the viscosity of the injected fluid, adjust the water-oil mobility ratio in the reservoir, and expand the sweep volume. At the mesoscopic level, firstly, injected polymer flooding can effectively adjust the water-oil mobility ratio and expand the wave capacity. Adding polymers to water will make it more difficult for the solution to penetrate into the formation, increase its viscosity, and reduce the flow of the solution. If the flow of crude oil is enhanced relative to the solution, the sweep range of the solution will be increased, thus making the water flooding effect better. Secondly, the flow resistance of highly permeable water in the reservoir can be increased, which is conducive to improving the permeability of water flooding. Adding polymers to water can increase the viscosity of water. Third, under this condition, a stable oil filament flow path can be formed. Reduce the proportion of participating in water flooding and improve the effect of water flooding. Fourthly, the polymer flooding effect is obvious. Under the action of polymer injection, the low permeability layer can be activated, the vertical sweep effect of the oil layer can be increased, and the flooded area of the oil layer can be increased, which is conducive to improving the exploitation effect [2].

2.2 Current situation of polymer flooding in oilfield

This oil area is located in the front facies - estuary area of lake basin. There is only one set of Lower Cretaceous Yaojia member oil layer in the whole oil area. The fluvial sand body has a small development range and is discontinuous in plane, mainly manifested as intermittent bands or scattered distribution. The reservoir has low porosity and permeability. After more than 20 years of development, the oilfield has entered a period of high water cut on the whole. At present, it mainly relies on the adjustment of water injection structure, which plays a certain adjustment effect on the water injection volume and liquid production of high water content Wells and formations, but has little effect on the adjustment within the formation [3].

The riverbed sandbody of Block 1 and Block 2 is flood plain facies with a large amount of riverbed sand. The riverbed body of Block 3 is distributary plain facies with small scale. The first group of bed of Block 3 is a narrow sand body with an effective thickness of 9.71 meters. Compared with Block 1, the drilling proportion of bed sand body on the plane is very different. The drilling proportion of bed sand body of Block 3 is 41.3%, and that of bed sand of Block 1 is 29.4%.

In 1995, polymer flooding was developed in Block 1, which had an oil bearing area of 15.35 m3. 123 Wells were drilled, 98 of which were injected, resulting in 765 PV.mg /L polymer and a 16 % increase in oil recovery. It is now entering the late stage of water drive development. Sanjie Puyi Formation was flooded in 2000. The reservoir contains 6 square kilometers and has a geological reserve of 762×10* t. There are 46 injection Wells and 45 production Wells in the five-point area well pattern. In December 2005, in the middle and late stages of polymer injection development, a polymerization method of 583PV-1/L was used to increase oil recovery by 13 percent.

3. Improve methods for improving polymer displacement efficiency

3.1 Stratified injection

In the case of polymer injection production in reservoirs, due to the large difference in lithology of reservoirs, the effect of polymer flooding as a whole cannot be guaranteed. In order to improve the oil displacement effect, it is necessary to take the initiative to use the method of stratified injection to effectively solve the contradictions between different layers of the reservoir and avoid the problem of mutual influence between different layers. In order to control polymer flooding accurately, we must work out a set of scientific and effective polymer flooding scheme according to the specific conditions of each reservoir. According to the study, the stratified injection method can increase the oil utilization efficiency of the reservoir by 10%, control the rate of increase of water injection pressure in the reservoir formation and reduce the water cut in the producing well during the oil development process. At the same time, in the process of injection molding, technicians should scientifically control the time of injection molding. For the reservoir to be stratified injection molding, the injection molding time will become longer, and the effect can become worse. The greater the permeability difference between different reservoirs in the reservoir, the earlier the stratified water injection time of polymers. For reservoirs with very different permeability between the two reservoirs, stratified injection of polymerization agent can achieve more ideal oil displacement effect compared with no stratified injection [4].

3.2 Alternate injection

At present, in the process of polymer flooding, there are often some difficulties or high costs. In order to effectively solve these problems, it is necessary to use alternate injection to develop heterogeneous reservoirs, so as to effectively reduce the impact of heterogeneous characteristics of the reservoir, so as to achieve the purpose of saving the economic cost of polymer flooding. The use of scientific compatibility mode and technology is the fundamental and precious treasure to ensure the compatibility efficiency can be fully played. The data show that polymer flooding by alternate water flooding can significantly improve the liquid absorption capacity of heterogeneous reservoirs under low permeability conditions, reduce the heterogeneity of heterogeneous reservoirs, improve the use level of each reservoir, and promote the polymer flooding effect. In the process of oilfield development, the alternate water injection method can obviously prolong the operation time of the oilfield under the condition of low water cut, so as to improve the oilfield development effect and increase the oilfield production. The effect of polymer flooding can be maximized by appropriately setting different injection times under different injection times. If the total amount of polymer injected remains unchanged, and the cycle set by the alternate injection is relatively short, the number of alternate injection will increase, resulting in the reduction of the amount of polymer injected each time, thus restricting the effect of the alternate injection, and thus increasing the difficulty of water injection profile adjustment, thus causing an adverse impact on the improvement of oil displacement efficiency. The results show that the effect of polymer flooding can be significantly improved and the oil production can be significantly increased by alternating polymer flooding at the later stage of polymer flooding. Compared with the conventional injection method of polymer, the injection of alternating polymer can not only effectively enhance the oil displacement effect, but also reduce the amount of polymer and maximize the oil displacement efficiency of the reservoir polymer, which is of great significance to the production increase [5].

4. Analysis of dynamic change characteristics and optimization measures

4.1 Characteristics of reservoir agglomeration driving state in narrow channel sand body

（1）Injection pressure changes

After polymer flooding, due to the high viscosity of polymer solution, the retention of polymer solution in the reservoir leads to the decrease of permeability, the increase of permeability resistance, and the increase of water injection pressure. The initial water injection pressure and growth range of polymer pools in Block No. 3 and Donggou No. 1 are different in development degree. The initial water injection pressure of polymer reservoirs such as Block 3 and Donggou Strip 1 was 10.5-11.5MPa, with an increase of about 3.2MPa. In the pure oil area, the initial water injection pressure of Xigou No.1 and No.2 blocks is only 9.0 MPa, with an increase of 3.8 MPa and 3.6 MPa.

(2) Changes in water content

In the process of increasing the amount of polymerization agent added, the comprehensive water cut first increased, then gradually decreased, and then continued to increase after reaching the lowest level. In the first and second blocks, the water cut in the central well of the first and second blocks with river development decreased by 25-30% within 12 months, while the water cut in the third block with narrow river development decreased by 20%. There was only a small change in six months.

(3) The change of polymer flooding concentration

According to the polymer injection practice in the pilot test and industrial extension block, after polymer injection, the polymer content in the produced fluid of the well has been increased to a certain extent, and the first and second blocks take effect first and then polymerization, but in the third block, a narrow river channel, it takes effect first and then polymerization, and the increase of the concentration is smaller than that in the first and second blocks.

4.2 Adjustment measures of narrow and small channel sandstone reservoir polymer flooding

(1) Ensure injection quality

By observing the dynamic changes of polymer flooding, the pressure tends to rise throughout the underground mining site. In the case of pressure rise, attention was paid to recording the injection rate of the well during the design operation. The real-time injection rate of the sand body was observed at all times, and the injection amount and method of polymer were adjusted according to the injection rate. In view of the problems existing in the development of wellhead in the riverbed, the method of high density rapid injection can be adopted to solve them. Polymer flooding is an enhanced oil recovery method that features high density rapid injection as follows: Polymer flooding High density rapid injection is a rapid injection method that can inject large amounts of polymer solution into the reservoir in a short period of time. This reduces the retention time of the polymer solution in the reservoir and increases the viscosity of the polymer solution, resulting in improved oil recovery. Polymer flooding also has a relatively low cost of high density and rapid injection because it can deliver large amounts of polymer solution to the reservoir in a short period of time and can be calculated by using efficient oil recovery calculation methods. Polymer flooding is also safe with high density and rapid injection. Because it can inject large amounts of polymer solution into the reservoir in a short period of time without causing damage to the reservoir. In addition, this method of injection ensures the required polymer concentration in the reservoir, thus enhancing oil recovery. However, to ensure the safety of high density and rapid injection in polymer flooding, strict controls and monitoring are required during implementation to ensure that the polymer solution does not cause damage to the reservoir. The low permeability of Wells and the poor connectivity of subsurface Wells can lead to reservoir failure. To prevent this from happening, the polymer injection rate and concentration in these Wells must be reduced to ensure the continuous supply of polymer, thus achieving polymer flooding in the narrow riverbed.

(2) Pay attention to the initial injection volume

Both theoretical and numerical simulation show that the larger the molecular weight of polymer, the greater the increase of the amount of crude oil, but the selection of molecular weight should also take into account the matching with the reservoir. According to the calculation, block 3 is based on the polymer of medium molecular weight, and its pore volume is the smallest, so I400×10 polymer should be selected at the beginning of polymer injection.

(3) Alleviate the contradictions between layers

The numerical simulation results show that when the reservoir is relatively stable and the permeability difference of uneven reservoir is 5, the recovery rate of forward jointed reservoir can be increased by 5% than that of cage injection, and the earlier the injection process, the better. Due to the short existence time of low-value reservoir in Block 3, in the early stage of reservoir rise, the reservoir with separated layers is selected, and the annular decompression tank injection technology is used to control the injection of high permeability layer and increase the injection of thin layer in the reservoir, so as to achieve the purpose of controlling the rise rate of the reservoir. On this basis, the developed oil fields are injected 19 times, and some oil fields are divided into different layers to achieve the purpose of the contradiction between the liquid production of high water cut section and the regulation zone. On this basis, the corresponding adjustment is made to maintain the suction profile at a higher level while the water content is restored. After the injection of well I0, the injection rate of Portuguese II_4 in the control zone decreased from 6.3 m/m to 7.3 m/m, while the injection rate of Portuguese I5.7 in the reinforcement zone increased from 7.9 m/m to 15.9 m/m. The combined water cut growth rate of the surrounding 10 producing Wells was relatively slow.

(4) Expand the sweep volume

In Block 3, the rate of increase in injection pressure slowed after the water cut was restored. Based on the pressure distribution, about 30 percent of the reservoir was not utilized in 50.5 percent of the Wells during this period. However, in this period, if the injection volume of macromolecular polymer is used, the wave volume can be larger. Therefore, when the injection volume is 14OPV-mg/L, the polymer is used. When the injection volume is 14OPV-mg/L, the polymer of I900×10 is used. The injection pressure in this block increased from 111.44MPa to 102.06MPa, while the pressure only increased by 0.62MPa. After the fourth month of injection, the pressure remained stable. In the condition of reservoir and reservoir, the reservoir of the reservoir and the reservoir of the reservoir are stable.

After the above adjustment, the water content of Zone 3 is still 88.53% at 439 PV.mg /L, and the recovery rate is 1% higher than that of Zone 2.

5. Conclusion

In the development of narrow reservoirs, the application of polymer flooding technology can effectively increase the development output of oil fields and increase the national crude oil reserves. In view of the relevant

https://doi.org/10.1051/e3sconf/202339401021

problems encountered in the current development, some adjusting measures are put forward to ensure that the "polymer flooding" technology can be successfully applied to the development of narrow and miniaturized sandstone reservoir.

References

1. Xue Guoqin, Kong Bailing, Li Xiyu, Long Weijiang, Wang Rui. Henan oilfield polymer flooding technology [J]. Journal of petroleum geology and engineering, 2009023 (003) : 50 to 52.
2. ZHANG Zhiyuan. Polymer flooding oil recovery technology [J]. Journal of Daxian Teachers College,2003,013(02):19-21.
3. SUN Zhonghua. Feasibility study on local polymer injection in narrow and small channel sand bodies [J]. China New Technology and New Products,2009(03):11-12.
4. YI Zhenlin. Study on sand body prediction method of narrow channel combined with well and earthquake [J]. Journal of Southwest Petroleum University (Science & Technology Edition),2012,034(01):19-24.
5. LI Qian. Analysis on the characteristics and adjustment measures of accumulation driving state of narrow channel sand reservoir [J]. Science and Technology and Enterprise, 2014(10):1.

https://doi.org/10.1051/e3sconf/202339401022

Productivity dynamic analysis of fractured horizontal well in ultra-low permeability reservoir

YuLong Xing

Institute of Geology, No.4 Oil Production Plant, Daqing Oilfield Co., LTD., Daqing Heilongjiang 163500, China

Abstract: Daqing Oilfield is rich in low permeability reservoirs. In order to fully understand the development dynamics of ultra-low permeability reservoirs, this paper takes Daqing Oilfield as an example and studies the productivity dynamics of fractured horizontal Wells by numerical simulation method. The results show that the productivity of fractured horizontal Wells decreases exponentially with the decrease of formation permeability. The productivity of fractured horizontal well increases with fracture length and decreases with fracture spacing. With the increase of formation coefficient and reservoir thickness, the productivity increase of fractured horizontal Wells decreases. In the case of small well spacing, the smaller the formation coefficient and reservoir thickness, the higher the productivity of fractured horizontal Wells. Compared with conventional vertical Wells, fractured horizontal Wells have better production performance.

Key words: Fractured reservoir; Fractured horizontal well; Well test analysis.

1. Introduction

With the development of economy, low permeability reservoir reserves are increasing. Daqing Oilfield is the first oilfield of ten million tons built in China, but its low permeability reservoir reserves account for a small proportion of the total reserves. For low permeability reservoir, horizontal well exploitation is one of the effective means to achieve high and stable production [1]. In Daqing oilfield block of Daqing Oilfield, the production of oil well has been greatly increased by using horizontal Wells. Therefore, it is of great significance to fully understand the dynamic characteristics of productivity of low permeability reservoir to effectively guide the development of this kind of reservoir.

2. Basic conditions of ultra low permeability reservoir

Daqing oilfield is located in the central and western part of Heilongjiang Province. It is a structural lithologic reservoir with mudstone and sandstone, followed by fine sandstone, siltstone and limestone. According to mercury injection data, core analysis and core photos, the average porosity and permeability of Daqing oilfield are 9.6% and 0.049×10-3μm2, which belong to extremely low permeability reservoir. According to the regional structural analysis, Daqing oilfield belongs to large complex anticline structure. The Daqing oilfield is mainly composed of braided river deposits with sand bodies developed between channels. The main reservoir group developed shore-shallow lacustrine - delta plain subfacies deposits, with small scale, poor continuity, limited distribution range, large thickness variation, and transverse superimposed multiple layers. According to the analysis of core and mercury injection data, the average pore throat radius of Daqing reservoir group is 0.059μm2 and the average throat radius is 0.122μm2. The natural energy of Daqing oilfield is poor and the formation pressure drops fast. According to the geological characteristics of Daqing oilfield and the results of reservoir numerical simulation, it can be seen that the oil reservoir in Daqing oilfield has the characteristics of low starting pressure gradient and uneven reservoir pressure distribution.

3. Physical model of fracture network penetration in liquid phase of fracturing

For the special low permeability reservoir, the method of horizontal drilling and stage fracturing must be used to produce oil. Due to reservoir lithology, if hydraulic fracturing is carried out in a fractured low permeability reservoir, natural fractures of the reservoir will be opened and connected to each other when the main fractures are formed, and a complex fracture network structure and SRV area R will be generated around them (as shown in Figure 1) [2]. In order to simplify the original fractured reservoir and its internal fracture network structure, the Warren-Root dual medium model is proposed. The one-quarter penetration zone of a single fracture was analyzed, and the permeability in the reservoir was divided into three parts (Figure 2) : Zone O (0< x< x_(f,)0< y< y_e). The reservoir was an initial, fracture-free, secondary medium, Darcy free, slow, secondary permeability reservoir; The study area I (0< x< x_f,0< y< y_e) refers to a two-layer reservoir with quasi-stable permeability centered on the big fault. Where, F (0< x<x_f) is the linear flow of the reservoir in the simulated fracturing process, which is influenced by factors such as reservoir and skin action. The fluid moves from zone 0 to Zone I, then from zone I to the main fracture zone, and eventually from the main fracture zone into the horizontal wellbore.

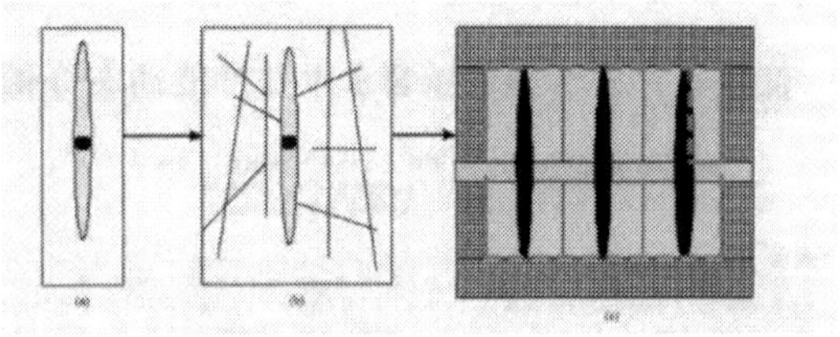

Figure 1. Schematic diagram of fracture morphology of fractured well

The main assumptions of the model are: (1) reservoir up and down as the closed boundary; (2) There are both lithologic and liquid micropressures with constant compressibility; (3) The fracture is distributed vertically through the horizontal wellbore with equal spacing. The half-length of the fracture is x_f, the width is w_f, and the height is equal to the thickness of the reservoir h. The fracture has limited conductivity. (4) the fracture closed, no flow through, closed the border between adjacent cracks, ignoring the influence of the interference of the seam; (5) regional O and I are double medium reservoir, ChuRong ratio and flow coefficient respectively.

4. Case analysis

Daqing Oilfield is a typical low permeability reservoir. The main development layer of this oilfield is Daqing reservoir with an average permeability of 0.81 mD and an average effective thickness of 14.4m. Through the analysis of the geological characteristics of the oilfield, it can be seen that the Daqing oil reservoir is mainly characterized by low porosity, low permeability and low pressure [3]. In addition, there is a certain degree of water sensitivity in Daqing oil reservoir. The oilfield has entered the middle and late stage of development, but because of the development of fractured horizontal Wells, the reservoir quality has been improved to some extent, and the output of single well has increased significantly. The development effect of fractured horizontal well in this oilfield is studied by numerical simulation method. The geological model is established according to the actual geological parameters, and the fine three-dimensional geological model is established by Eclipse software. On this basis, it is divided into well area and non-well area, and grid division is carried out respectively. Finally, Eclipse software is used for numerical simulation calculation to obtain the distribution of reservoir pressure field under different well spacing. When the well spacing is 500m, the pressure distribution of the oilfield is more uniform, indicating that the development effect of fractured horizontal Wells is better.

Table 1 Relationship between each fracture of Horizontal well and Production (a)

Fracture half-length scheme	crack 1	crack 2	crack 3	crack 4	crack 5
scheme 1	150	110	70	110	150
scheme 2	140	110	70	110	140
scheme 3	130	110	70	110	130
scheme 4	150	70	110	70	150
scheme 5	140	70	110	70	140
scheme 6	130	70	110	70	130
scheme 7	150	150	150	150	150
scheme 8	70	70	70	70	70

Table 2 Relationship between each fracture of Horizontal well and Production (b)

scheme	Production per crack (m^3/d)					Chief ginseng (m^3/d
	1		3	4	5	
scheme 1	3.93	1.67	0.96	1.67	3.93	12.16
scheme 2	3.74	1.73	0.97	1.73	3.74	11.91
scheme 3	3.55	1.8	0.98	1.80	3.55	11.68
scheme 4	4.12	0.93	1.90	0.92	4.11	11.98
scheme 5	3.92	0.98	1.92	0.98	3.92	11.72
scheme 6	3.73	1.04	1.94	1.04	3.73	11.48
scheme 7	3.68	1.88	1.73	1.88	3.68	12.85
scheme 8	2.58	1.57	1.42	1.57	2.58	9.72

5. Productivity dynamic analysis

5.1 Influence of starting pressure difference on horizontal well output

Under the condition that the non-factor starting pressure gradient is 0, 0.05, 0.1, 0.3 and 0.5, and under different starting pressure differences, the non-factor production changes over time are shown in Table 3. As can be seen from the chart, the production process of the fractured reservoir is characterized by two steady intervals of decline compared to the homogeneous reservoir. According to the shape of the characteristic curve, the production decline process of low permeability fractured reservoir can be divided into seven periods. The first period is the straight flow from the main fracture to the well hole. In this period, there are no factors, and the production curve is a straight line with slope of 0.5, which exists for a long period of time. Phase II was mainly caused by fissure edge action and initial water injection in the improvement area. Its production law was a straight line, and its slope was close to 1. Phase III is a line formed within the treatment zone, and a line with a slope of 0.5 is formed within the treatment zone. In the reformed zone I, the first stage is the surge from bedrock to micro-cracks. In this period, the seepage flow tends to be stable and the output curve tends to be stable for a long time. Phase V is a reservoir in which the linear flow period of the reservoir is not significant, the initial pressure difference of the reservoir and the permeability of the reservoir are small, the production curve of the reservoir is one line, and the reservoir is a reservoir in a reservoir. The V stage is the channeling period from bedrock to small fissure, and its output variation rule is two relatively stable intervals. Period V is the marginal response period of region 0.001, during which output decreases sharply. In the early stage, starting pressure difference has no obvious effect on oil recovery of oil-water two-phase flow. With the increase of starting pressure difference, the fluid flow resistance in the reservoir increases, the supply capacity of the reservoir weakens, and the productivity of the reservoir in the middle and late period decreases.

5.2 Influence of the number of fractures on productivity during horizontal well fracturing

The results show that the degree of fracturing of artificially fractured horizontal Wells is closely related to the degree of fracturing. As can be seen from the figure, with the increase of the number of fractures, dimensionless production of fractured horizontal Wells is higher, but the increase is smaller. When the increase exceeds a certain range, the increase changes little. At the same time, the number of cracks increased, increased the construction cost, improve the construction difficulty. The results show that when the length of horizontal well is fixed, the number of fractures in horizontal well hydraulic fracturing has an optimal range in terms of economy and construction cost.

5.3 Influence of fractures on horizontal well production during fracturing

For horizontal Wells, assuming that five major fractures are opened longitudinally, the production decline trend of fractured horizontal Wells is shown in the table below when the adjacent fractures (d,) are 100,200, and 300 m respectively. As can be seen from the chart, as the proximity increases, the output increases in the middle and middle stages of production, and the downward trend of output flattens out. Under the condition that the number of main fractures in artificial fracturing remains unchanged and the distance between fractures is larger, the scope of reconstruction region I is larger, and the supply range and capacity of the peripheral region to region I is larger.

5.4 Effects of major fractures on the production of fractured horizontal Wells

The following figure shows the variation law of fractured horizontal well productivity under artificial fracturing conditions without affecting production. The results show that the permeability after fracturing has a great effect on the initial production. The higher the permeability, the higher the initial production and the smaller the permeability. Its effect on fracturing effect decreases with the extension of production cycle. Under fixed pressure, both the output and supply of the main fracture are larger, and the linear flow of the fracture is maintained for a long time.

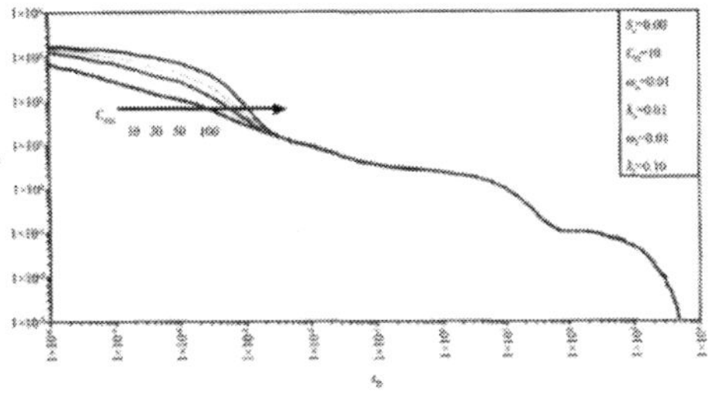

Figure 2 Influence of diversion capacity on productivity decline curve of fractured horizontal Wells

5.5 Influence of storage capacity ratio and micro-fracture channeling coefficient on productivity of the reconstruction area

Under the condition that other parameters remain unchanged, the output decline curve of fractured horizontal Wells is shown in the figure [4] after the storage capacity ratio and channeling coefficient are adjusted to different values in the reconstructed region I. Results: With the increase of o value, the downward trend of output in the middle and late period was gradually moderated, while the linear flow state lasted longer. The larger the input ratio, the larger the medium-term output, but the smaller the decline. The higher the elastic storage capacity ratio o, the higher the elastic capacity of the fracture system and the ability to supply the main fracture system, and the longer the linear flow cycle, so the medium-term output is larger, but the downward trend is relatively stable. The larger the channeling coefficient in the inner zone, the smaller the permeability difference between the inner zone, the matrix and the micro fracture, the stronger the matrix supply to the micro fracture, the larger the medium-term yield, the gentler the attenuation, and the earlier the inner zone boundary appears [5].

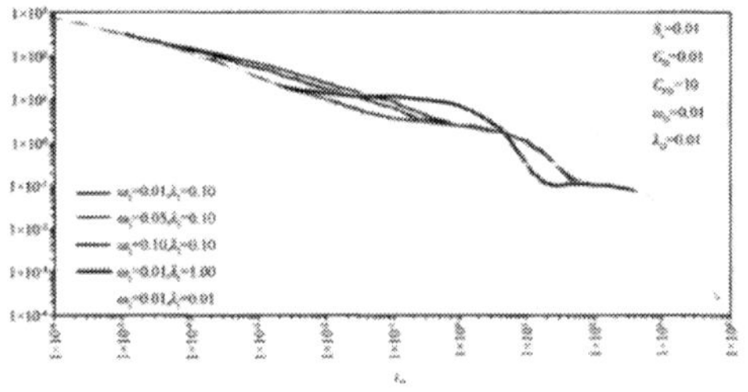

Figure 3. ω_f、λ_f Effect on productivity decline curve of fractured horizontal well

6. Conclusion

Based on the stress sensitive horizontal well productivity equation, this paper proposes a method suitable for fractured horizontal well productivity analysis in low permeability reservoir, and takes Daqing oilfield as an example. When the permeability of the fractured horizontal well is reduced to 0.1×10-3 mD, the production of fractured horizontal well will show an exponential decline law. When the fracture spacing is less than a certain value, the production of fractured horizontal well decreases with the increase of fracture length. When the fracture spacing is greater than a certain value, the production of fractured horizontal well increases with the fracture spacing. Compared with conventional vertical Wells, the lower the formation coefficient and reservoir thickness, the higher the productivity of fractured horizontal Wells.

References

1. Frances Ding. Productivity and Numerical Simulation of stage-fractured horizontal well in offshore low permeability Gas Reservoir [D]. Southwest Petroleum University, 2018.
2. Deng Xuefeng. Optimal design of rational production pressure difference for fractured horizontal Wells in tight low permeability reservoir [J]. Lithologic Reservoirs, 2017, 29(1):5.
3. ZHOU Wenwu. Productivity Simulation of fractured horizontal well in low permeability reservoir [D]. Southwest Petroleum University, 2018.
4. Zhu Mingxia, Tang Qiang. Analysis on Main controlling factors of productivity of fractured horizontal well in low permeability reservoir [J]. China Petroleum and Chemical Standards and Quality, 2020(5):2.
5. CHENG Ziyue. Study on Influencing Factors of productivity of multi-stage fractured horizontal well in M block of ultra-low permeability reservoir [J]. Contemporary Chemical Industry, 2019, 48(8):4.

Adaptability Analysis of Three or Four Strips with Small Well Spacing in Zone A Transition Zone

Xiaojuan Liu

Geological Research Institute of Daqing Oilfield Third Oil Production Plant, Daqing, China

Abstract: Small well spacing tests were carried out in three and four zones of the transitional zone to solve the problems of poor effect of water flooding and low reserves utilization. In order to improve the effect of water drive development in the transitional zone of Area A, the two test zones in area a have been reopened and utilized, well pattern is changed from anti-nine-point method to five-point method. Through strengthening the tracking adjustment of oil and water wells, the variation law of production and water cut in the test well area with small well spacing in the development process is understood, and the adaptability of different well patterns and well spacing in three or four zones is verified, it provides a technical reference for the adaptability of water injection in three or four zones of the transition zone.

Key words: Transition zone, small well spacing injection production system adjustment adaptability.

1. General situation of three or four zones in Area A

In Area A, the main depositional type of the transitional zone is the delta inner front, and the large and medium channel sands are less developed, the drilling bed number ratio is 23.0% , the ratio of drilled layers is 77.0% , which is distributed as narrow strip, lump and thick lump. The development of oil reservoir is gradually worse from one belt to four belts. The average single well sandstone and effective thickness of three or four belts are obviously lower than one or two belts.1.2 Physical Property Characteristics of Crude Oil in Three or Four Zones

2. Analysis of well network adaptability

Since entering the later stage of high water cut development, the adaptability of the injection production system in the three or four zones has gradually deteriorated, mainly manifested in the following aspects:

2.1 large injection-production well spacing, low number of injection-production wells, and low control of well pattern on sand body

Three zones of transition zone in area a adopt 350m well spacing four-point method area water injection well pattern, four zones adopt 250m well spacing inclined linear water injection well pattern. The channel sand bodies in transition zone of Area A are small-scale, discontinuous, narrow-strip and sporadic, and the channel sand is mainly controlled by single well. Under the current well pattern, the proportion of imperfect sand body below 4 well points is high, and the control of water flooding in channel sand is low. In Area A, the density of well pattern is small, the distance between injection and production wells is 250-350m, the rate of injection and production is low, the ratio of effective thickness is high in one-way and two-way communication, but the ratio of effective thickness is low in three-way communication, far below one or two bands. At present, it is difficult to control narrow and small sand bodies with small development area by well spacing and well pattern density.

2.2 The injection pressure is high, the injection condition is bad, the water injection adjustment is difficult

The injection condition of the three or four zones in area A is obviously worse than that of the one or two zones due to the development of oil layers and the characteristics of well patterns. The average injection pressure of single well in three or four zones is higher than that of the whole zone. The average intensity of single injection and actual injection was lower than that of the whole region. The completion rate of three or four strip injection is obviously lower than that of the whole region. The average water injection interval of single well is lower than that of the whole area, and the water injection adjustment is difficult.

2.3 Large proportion of low production and low efficiency wells, low recovery rate

From the perspective of single well fluid and oil production intensity, the three or four zones are far lower than the level of the entire region. From the daily oil production classification table of different zones in Zone A, it can be seen that the proportion of low production and low efficiency wells in the three to four zones is relatively large, with the proportion of low production wells with oil production less than 1.0t being 31.03% and 49.18%, respectively, which are significantly higher than those in the first and second zones. In terms of extraction degree, the extraction degree of the three and four belts is lower than that of the first and two belts. The extraction degree of the three belts is 33.46%, while the extraction degree of the four belts is only 16.36%.

2.4 Scattered distribution of remaining oil, making it difficult to tap potential

Based on the results of fine geological research, the paper systematically analyzes the remaining oil wells in three or four zones of the transitional zone in SA a area, and determines several kinds of remaining oil: (1) the well spacing is large, the controlling area of sand body is small, and the controlling degree of water flooding is low; Formed the remaining oil between wells. The remaining oil is scattered, mainly distributed in the front facies of the thin difference layer and the surface of the outer layer, the sand body distribution area is large, the oil layer is thin, poor physical properties, this type of remaining oil is more than 50% . (2) the remaining oil is distributed in the imperfect injection-production zone and some sand bodies in the stagnant zone have not been injected, resulting in imperfect injection-production systems such as injection-production without injection, production without injection or production without injection, in some wells, the remaining oil is formed because the water injection has not been driven to or bypassed. (3) there is some residual oil on the edge and top of the river course, because water passes through the center of the river course, the water can not drive it, or the oil layer becomes worse or even shrinks out at the edge of the river course, or the oil layer has not perforated to form the pressure relief point, the remaining oil is formed because of the poor water flooding effect and the serious heterogeneity in the thick oil layer, but the top is not flooded or the degree of flooding is low. In the late stage of high water cut production, although there is still a certain amount of remaining oil, the well pattern has a small control area on sand body.

3. Application of small well spacing test area

In response to the problems of poor development of oil layers and low well density in Zone A, such as difficulty in injection and production, poor reservoir utilization, low recovery rate, and poor development effect, combined with the distribution characteristics of remaining oil, a method of appropriately reducing the injection production well spacing was adopted for the three and four transitional zones in Zone A. The two experimental zones were reopened and utilized. After a series of adjustments to the injection production system such as sealing, hole filling, and injection conversion, the experimental zone achieved good development results.

3.1 Reuse of experimental area and increase in well network density

After the trial area is put into operation, it is equivalent to a one-time encryption of the original well network. Compared with the old well network, the well spacing is reduced by twice and the well network density is increased. The density of the 125m well spacing well network in the central experimental area has increased from 17.4 to 65.8 wells/km2, while the density of the 175m well spacing well network in the eastern experimental area has increased from 6.8 to 30.6 wells/km^2.

3.2 injection production adjustment, oil and water well drilling, and tapping remaining oil potential

After the trial area was put into production, the original well network was adjusted in the injection production system, from the original inverse nine point method to the five point area well network. In response to the single mining layer, oil and water wells were drilled to effectively excavate remaining oil. A total of 727 small layers were added to the oil and water wells in the two experimental areas, with a sandstone thickness of 913.1m and an effective thickness of 530.7m.

4. Effect analysis of small well spacing test area

4.1 Increasing the proportion of multi-directional connectivity and improving the degree of water drive control

After infilling the well network, adjusting injection and production, and supplementing oil and water wells, the multi-directional connectivity ratio of the 125m and 175m well spacing sets of five point method area well network has reached over 45%, and the degree of water drive control has both reached over 90%.

4.2 Injection pressure decreases and injection conditions significantly improve

After the trial area was put into production, the start-up pressure decreased and the injection pressure decreased, with an average single well injection pressure of 12.22 MPa, a decrease of 1.16 MPa compared to the original well network. Apparent water injectivity index and injection completion rate increased, and the injection completion rate reached 92.1%, which is 20.4% higher than the original well network, indicating a significant improvement in injection conditions.

4.3 The proportion of low production and low efficiency wells has decreased, and the recovery situation has improved

From the production end, the average single well fluid production intensity and oil production intensity increase 0.34 t/D respectively. M and 0.06 t/d. M, the ratio of low production and low efficiency wells is 26.19% less than the original well pattern, and the recovery condition is better than the original well pattern. After well pattern infilling, the problem of"Difficult injection and difficult production" caused by three or four well pattern with small density and small sand body is improved.

4.4 Oil reservoir utilization is better than the original well network

From the water injection profile of the water injection well, the effective thickness utilization ratio of the two experimental areas increased by 18.51% and 23.73% compared to the original well network. From the production profile of the oil well, the effective dynamic thickness ratio of the two experimental areas increased by 4.97% and 14.16% compared to the original well network. The profiles of oil and water wells are better than the original well network. From the perspective of the oil layer group, both experimental areas show the highest proportion of S2 utilization, with significant improvement in the S2 and S3 groups, with a well spacing of 175 meters being better than 125 meters in utilization.

Table 1 Comparison of water absorption status between the experimental area and the original well network

Oil reservoir group	Original well network (%)			125meter (%)			175meter (%)		
	Number of layers	sandstone thickness	effective thickness	Number of layers	sandstone thickness	effective thickness	Number of layers	sandstone thickness	effective thickness
S1	41.7	33.6	37.2	23.5	31.4	22.6	30.8	33.3	38.8
S2	50	62.5	62.3	61.3	72.2	78.3	73.3	81.1	86.3
S3	0.0	0.0	0.0	12.5	17.6	19.1	38.5	22.4	24.2
subtotal	43.3	50.9	50.6	54.6	68.1	69.2	33.1	69.7	74.3

4.5 Significant increase in oil recovery rate

After the trial area was put into production, the oil recovery rate significantly increased, with the two sets of oil recovery rates increasing by 0.23% and 0.17% for the 125m and 175m well spacing, respectively. Through the analysis of development index prediction curves, the 125m and 175m well networks are expected to increase oil recovery by 1.62 and 2.80 percentage points.

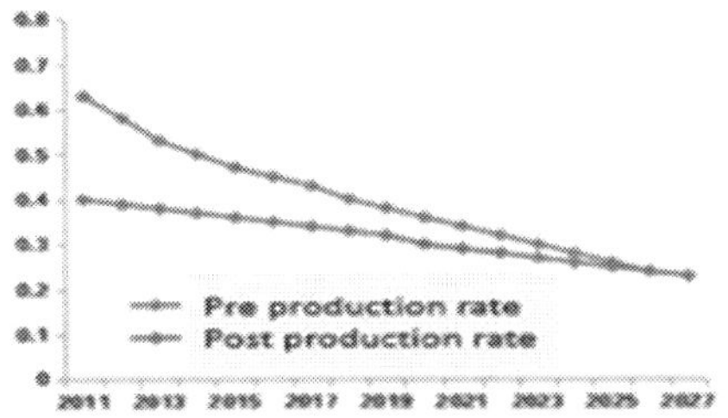

Figure 1 -Production rate curve before and after infill of 175m well network

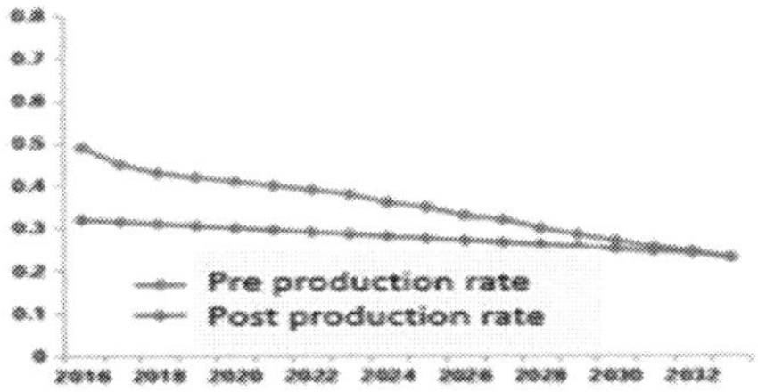

Figure 2 -Production rate curve before and after infill of 125m well network

5. Conclusion

1. The three or four zones are affected by the development of oil reservoirs and the characteristics of well networks, resulting in low water drive control, poor reservoir utilization, and low recovery. Entering the later stage of high water cut production, although there is still a certain amount of remaining oil, the control area of the well network on the sand body is small, and the difficulty of tapping potential under the current well network conditions is increasing
2. Combining the distribution characteristics of remaining oil in three or four zones, it is feasible to adopt the method of appropriately reducing the injection production well spacing to improve the development effect. After the trial area was put into operation, it was equivalent to a one-time densification of three or four belt wells, increasing the density of the well network and improving the degree of control. The trial area verified the feasibility of the small well spacing five point method area well network in the three or four belt wells.
3. After the trial area was put into production, the injection and production conditions were improved, the oil recovery rate was significantly increased, and the oil reservoir utilization was better than the original well network, indicating that small well spacing is beneficial for improving the development effect of three or four zones.

References

1. Zhang Yingzhi. Practice and Understanding of Oilfield Development Adjustment in Sabei Development Zone of Daqing Oilfield: Petroleum Industry Press, 2012
2. Liao Guangzhi, Niu Jingang. Industrial Polymer Flooding Effects and Main Practices in Daqing Oilfield [J]. Daqing Petroleum Geology and Development, 2004:2-3
3. Wang Demin. Developing New Theory and Technology of Tertiary Oil Recovery to Ensure the Sustainable and Stable Development of Daqing Oilfield (Part 1) [J]. Daqing Petroleum Geology and Development, 2001:3-7
4. Liu Xiang'e. Collection of papers on overall water plugging and profile control technology in oilfield blocks [M]. Petroleum Industry Press. 1994

Seismic Intelligent Prediction of Thin Interbedded Sand Body--Taking Heidimiao Reservoir in X43 Block as an Example

Cunsheng Wang [1], Ruishan Du [2,3], Pan Wang [4], Ruohui Yin [1]

[1] School of Earth Sciences, Guilin University of Technology, Guilin 541000, China
[2] School of Computer and Information Technology, Northeast Petroleum University, Daqing 163318, China
[3] FAPS Energy-Tech Ltd. of Heilongjiang Province, Daqing 163318, China
[4] Geological Research Center, Geophysics Research Institute, Bureau of Geophysical Prospecting INC., Zhuozhou, Hebei, 072750, China

Abstract: The Upper Cretaceous Nenjiang Formation in Songliao Basin is characterized by delta front subfacies, with a large area of thin channel sand and sheet sand, which is a typical problem of thin interbed identification. In this paper, based on the theoretical basis of seismic sedimentology and artificial intelligence algorithm, the application layer slicing technology and seismic data after 90 ° phase shift processing are used to carry out artificial intelligence prediction of thin interbedded sand bodies, which is in good agreement with wells.

Keywords: Artificial intelligence; 90 °Phase Shift; Stratigraphic slice; Thin interbed.

1. Introduction

With the exploration and development of the Songliao Basin for many years, the middle and shallow Heidimiao oil layer (the third member of Nenjiang Formation) has gradually attracted attention. The sedimentary model of the third member of Nenjiang formation in Changling sag is mainly underwater branch channel, estuary dam and sheet sand deposition of delta front subfacies [1]. The reservoir is mainly composed of thin interbeds and sand mudstone interlayers. It is difficult to quickly predict the distribution range of thin interbeds. For this reason, Zeng Hongliu proposed the concept of ' seismic sedimentology ' to improve the lateral resolution of seismic data to make up for the lack of resolution of the original seismic data [2-5]. In this paper, taking the I sand group of Heidimiao oil layer in X43 block as an example, the boundary and morphological distribution characteristics of thin sand layer are quickly and finely characterized by stratum slicing, 90 ° phase shift technology and artificial intelligence method, and good results are obtained.

2. Seismic response characteristics

The I sand group of Heidimiao oil layer in the study area is a typical lithologic reservoir. The sedimentary period is dominated by delta front subfacies, some of which are pre-delta subfacies. Channel sand body and thin sand body are the main reservoir types. The thickness of channel sand body is 10-15 m, and the thickness of thin sand body is 2-5 m. The difference of sedimentary environment leads to the difference of sand body deposition, the change of lithology combination of sand body, which makes the seismic wave interfere with each other, the characteristics of seismic waveform combination are not uniform, and the phenomenon of ' cross axis ' appears in the seismic axis (figure 1).

 https://doi.org/10.1051/e3sconf/202339401024

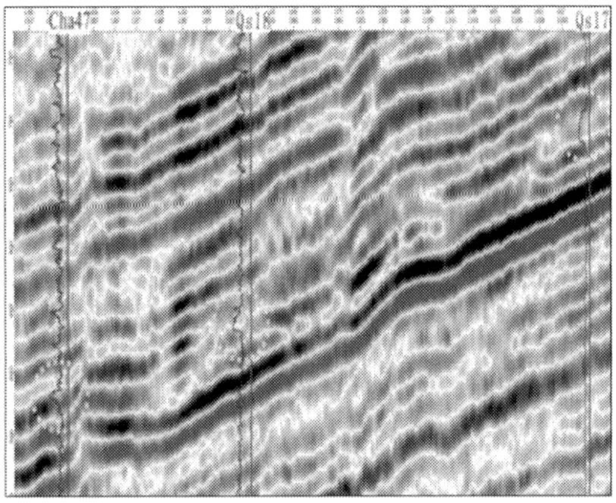

Fig.1 The ' cross-axis ' seismic profile of the I sand group of the third member of Nenjiang Formation

From the seismic profile, the wells drilled in the river have obvious seismic response characteristics, which are ' complex trough ' or ' lenticular ' seismic reflection characteristics. However, the boundary of channel sand body is vague, and the boundary of sand body needs to be further described in detail.

3. Seismic data analysis and processing

In seismic sedimentology, stratigraphic slice technology is one of the key technologies. A series of isochronous interfaces are realized by interpolating slices in equal proportion between two isochronous interfaces. Each isochronous interface is a comprehensive reflection of isochronous strata. The seismic attributes extracted by isochronous interface reflect the real stratigraphic attributes. Stratigraphic slices can reflect the distribution of sand bodies, which can be more effectively used to study the shape and distribution of sand bodies on the plane and the distribution characteristics of sand bodies in vertical direction [6-7].

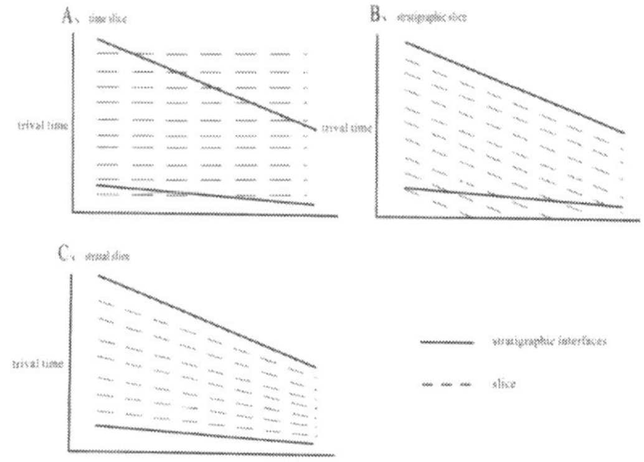

Fig.2 Time slice, layer slice and stratum slice technology (the difference and comparison of the three)

In the original seismic profile, the seismic axis of the sand group I of the Heidimiao oil layer in the Heidimiao oil layer appeared " through the axis " and other phenomena as shown in 3. After 90 ° phase shift processing, the seismic reflection characteristics of the corresponding trough are unified. The profile can clearly describe the thickness change of the sand body, and the corresponding relationship between the reservoir thickness and the trough reflection width is consistent : the seismic reflection characteristics of the wells with thicker reservoirs, such as Qs17 well, are wide trough characteristics ; the width of the trough in the seismic reflection characteristics of the thin reservoir, such as Cha47 well, becomes narrower (Fig.3). The thin sand body, morphology and distribution range can be described more accurately by fine interpretation of the processed seismic data and extraction of wave impedance attributes(Fig.4).

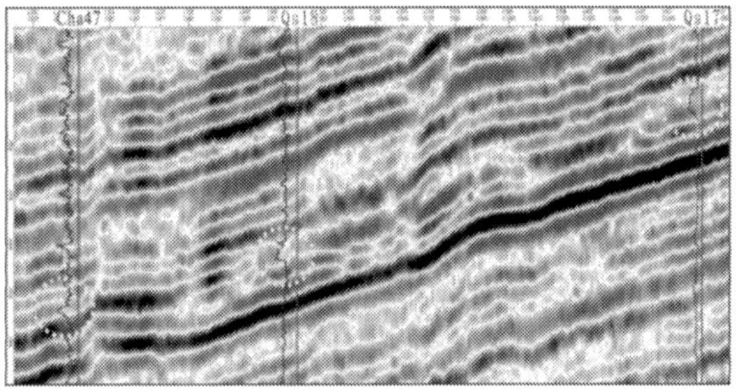

Fig.3 Seismic profile after 90-degree phase shift through the I sand group of the third member of Nenjiang Formation

https://doi.org/10.1051/e3sconf/202339401024

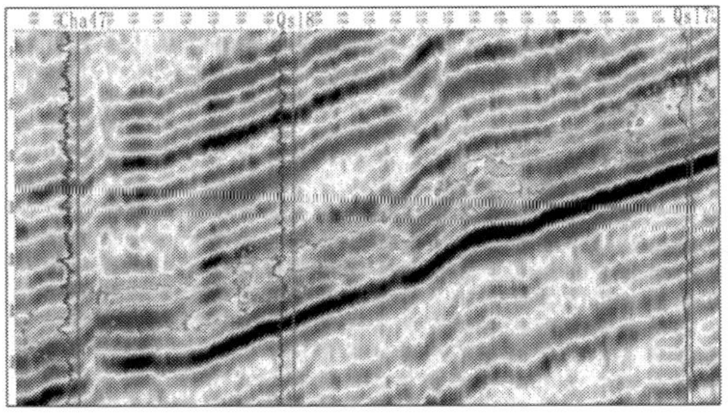

Fig. 4 wave impedance attribute profile and seismic profile superposition diagram

Reservoir prediction using seismic attributes is an important means for people to understand and detect reservoirs. Due to the complex nonlinear relationship between seismic attributes and reservoir parameters, traditional linear prediction methods (regression analysis, discriminant analysis, etc.) are difficult to achieve satisfactory prediction results. In contrast, neural network has better self-learning ability, associative storage ability and fast optimization ability. Therefore, in recent years, neural network technology has been increasingly applied to the field of prediction, and has better performance in reservoir prediction. The topology of BP neural network with a single hidden layer is shown in (Figure 5), including input layer, hidden layer and output layer. The input layer and output layer are generally one layer, and the hidden layer can be set to one or more layers according to the actual situation. The neurons between adjacent layers are fully connected, and there is no connection between neurons in the same layer.

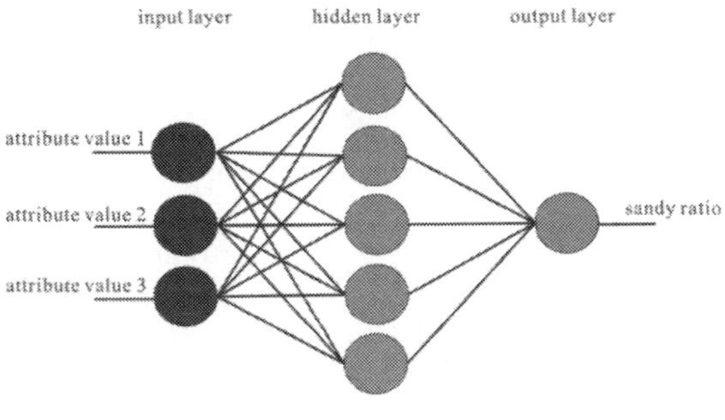

Figure 5 BP neural network topology structure

4. Conclusion and analysis

Based on the seismic data volume after 90 ° phase shift processing, the impedance attribute is calculated to obtain the wave impedance body, and the attribute extraction is carried out by the formation slicing technology. The BP neural network is used to model and predict the thin sand body information of the well point. The prediction results show that the main diversion channels in the study area are mainly developed in the bands of Qs17 and Qs18 wells, which are the red areas in the figure. The thin-layer river channel is mainly developed in the Qs16 and Cha13 well areas, which are the yellow areas in the map. The sheet sand reservoir is mainly developed in the Cha28 and Cha29 well areas, which is the blue area in the map (Figure 6), which is in good agreement with the well point information.

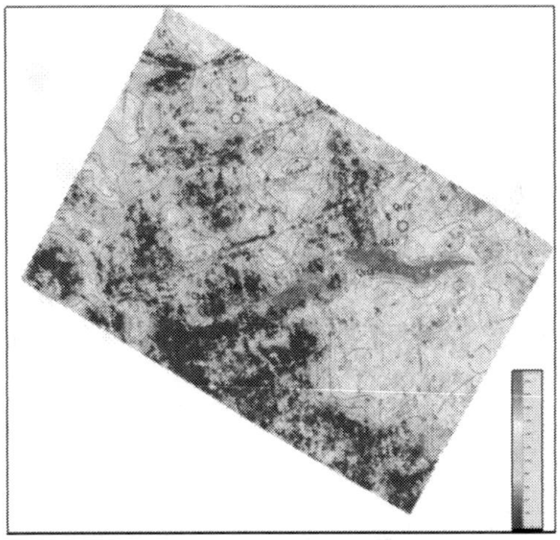

Fig. 6 River sand and thin sand superposition diagram

5. Conclusion

①The channel boundary can be quickly and finely characterized by stratum slicing and 90 ° phase shift technology.

②Combined with the calculation results of wave group antibody and well point data, the prediction of artificial intelligence thin sand body can better describe the boundary and morphological distribution characteristics of thin sand body.

References

1. Liu Haibin.Sedimentary characteristics and lithologic reservoir types of the third member of Nenjiang Formation in Changling Sag [J].Inner Mongolia Petroleum, 2015,41 (11) : 140-142.
2. ZENG H L, BACKUS M M, B ARROW K T, et al. Stratal sli cing;Part 1, realistic 3-D seismic model[J]. G eophysics, 1998, 63(2): 502-513.

https://doi.org/10.1051/e3sconf/202339401024

3. ZENG H L,HENRY S C, RIOLAJ P. Strata slicing: Part II, real 3-Dsei smi c data[J]. G eophysics, 1998, 63(2): 514-522
4. ZENG H L, B ACKUS M M. Interpr etive advantages of 90° -pha sewavelets: Part 1: Modeling[J] G eophysics, 2005, 70(3): C7-C15.
5. ZENG H L, B ACKUS M M. Interpr etive advantages of 90° -phasewavelets: Part 2: Seismic appli cations[J]. Geophysi cs, 2005, 70(3):C 17-C24.
6. Xiang Nian.Application of seismic sedimentology in the characterization of unconventional thin sand bodies [J].Petrochemical technology, 2020,27 (06) : 152-153 + 186.
7. Li, Jiang, Fan, Liu, Wang, Qi.Research on seismic sedimentology of Daqing placanticline oilfield development [J].Petroleum geophysical exploration, 2022,57 (04) : 916-925 + 742-743.

Research on infill adjustment method for XX narrow and thin sand body oilfield

Hongliang Gao

The Seventh Oil Production Plant of Daqing Oilfield Co., Ltd., Daqing, China

Abstract: The sedimentation of XX oilfield is mainly composed of underwater distributary channel sedimentation, and the sand body has the characteristics of "narrow, thin, and scattered". Although basic well network development, primary and local secondary infill, edge expansion adjustment, and corresponding injection production relationship adjustments have been carried out, there are still problems in current development, such as poor well network adaptability, low utilization degree of thin and poor layers, fast production decline rate, long shut-in wells, and many inefficient wells. In order to explore the potential of infill adjustment in narrow and thin sand body oilfields and further improve oilfield recovery efficiency, well pattern infill adjustment work has been carried out, achieving good results. A supporting infill adjustment technology has been formed, which includes "well seismic joint fine reservoir prediction, multi-disciplinary research on remaining oil quantification, multi-parameter evaluation of infill well location optimization, and multi information fusion perforation layer optimization", effectively improving the development effect.

Keywords: narrow and thin sand body oilfield; Non uniform encryption; Remaining oil; Reservoir Description

1. Introduction

XX Oilfield is located in the southern part of XX structure, with a main development of one set of XX oil layer. It belongs to delta front facies sedimentation with low permeability and is mainly composed of underwater distributary channel sedimentation. The sand body is narrow and thin, with obvious banding characteristics, generally between 100-200m in width. The sand body is small in scale and scattered in distribution. Although it has entered the stage of ultra-high water cut development through basic well pattern development, primary and local secondary infill, expansion adjustment, and corresponding injection production relationship adjustment, there are still problems such as poor adaptability of the well pattern, low degree of utilization of thin and poor layers, fast production decline rate, long shut-in, and many inefficient wells, resulting in significant technical difficulties in readjustment. In order to explore the potential of infill adjustment in XX narrow and thin sand body oilfield and further improve oilfield recovery efficiency, infill adjustment work has been carried out, achieving good results. A supporting infill adjustment technology has been formed, which includes "well seismic joint fine reservoir prediction, multi-disciplinary research on remaining oil quantification, multi-parameter evaluation of infill well location optimization, and multi information fusion perforation layer optimization".

2. Fine reservoir prediction combined with well seismic analysis

Through precise reservoir description, precise understanding of structures and reservoirs can effectively guide reservoir prediction and well location optimization for infill wells.

2.1 Fine characterization of faults

Using multiple methods such as breakpoint data guidance, coherent body, ant body recognition, etc. to comprehensively identify small faults with a fault distance of more than 5m and an extension length of more than 100m. By accurately identifying faults, it effectively guides the deployment of internal well locations and the adjustment of injection production structures in the well network.

 https://doi.org/10.1051/e3sconf/202339401025

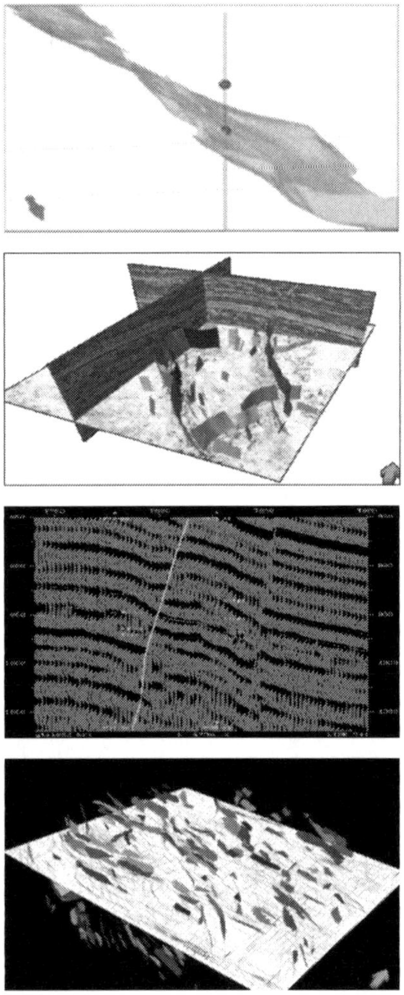

Fig. 1-1 Fault identification method

One is to newly identify small fault occlusion, change the injection production relationship, and deploy infill wells at positions with adjustable thickness increases. As shown in Figure 1-2, due to fault occlusion, residual oil is formed between the three oil wells, and one infill well can be deployed.

The second is that the extension length of the fault has changed, the injection production relationship has changed, and the adjustable thickness has been increased to deploy infill wells. As shown in Figure 1-3, the extension length of the fault has changed, and due to the obstruction of the fault, residual oil is formed between the three oil wells. One infill well and one injection well can be deployed.

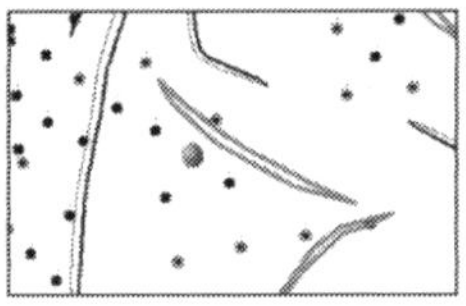

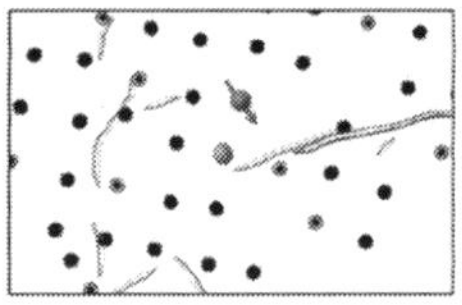

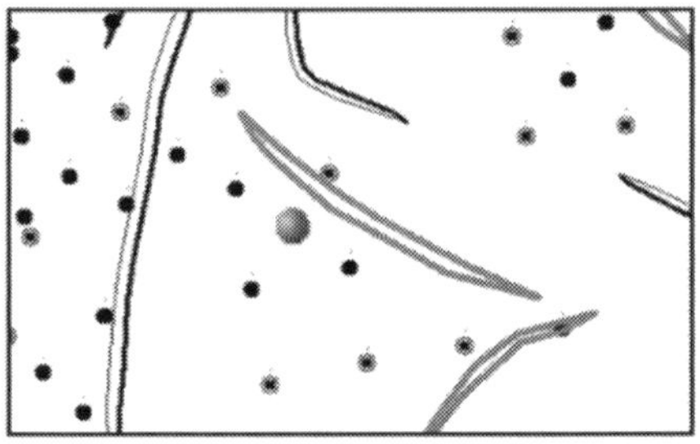

Fig. 1-2 Newly identified faults

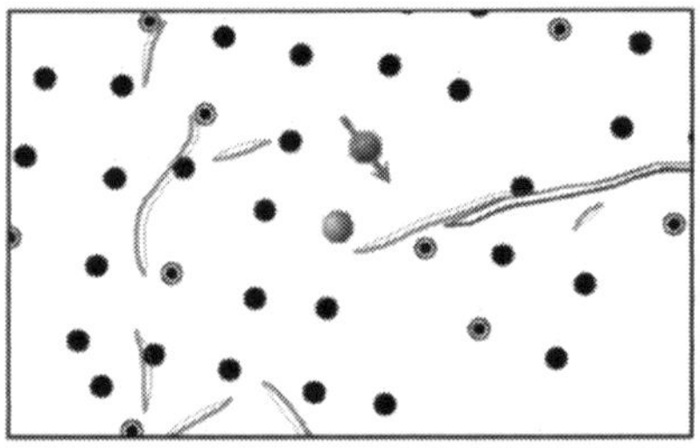

Fig. 1-3 Change in fault extension length

2.2 Fine characterization of reservoirs

By fine reservoir description, we can improve our understanding of sand bodies, accurately predict reservoirs, and effectively guide reservoir prediction and well location optimization for infill wells.

One is to control the time window according to sedimentary units under dense well network conditions, and apply a multi type attribute fusion method to improve the correlation between seismic attributes and sand bodies, with a correlation coefficient of 0.46, an increase of 0.03 (Figure 1-4).

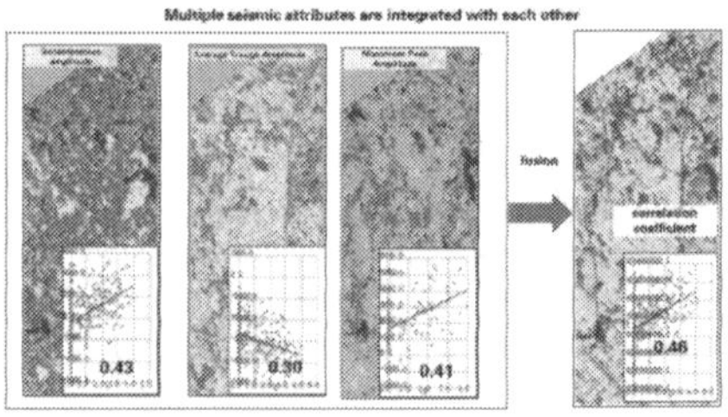

Fig. 1-4 Seismic Attribute Optimization

Secondly, based on the sedimentary environment and sand body characteristics, the seismic attribute control of the main channel and the indication control of seismic waveform in thin and poor layers are adopted. The inversion method is optimized by unit, and the prediction accuracy of sand bodies above 2m is 80.9%, an increase of 1.1%, and that of thin sand bodies between 1m and 2m is 63.1%, an increase of 12.8% (Figure 1-5).

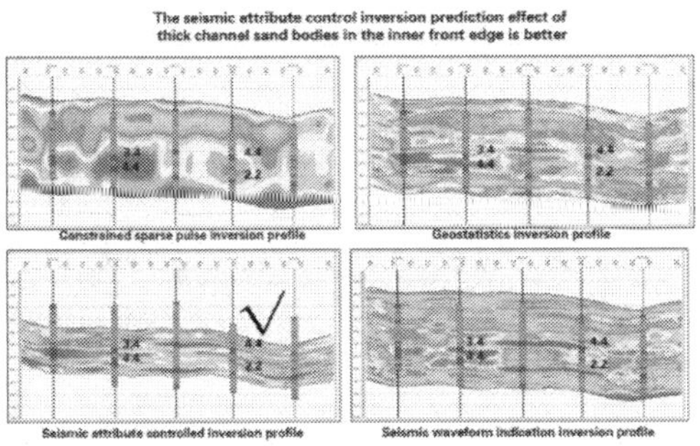

Fig. 1-5 Fine reservoir prediction

Thirdly, based on the prediction results of sand bodies, combined with the sedimentary understanding of each unit, the characterization method of "attribute control, inversion boundary, and logging facies" is adopted to complete the fine characterization of well seismic combined sedimentary facies.By comparing the new and old facies belts, the development characteristics of sand bodies are confirmed, and there are four main types of sedimentary microfacies changes, which effectively guide the optimization of infill well locations and increase the adjustable thickness of infill wells.

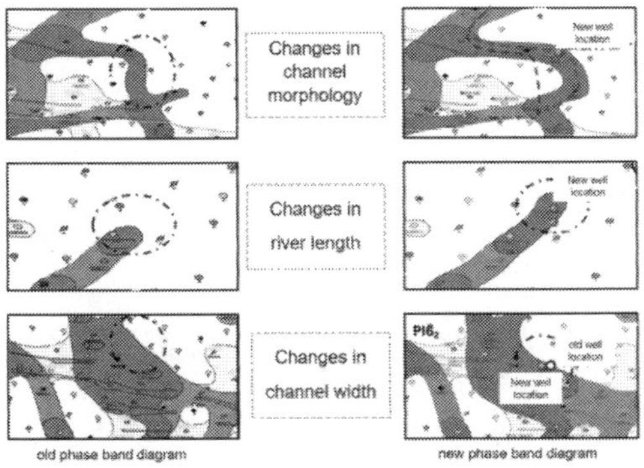

Fig. 1-6 Fine reservoir characterization guides well location deployment

3. Multidisciplinary research on quantification of remaining oil

By using multi-disciplinary fine description as a means, on the basis of fine geological research, high-precision phase control modeling technology and fine numerical simulation technology are applied to quantify the distribution of remaining oil and select potential areas for infill and adjustment.

3.1 Combining Well and Earthquake to Establish a Fine Geological Model

Structural modeling: Based on the combination of "well seismic" and structural achievements, establish precise fault and layer models; Sedimentary facies modeling: Based on the results of unit sedimentary microfacies, establish a sedimentary facies model; Attribute modeling: Using sedimentary microfacies as constraints, establish a reservoir attribute model with a planar grid step size of 20m.

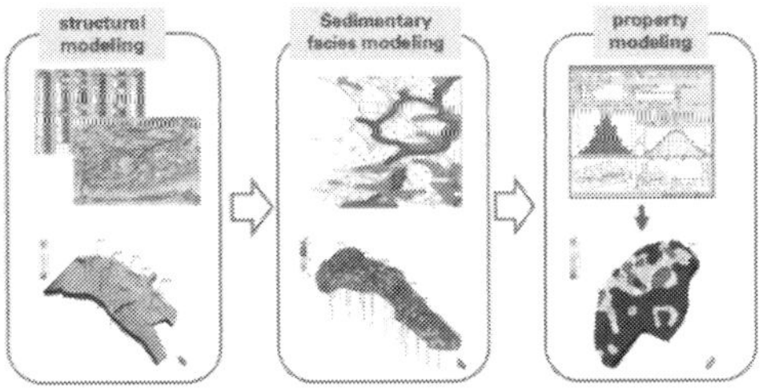

Fig. 2-1 Fine Geological Modeling

3.2 Numerical simulation of phase separation zone to quantify overall remaining oil

Based on the heterogeneity of the reservoir, lithofacies zoning was carried out, and different permeability curves were selected for numerical simulation based on the differences in microfacies properties. The single well fitting rate was increased to over 83%, quantifying the distribution of remaining oil in the entire area.

On the plane, the distribution of remaining oil is controlled by factors such as structure, reservoir, sedimentary microfacies, and well network, and there are mainly five types. The remaining oil is mainly composed of incomplete injection and production, uncontrollable well network, and thick oil layers, accounting for 69.34% of the total remaining reserves.

Fig. 2-2 Residual oil plane distribution map

Vertically, due to interlayer interference, there is a significant difference in the utilization of the XX Formation oil reservoir, with the remaining oil mainly concentrated in the XX1-9 sandstone formation, accounting for 70.11% of the remaining reserves. Among them, XX23, XX62, XX82, and XX92 have the most abundant remaining oil, accounting for 48.87% of the remaining reserves. Entering the ultra-high water cut development period, although the remaining oil distribution in XX Oilfield is scattered, there is still potential for infill adjustment in some areas.

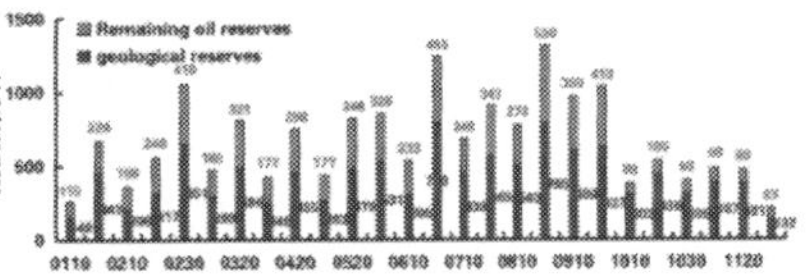

Fig. 2-3 Vertical residual oil distribution map

4. Multi parameter evaluation and optimization of infill well location

In response to the characteristics of narrow, thin, and scattered sand bodies in narrow and thin sand fields, as well as the high proportion of inefficient wells with uniform well distribution, a non-uniform infill adjustment principle of "uniform well distribution and selective drilling" is proposed. The multi-parameter comprehensive evaluation method is used to optimize the infill well location to improve the effectiveness of infill wells.

4.1 Well network encryption method

By using numerical simulation technology for comprehensive evaluation, the original well pattern is a reverse nine point area injection production well pattern, and the best adjustment method is diagonal center encryption. This method has a slow water cut increase speed and high recovery degree, which is conducive to later adjustment (Table 3-1, Figure 3-2).

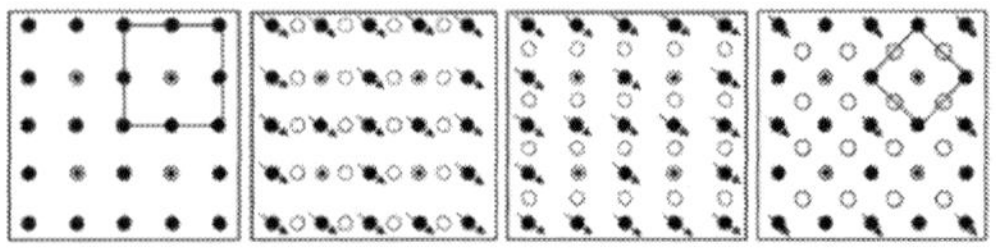

(a) (b) (c) (d)
(a) Original well network
(b)Add columns between columns
(c)Adding rows between rows
(d) Diagonal center encryption

Fig. 3-1 Optimization of encryption adjustment methods

Table 3-1 Prediction Results of Different Well Pattern Indicators

Well network adjustment method	Initial oil recovery rate	Adjust for five years		Adjust for ten years		Adjusting for fifteen years	
		Extraction degree	to contain water	Extraction degree	to contain water	Extraction degree	to contain water
	(%)	(%)	(%)	(%)	(%)	(%)	(%)
Original well network	0.46	32.70	95.91	33.51	96.91	34.21	97.31
Add columns between columns	0.75	33.92	96.51	35.41	98.01	36.17	98.76
Adding rows between rows	0.73	33.65	96.16	35.32	97.66	36.23	98.16
Diagonal center encryption	0.69	33.53	95.66	35.73	97.41	37.26	97.91
Adjustment of injection and production system	0.48	32.89	96.01	34.42	97.26	34.95	97.76

4.2 Optimization method for infill well location

Based on the characteristics of "overall scattered and local concentrated" distribution of remaining oil, the method of "uniform well distribution and selective drilling" was adopted to determine seven main control indicators. Combined with the research results of technical and economic boundaries, the principle of selecting infill wells was determined. In order to effectively reduce the proportion of inefficient wells, a multi-parameter comprehensive evaluation method is introduced to achieve multidisciplinary intervention analysis and avoid errors caused by using adjustable thickness as a single evaluation indicator.

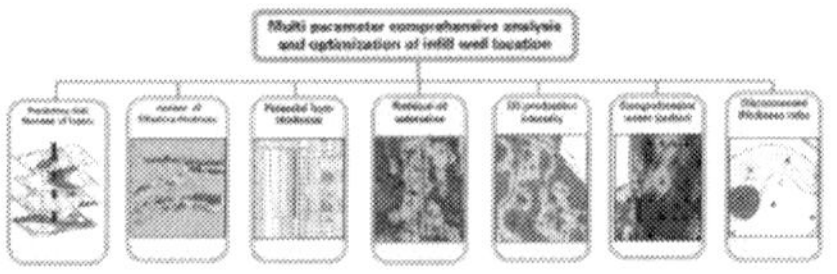

Fig.3-2 Multiparameter comprehensive evaluation method

Taking the XX encryption area of XX oilfield as an example, 61 encryption wells were uniformly deployed using diagonal center encryption (Figure 3-4) as the preferred object. By establishing parameter relationship diagrams such as layer number effective thickness, remaining oil saturation adjustable thickness, and comprehensive water content daily oil production (Figure 3-3), 13 encryption wells were finally selected through comprehensive screening. In the initial stage, an average of 10 small layers were encountered during single well drilling, with an effective thickness of 4.3m. The daily oil production in the initial stage was 3.1t, and the water content was 75.6%.

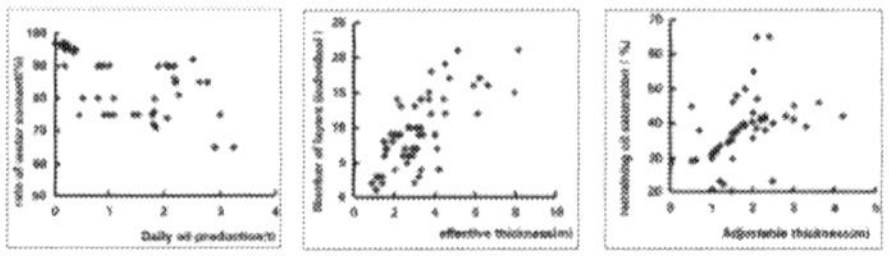

Fig.3-3 Encryption Adjustment Parameter Relationship Chart

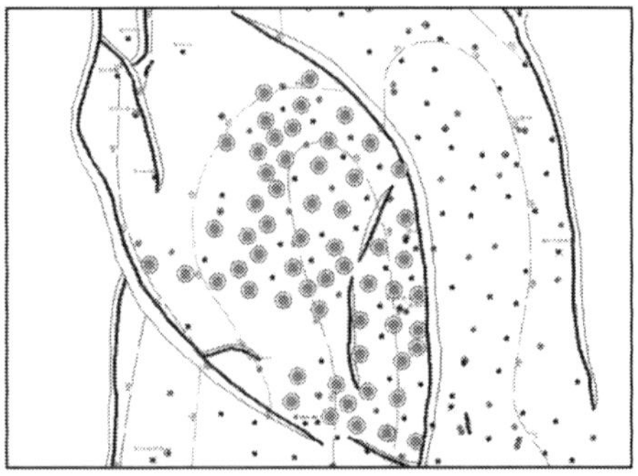

Uniform well distribution

https://doi.org/10.1051/e3sconf/202339401025

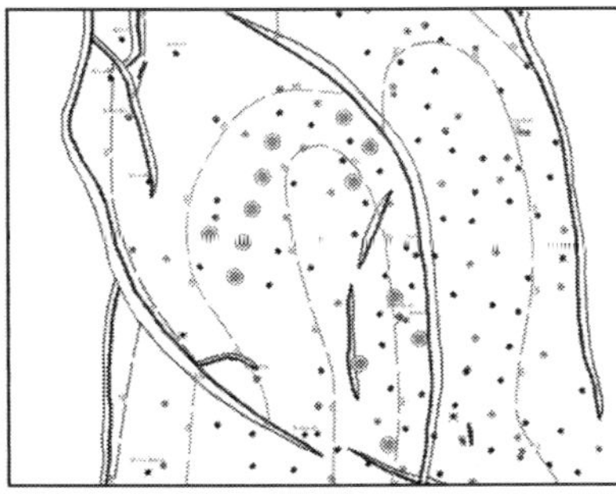

Optimization results

Fig. 3-4 Optimization Example of Multi parameter Comprehensive Evaluation Method

5. Optimization of Perforation Horizon Based on Multiple Information Fusion

Using various data such as logging, means of production and sedimentary facies belt, the water flooded layer can be accurately identified, the perforation horizon can be optimized, and the initial water cut of infill wells can be controlled.

5.1 Comprehensive Determination of Water Flooding Degree

Based on logging interpretation, combined with various data such as remaining oil distribution, production situation, and testing, comprehensively analyze the degree of water flooding in the encrypted well reservoir, laying the foundation for perforation layer selection.

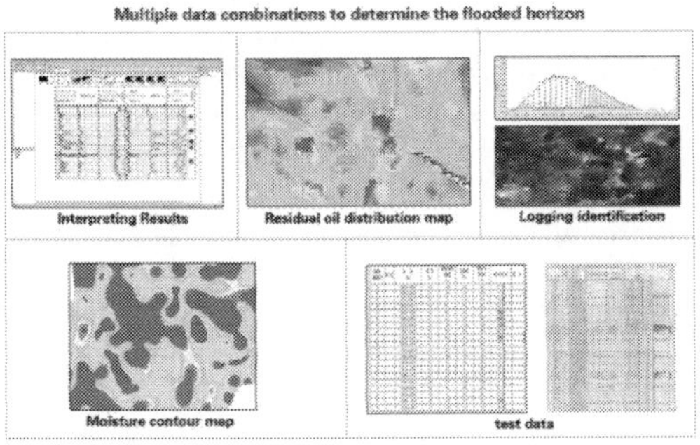

Fig. 4-1 Multiple data combinations to determine the flooded horizon

5.2 Optimization of perforation layer selection principles

Due to the high degree of water flooding, poor sand body thickness, scattered distribution of remaining oil, and imbalanced water flooding conditions in the Putaohua Oilfield, the selection of perforation layers is optimized based on the principle of "classification evaluation and control of water content". In the initial stage, in order to ensure the production effect, the optimization of perforation layers was mainly focused on low water flooded and non water flooded layers. With the continuous understanding of the reservoirs in the infill adjustment zone, the optimization method was gradually improved, and supplementary and modified perforation modes were added.

The perforation and completion modes of reservoirs are mainly divided into the following four categories (Table 4-1):

One is the selective perforation mode. Under the condition that the adjustable thickness can meet the production capacity needs, strictly control the perforation thickness and only inject reservoirs with low water flooding degree;

The second is the extended perforation mode. In response to the high degree of water flooding in the entire well and the lack of development of thin and poor reservoirs, due to the very small adjustable thickness, it is considered to appropriately relax the water content limit to increase single well production and open up medium to high water flooded risk reservoirs;

The third is the supplementary perforation mode. In response to the situation of high water flooding but relatively developed thin and poor reservoirs, due to the small adjustable thickness, selective expansion of mud and sandstone layers can be used to increase the adjustable thickness of a single well while shooting out oil layers with lower water flooding levels. Depending on the specific situation, the completion method should be considered and determined;

The fourth is the modified perforation mode. Considering the poor development of reservoirs and the difficulty in achieving the designed production capacity of the plan, based on the thickness of the reservoir, the properties of the oil layer, and the connectivity relationship on the sand body plane, consideration is given to adjusting the potential through fracturing transformation.

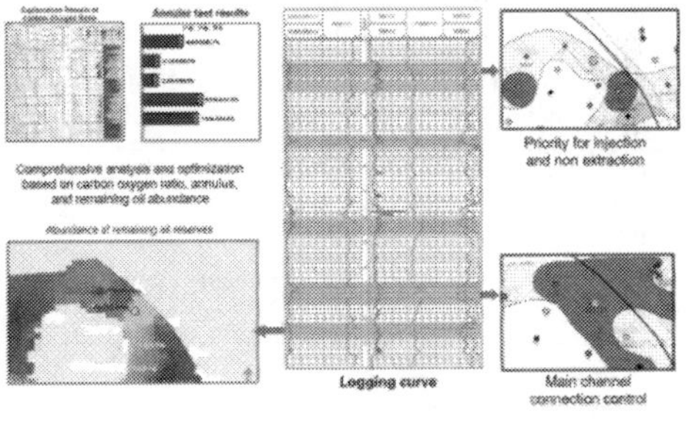

Fig.4-2 Example of perforation layer optimization

6. Encryption adjustment effect

A total of 425 infill wells were put into operation in XX Oilfield. In the initial stage of infill wells, the average daily liquid production of a single well was 12.5 tons, daily oil production was 2.7 tons, and the comprehensive water content was 78.40%. The decline in production slowed down by 2.5 percentage points, achieving good results. After the encryption adjustment, the degree of water drive control has increased to 77.6%, an increase of 2.6 percentage points. The application of set encryption adjustment technology can effectively improve the encryption adjustment effect and improve the oilfield development effect.

Table 5-1 Comparison of Initial Production of Infill Wells

region	Design			initial stage		
	Nissan liquid (t)	Daily oil production (t)	to contain water (%)	Nissan liquid (t)	Daily oil production (t)	to contain water (%)
XX Oilfield	8.0	1.6	80.00	10.9	2.2	79.82

6.1 Analysis of Typical High Yield Wells

According to statistics, there are 12 infill wells with a daily oil production of more than 10t in the initial stage of production. The average single well encountered 10 small layers, with a sandstone thickness of 13.1m and an effective thickness of 7.4m. In the initial stage of production, the average daily liquid production of a single well was 16.9t, with a daily oil production of 13.7t and a comprehensive water content of 18.97 (Table 5-2). From the distribution area, it is mainly concentrated in areas with abundant residual oil at the fault edge, areas with imperfect injection production relationships, and areas with newly added sand bodies.

Table 5-2 Production Statistics of High Yield Infill Wells

Serial number	Well number	Number of layers (piece)	Sandstone thickness (m)	effective thickness (m)	Number of shooting layers (piece)	Effective shooting (m)	Nissan liquid (t)	Daily oil production (t)	to contain water (%)
one	Well 1	6	7.3	4.0	6	4.0	14.4	13.4	6.94
two	Well 2	8	17.4	12.3	4	7.6	27.2	10.0	63.24
three	Well 3	7	13.4	9.5	3	5.4	28.1	18.5	34.16
four	Well 4	10	13.4	10.9	8	10.0	17.5	13.5	22.86
five	Well 5	11	15.0	3.2	5	2.0	12.5	10.8	13.60
six	Well 6	11	13.2	5.6	8	3.4	17.2	15.0	12.79
seven	Well 7	10	14.9	10.0	4	6.3	25.4	13.0	48.82
eight	Well 8	9	13.5	7.5	5	8.9	14.3	13.0	9.09
nine	Well 9	10	9.9	5.0	15	7.5	13.0	13.0	0.00
ten	Well 10	10	11.2	5.4	9	5.4	11.1	11.1	0.00
eleven	Well 11	11	15.6	10.3	7	8.7	13.6	13.6	0.00
twelve	Well 12	11	14.6	10.8	10	10.8	15.2	15.2	0.00
average		10	13.3	7.9	7	6.7	17.5	13.3	23.58

7. Conclusion

(1) The application of fine reservoir structure interpretation and fine reservoir prediction can effectively guide the adjustment of well network encryption, providing guarantee for improving the production effect of encryption wells;

(2) Entering the ultra-high water cut development period, the remaining oil distribution in XX Oilfield is scattered, but there is still potential for infill adjustment in some areas;

(3) The gradually improved multi information fusion optimization perforation layer technology has played an important role in ensuring production and controlling water content;

(4) The application of the supporting encryption and adjustment technology of "well seismic joint fine reservoir prediction, multi-disciplinary research on remaining oil quantification, multi-parameter evaluation of encryption well location optimization, and multi information fusion perforation layer optimization" can effectively improve the encryption and adjustment effect.

References

1. Chen Yuanqian. Determination of well network density under different well layout methods [J]. Petroleum Exploration and Development, 1986, 13 (1): 60-62
2. Wu Jingchun et al. Research on Well Pattern Encryption Methods in Block L of Daqing Low Permeability Oilfield [J]. Science and Technology and Engineering, 2011, 11 (34): 8590-8593
3. Wei Jide, Du Qinglong, Lin Chunming, et al. Factors and Classification of Residual Oil in Daqing Oilfield [J]. Petroleum and Natural Gas Geology, 2001 (01)
4. Jiang Xiufen, Zhang Yunxiu, Ren Jirong, et al. Analysis of the effect of secondary well network encryption adjustment [J]. Daqing Petroleum Geology and Development, 1995, 03:53-55
5. Cao Renyi, Cheng Linsong, Xue Yongchao, et al. Research on Optimization and Adjustment of Well Patterns in Low Permeability Reservoirs [J]. Journal of Southwest Petroleum University, 2007, 04:67-69

Research on water driving state evaluation and adjustment technology based on streamline

Yue Hu

Institute of Geology, the Third Oil Production Plant, Daqing Oilfield Company, Daqing 163113, China

Abstract: The general reservoir engineering method is the injection volume and production between the regular injection-production well pattern (five-point method, seven-point method, nine-point method, etc.), geometric average calculation or kh method. The accuracy of the calculation results is low and it is difficult to reflect the political enemy dynamics of reservoir heterogeneity and correctly evaluate the water characteristics of reservoir development and characteristics. Problems of water injection, distribution and well efficiency are not solved. We can intuitively understand the dynamic flow characteristics of fluid in underground by using the oil-first device technology, and combined with the basic principles of reservoir engineering, a method of hydraulic reservoir dynamic evaluation is proposed. Using this method, many variables such as drainage area of oil well, flooding velocity, flooding swept area between oil Wells can be evaluated quantitatively, and the development effect of oil field can be evaluated accurately. According to the analysis of the characteristics and factors of the distribution of the wire, the corresponding adjustment measures are taken to make the dominant wire weaker and the weak wire stronger, and finally achieve the purpose of improving the recovery rate.

Key words: water drive evaluation; Streamline numerical simulation; Development adjustment

1. Introduction

In order to accurately understand the law of underground fluid in water drive reservoir and improve the ultimate oil recovery, the evaluation model of water drive state was established on the basis of flow line numerical simulation technology, and the evaluation method of water drive state was proposed. On this basis, the corresponding adjustment and optimization technology was studied. Through the dynamic evaluation of flow line, the relationship between injection-production Wells can be evaluated quantitatively, which can guide the development and adjustment of reservoir. Using low cost adjustment techniques such as hierarchical subdivision recombination, dynamic deployment and well pattern thinning, good development results have been achieved.

2. Research on water driven state evaluation technology

2.1 Use oil - well distribution coefficient

Well Allocation Factors are the amount of fluid that a well receives (or contributes) to other Wells [1]. For an oil well, the oil well distribution coefficient is the relative contribution of the surrounding Wells to its liquid production, oil production and water production.

2.2 Establishment of water-driven state evaluation model

The traditional methods in reservoir engineering adopt regular well pattern design, and the production distribution is usually fixed geometric average method. The percolation area of the reservoir is limited in the specified well pattern unit, and the boundary is artificially fixed, which has great blindness and does not conform to the actual seepage condition of underground reservoir. Therefore, a new calculation method is proposed, which takes the oil well as the center and establishes the dynamic well pattern between the oil well and the surrounding water injection well by streamline simulation, and realizes the distribution coefficient calculation by the variable well pattern flow method. In this way, the distribution of the surrounding water injection well and the actual dynamic injection volume of the water injection well are taken into account. The streamline model is realized by the "five-step method". The first step is to clarify the cumulative production, cumulative injection and related production and water injection time of oil and water Wells. The second step marks the existence of injection-production relationship and establishes the historical injection-production flow line; The third step is to use the superposition analysis of production injection, new well, measures and monitoring data to evaluate the nature of streamlines and judge the strong, sub-strong and weak streamlines. The fourth step is to remove the weak streamlines such as short corresponding time, ineffective injection and production, weak flooding of new Wells and less cumulative production and cumulative injection, so as to dilute the complex streamline network, and analyze and calculate the streamline water flooding swept area. The fifth step is to describe the flooded area and the remaining oil-rich area by synthesizing the strength of various flow lines, corresponding time, injection and production efficiency, and cumulative production and cumulative injection.

2.3 Dynamic evaluation method

Since the established streamline model fully considers the heterogeneity of the reservoir, the irregular form of well pattern and the dynamic variation characteristics of injection-production Wells, it can realize the dynamic evaluation of the water drive characteristics of the reservoir, the water injection efficiency, the connectivity degree of injection-production Wells and the effectiveness of the well. Main evaluation indexes: (1) Cumulative water drive contribution oil production-cumulative water injection relationship curve; (2) The relation curve of current water drive contribution to oil production-current water injection; ③ Flow distribution diagram of well pattern unit. After selecting relevant parameters, the corresponding dynamic evaluation method can be established. Firstly, the cumulative deficit diagram of the oilfield or block is drawn to analyze the overall injection-production situation within the oilfield or block, including the injection-production ratio of the oilfield, formation pressure maintenance level, injected water loss, etc., so as to preliminarily evaluate the water flooding effect of the oilfield. Secondly, a typical well group is selected for detailed analysis, and the flow distribution diagram within the well pattern unit is drawn to evaluate the contribution of water flooding in a single well, as well as the connectivity between oil and water Wells and zones. Finally, the Wells with poor development effect (injection well and production well) are selected and corresponding adjustment schemes are formulated to improve the water flooding effect.

2.4 Quantify the relationship between injection-production Wells

For the whole oilfield, the key of water driving state evaluation should be specific to a single well. By calculating the distribution coefficient of each well in the well pattern unit, the connectivity degree of each well and the actual injection-production distribution can be described quantitatively. Similarly, the distribution coefficient of each phase fluid in a single well can be calculated to clarify the flow characteristics of each phase fluid in the reservoir, so as to guide the dynamic allocation of oil and water Wells [2].

The larger the distribution coefficient of the injection well corresponds to the well, the more water injected into the well. If the flow field is not adjusted in time, the premature flooding of the well will be caused.

As can be seen from the flow line distribution of the one-injection and two-production theoretical model, a large amount of injection water flows to well P2 from the injection well I1, and it can be seen from the pie chart of distribution coefficient that the water volume accounts for 82%, forming the main channeling channel. However, in the direction of well P1, the water volume is less, resulting in poor or even ineffective water flooding effect.

Similarly, if the distribution coefficient of an oil well to a certain injection well is larger, it indicates that the production volume contributed by the oil well is mainly from the direction of the injection well. According to the flow line distribution of well P2 in the two-injection and one-production theoretical model, it can be seen that the fluid production of the well mainly comes from the contribution of injection well I1, and the pie chart of distribution coefficient shows that the contribution rate reaches 86%.

3. Study on water drive adjustment and optimization technology

Based on the evaluation results of water driving state and the analysis of the characteristics and factors of flow line distribution, targeted adjustment measures are taken to change the dominant direction of underground fluid, increase the water drive intensity in the direction of weak flow line, improve the overall sweep coefficient, and improve the oilfield development effect.

3.1 Increase the sweep volume of injected water

The injection water in the early stage is used to advance evenly outward in the formation around the bottom hole. First, the injection water advances along the high permeability section with low seepage resistance. When the formation pressure in the higher permeability section increases, the injected water is forced into the very low permeability section, which increases the formation pressure in the low permeability section, reduces the pressure difference between the two, and effectively improves the swept volume of the injected water. See Figure 1.

$$q_o = \frac{2\pi k_o h_o \Delta p}{u_o\left(\ln\frac{r_e}{r_w} + S\right)}$$

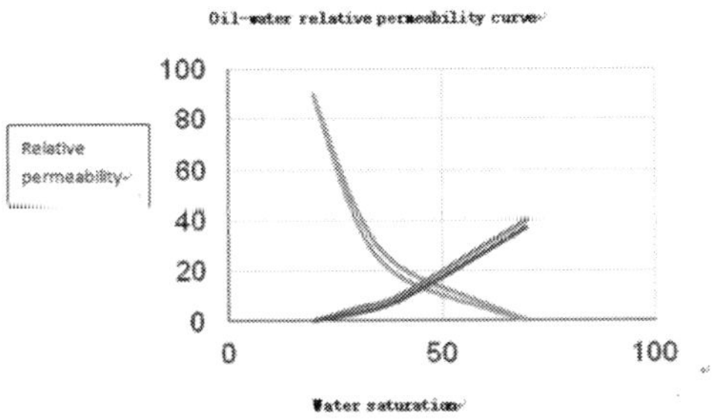

FIG. 1 Relation between injected water and volume

3.2 Dynamic vector allocation, optimize fixed streamline

The vector dynamic deployment is carried out by streamline numerical simulation. Combining with the results of streamline dynamic simulation, the following technical ideas of dynamic deployment are formulated for the purpose of improving the waterflood sweep sweep: (1) Based on the water-driven state evaluation model, the waterflood sweep sweep coefficient of a single well and the average sweep coefficient of a block are calculated. ② Optimization of water injection and liquid extraction. By comparing the sweep coefficient of single-well waterflooding with the average sweep coefficient of block waterflooding, if the sweep coefficient of single-well waterflooding is small, the water injection will be increased, and the oil well in the direction of weak flow line will be extracted, while the oil well in the direction of strong flow line will be shut down. On the contrary, the water injection should be reduced and combined with measures to control the rise of water in the well. ③ Use the optimized water injection to simulate the production effect after a certain period of time. Repeat the above steps until you optimize the best solution for improving the sweep efficiency of water drive.

3.3 Improve the injection-production well pattern and add new flow lines for potential exploitation

By perfecting the injection and production well pattern in the area where reserves are out of control, a new flow line is established to excavate the remaining oil. There are two ways to establish a new flow line: one is to deploy a new well, and the other is to adjust the flow field (including diversion or diversion). The main forms include: (1) deploying new Wells in uncontrolled well pattern areas or using old Wells to sidetrack and establish new injection-production well pattern; (2) The original strong flow line direction of the oil well after flooding, forming a row or cut water injection, change the original water drive direction; (3) The Wells with low production are diverted to form multi-directional water flooding, and the Wells with less cumulative water injection are diverted to pumping combined with the optimization of well pattern. (4) Combined with profile control, drive control, water sticking and other measures, weaken the strong streamline, promote the addition of other directions of the streamline.

3.4 Optimize injection-production well spacing and control plane streamline

Based on a clear understanding of reservoir heterogeneity and sedimentary microfacies, the spacing of injection-production Wells is changed to establish a reasonable displacement pressure gradient between the injection well and the well, and control the flow line is too strong or too weak, so as to maximize the development of the reservoir, improve the development effect and increase the water drive recovery factor. The main methods are as follows: (1) In the main river channel or the area with good physical properties, dilute the well pattern, widen the injection-production well spacing, reduce the displacement pressure gradient, and control the strong flow line. (2) In the lateral edge of the river channel or the area with poor physical property, well spacing should be filled, the distance between injection and production Wells should be reduced, and the displacement pressure gradient should be improved, so that appropriate flow lines can be established between oil and water Wells, and water injection can be effective. (3) In the joint area of the main river channel and the side edge, change the well pattern, water injection in the river channel, oil recovery in the side edge, combined with the measures of increasing injection and extraction, increase the flow line, expand the water drive sweep.

3.5 Subdivide and reorganize layers to improve three-dimensional streamline

Based on Darcy's formula, the seepage resistance term R 'is defined as pseudo seepage resistance coefficient, representing the oil-water two-phase seepage capacity varying with saturation. The larger the pseudo drag coefficient difference is, the more serious the interlayer interference is.

Aiming at injection Wells with prominent interzonal contradictions, the four-stage optimization method of layer subdivision and recombination is adopted to improve injection profile and balance interzonal operation. The four-stage optimization method of hierarchical subdivision and recombination is mainly as follows: the first step is to screen the pseudo-seepage resistance level difference. The pseudo-seepage resistance level difference limit in the adjustment zone is determined to be 5.5. According to the calculation method of the pseudo-seepage resistance coefficient, the pseudo-seepage resistance level difference of each scheme is calculated and the scheme with the pseudo-seepage resistance level difference less than 5.5 is screened out. In the second step, schemes with single control reserves less than 4.5×104t are excluded. The third step is to predict the development index and select the higher degree of recovery at the end of 15 years after the implementation of the program. The fourth step is to optimize the economic indicators, to optimize the scheme with the highest economic indicators. Through the stepwise screening of the four-stage optimization method, the hierarchical division scheme with small

https://doi.org/10.1051/e3sconf/202339401026

pseudo seepage resistance difference, a certain reserve base and the best development economic index is optimized.

4. Application Effect

The technology has been applied to W2 and G7 blocks in Jiangsu oilfield, and the recovery efficiency has been increased by 1.8% and 1.3% respectively, while the natural decline rate has decreased by 9.8% and 10.9%.

5. Cognition and Conclusion

(1) The conventional water drive reservoir evaluation usually considers the static geological model, which is unable to quantitatively evaluate the water drive state of the reservoir and the effect of the well. The reservoir dynamic evaluation method based on streamline simulation can realize the quantitative evaluation of the relationship between injection-production Wells, and can better guide the reservoir development adjustment.
(2) In the late stage of high water cut development, flow line adjustment and optimization is an important means to slow down production decline and further improve reservoir recovery efficiency. According to the different distribution characteristics of streamline, the corresponding streamline adjustment measures can effectively improve the development effect.

Reference

1. Wu Junlai. A new method for water drive reservoir dynamic evaluation based on 3D streamline simulation [J]. Journal of Xi 'an Shiyou University (Natural Science Edition),2011,26(2):43-47.
2. Wang Hongguang. Simulation study on characteristics of reservoir physical properties in high water cut period [J]. Acta Petrolei Sinica,2004,25(6):53-58.
3. Zhang Yingzhi. Practice and Understanding of Oilfield Development Adjustment in Sabei Development Zone of Daqing Oilfield: Petroleum Industry Press, 2012
4. Liao Guangzhi, Niu Jingang. Industrial Polymer Flooding Effects and Main Practices in Daqing Oilfield [J]. Daqing Petroleum Geology and Development, 2004:2-3
5. Wang Demin. Developing New Theory and Technology of Tertiary Oil Recovery to Ensure the Sustainable and Stable Development of Daqing Oilfield (Part 1) [J]. Daqing Petroleum Geology and Development, 2001:3-7
6. Liu Xiang'e. Collection of papers on overall water plugging and profile control technology in oilfield blocks [M]. Petroleum Industry Press. 1994

Spatio-temporal heterogeneity of transportation carbon emissions and its driving factors in China's main urban agglomerations

Shicheng Wei*

School of Life and Environmental Sciences, Minzu University of China, Beijing, China.

Abstract: The transportation sector in China has the characteristics of large total carbon emissions, high level, and unbalanced spatial distribution. For carbon emissions reduction, it is of great significance to study the carbon emissions of transportation in China in different regions. Focusing on China's three urban agglomerations: Beijing-Tianjin-Hebei Region, Yangtze River Delta Region and Pearl River Delta Region, this paper explores and compares the spatio-temporal heterogeneity of China's traffic carbon emissions from 2000 to 2019 by using methods such as GWR and ESDA. The results show that: 1) As for carbon emissions, the total carbon emissions and per capita carbon emissions of the three urban agglomerations have shown a significant growth trend. The high-value aggregation in Beijing-Tianjin-Hebei Region has weakened, the high-value and low-value aggregation in the Yangtze River Delta Region has increased, and the change in the Pearl River Delta Region is not obvious. 2) As for influencing factors, motor vehicle ownership has the greatest impact on regional carbon emissions, but the impact intensity of motor vehicle ownership on carbon emissions of the three urban agglomerations is different. 3) As for spatio-temporal heterogeneity, after 2010, the spatial correlation of carbon emissions of the three urban agglomerations was lower than that of the surrounding areas, and all of them were weakened.

Key words: carbon emissions from transportation; urban agglomerations; driving factors; spatio-temporal heterogeneity.

1. Introduction

In recent years, global warming has become one of the most serious environmental problems in the world [1], which undoubtedly becomes a major crisis facing mankind [2]. The greenhouse gases produced by the burning of fossil fuels such as coal, oil and natural gas caused by human activities such as industry and transportation have become the main culprits of global warming. In response to this thorny issue, the United Nations Intergovernmental Negotiating Committee adopted the *United Nations Framework Convention on Climate Change*, the world's first international convention to control greenhouse gas emissions and deal with global warming, in 1992, which clarified the responsibilities among countries [3]. The Pa*ris Agreement* also provides an action guide for climate change, striving to control the global temperature rise within 1.5℃. In 2018, IPCC (Intergovernmental Panel on Climate Change) reached a consensus to achieve the goal of peaking carbon dioxide emissions and carbon neutrality (hereinafter referred to as "double carbon") [4]. China is not an exception. As the world's largest carbon emitter [5], it has also taken corresponding actions in the field of dealing with carbon emissions. In September 2020, at the 75th United Nations General Assembly in China, China formally put forward the goal of achieving "peaking carbon dioxide emissions" in 2030 and "carbon neutrality" in 2060. The goal of "double carbon" is of great strategic significance to the development of China [5].

In recent years, with the rapid development of transportation industry, it has become the second largest carbon emissions industry [6], second only to energy power generation and heating industry. According to the forecast of the International Energy Agency, China's carbon emissions from transportation will account for more than one-third of the world's carbon emissions from transportation by 2035 [7]. Therefore, the research on carbon emissions from transportation is quite instructive for reducing carbon emissions. Many scholars have studied the carbon emissions of different industries and fields from the aspects of energy consumption scale and structure [8], population urbanization [9, 10] and construction land [11]; among which there are many explorations on carbon emissions of transportation industry. In previous studies, LMDI [12-13], STIRPAT model [14-15] and grey relational analysis [16-17] were used to combine carbon emissions from transportation

* Corresponding author: 1585720134@qq.com

with regional economic research, and explore the relationship between urban economic development and carbon emissions from transportation, revealing that regional carbon emissions from transportation are significantly affected by regional transportation industry development and economic development [18]. In addition, the degree of transportation and economic activities among different regions lead to a certain spatial correlation and heterogeneity of regional carbon emissions from transportation [19]. Combined with the spatial econometric model, the interaction between social development and human activities such as population size, passenger and freight turnover, regional GDP and carbon emissions from transportation in space can be explored, and the effects of different factors on carbon emissions from transportation can be revealed and predicted [20].
There are obvious spatial differences in China's economic development, population distribution and industry. The eastern region has the characteristics of relatively developed economy, large population, dense road network, large number of motor vehicles and frequent traffic activities, forming Beijing-Tianjin-Hebei Region, Yangtze River Delta Region and Pearl River Delta Region urban agglomerations. By the end of 2019, the total population of the three urban agglomerations of Beijing-Tianjin-Hebei Region, Yangtze River Delta Region and Pearl River Delta Region accounted for nearly 50% of the total population of the country, contributing more than 40% to the national economy, and the motor vehicle ownership exceeded 35%. Such a huge population base, coupled with frequent economic activities and traffic activities, is bound to cause greater carbon emissions. Therefore, it is still of great significance to focus on the analysis of the characteristics of carbon emissions of the three major urban agglomerations in eastern China and their impact on China. However, few previous studies have directly compared and analyzed the characteristics of carbon emissions from transportation of the three major urban agglomerations. In other words, there is still great research prospect to fully compare the influence of carbon emissions among and within the three major urban agglomerations and analyze the influencing factors and spatial heterogeneity of carbon emissions from transportation from a local perspective.
In this paper, 30 provinces, municipalities and autonomous regions in China (except Tibet Autonomous Region, Hongkong, Macau and Taiwan Province) are taken as the research objects, and the carbon emissions from transportation of the three major urban agglomerations, namely Beijing-Tianjin-Hebei Region, Yangtze River Delta Region and Pearl River Delta Region, are emphatically analyzed. Using ESDA and geographically weighted regression (GWR), the spatio-temporal distribution and heterogeneity of carbon emissions from transportation are explored, and their influencing factors are revealed [20], and corresponding opinions are put forward in terms of carbon emissions reduction from transportation.

2. Methods and Data

2.1 Overview of the research area

The research area of this paper is 30 provincial administrative regions in China, and on this basis, it focuses on the Beijing-Tianjin-Hebei Region, the Yangtze River Delta Region and the Pearl River Delta Region in eastern China (Figure 1). Although the area of the three major urban agglomerations only accounts for about 6% of the country's total area, the population accounts for nearly 50% of the country, the motor vehicle ownership exceeds 35% of the country, and the passenger and freight turnover accounts for about 40% of the country. Such a huge base means that energy consumption from transportation is large and carbon emissions from transportation are large.
Under the background of China's efforts to achieve the goal of "double carbon", it is of great strategic significance [21] and research value to study carbon emissions from transportation for sustainable development and promote the high-quality development of urban agglomerations

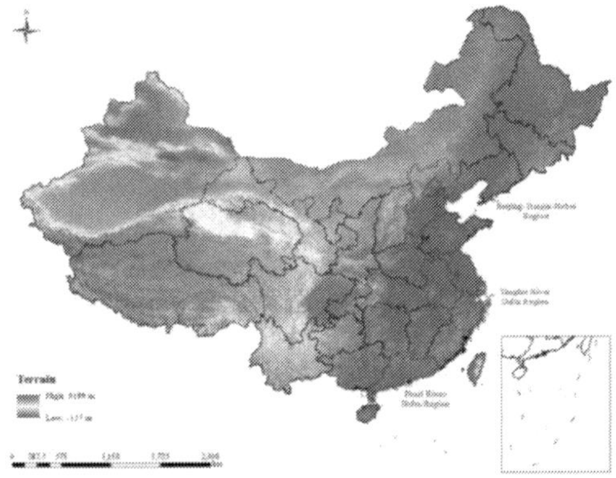

FIG 1 Profile map of Chinese provinces

2.2 Carbon Emissions Measurement Methods and Data Sources

2.2.1 Calculation of total carbon emissions

This study will use the "top-down method" to calculate the total carbon emissions from transportation of three major urban agglomerations in China and the carbon emissions from transportation of prefecture-level cities in each urban agglomeration.
The calculation formula of carbon emissions from fossil fuel consumption and transportation is shown in equation (1):

$$TCE_f=\sum_{i=1}^{9} FC_i \times ALC_i \times C_i \times O_i \times \frac{44}{12} \quad (1)$$

where TCE_f (total carbon emissions from fossil) represents the total carbon emissions from transportation (kg CO2) generated by fossil fuel consumption. i indicates the fuel type (in this study, there are 9 kinds of fossil fuels including crude oil, gasoline, kerosene, diesel oil, fuel oil, raw coal, coke, liquefied gas and natural gas). FCi represents the consumption of the first fossil fuel (kg). $ALCi$ represents the average low calorific value (kJ/kg) of the first energy source. Ci represents the carbon content (T/TJ) of the first fossil fuel. Oi represents the carbon oxidation rate (100%) of the ith fossil fuel.

The calculation formula of carbon emissions of electricity consumption and transportation is shown in equation (2):

$$TCE_e = EC \times EF \quad (2)$$

where TCE_e (total carbon emissions from electric power) represents the total carbon emissions from transportation (kg CO2) generated by power consumption. EC represents the electric power consumption (kWh). EF represents the factor that carbon emissions generated by electric power (kgCO2/ kwh), and the factors that carbon emissions generated by electric power of each province are shown in Table 1.

Table 1 Carbon emissions factors of electricity in different provinces in China

Region	***EF (kgCO2/ kwh)***
Beijing, Tianjin, Hebei, Shanxi, Shandong, Inner Monggolia	1.246
Heilongjiang, Jilin, Liaoning	1.096
Shanghai, Jiangsu, Zhejiang, Anhui, Fujian	0.928
Henan, Hubei, Hunan, Jiangxi, Sichuan, Chongqing	0.801
Shaanxi, Gansu, Qinghai, Ningxia, Xinjiang	0.977
Guangdong, Guangxi, Yunnan, Guizhou	0.714
Hainan	0.917

Therefore, the total carbon emissions (TE) (kg CO_2) of a certain research area is:

$$TE = TCE_f + TCE_e \quad (3)$$

Because the total amount of carbon emissions is of a large order of magnitude, all TCE involved in the followings are calculated in units of ten thousand tons.

2.2.2 *Per capita carbon emissions*

Per capita carbon emissions ($PCCE$) refers to the average carbon emissions of each person in a fixed area. Per capita carbon emissions are used to describe how much each person contributes to carbon emissions in this region. Per capita carbon emissions ($PCCE$) can be calculated by the following formula:

$$PCCE = \frac{TCE_x}{P} \quad (4)$$

where TCE_x represents the carbon emissions of the region x and P represents the number of permanent residents in the region.

2.2.3 *Carbon emissions intensity*

Carbon emissions intensity (CEI) refers to the carbon emissions generated by the growth of unit gross national product, which is mainly used to describe the relationship between the economy and carbon emissions in a certain region, and is of great significance to analyze whether the region can achieve a low-carbon development model. Carbon emissions intensity (CEI) is shown in the following formula:

$$CEI = \frac{TCE_x}{GDP} \quad (5)$$

where TCE_x represents the carbon emissions of the representative region x , GDP represents the gross national product of the corresponding region.

2.2.4 *Data sources*

In this study, it is necessary to collect the data of the Yangtze River Delta Region, Guangdong, Hong Kong and Macao, Beijing-Tianjin-Hebei Region and the provinces and cities included in the urban agglomerations from 2000 to 2019, involving GDP, car ownership, passenger turnover and freight turnover. These data are obtained from the official website of the National Bureau of Statistics and the *China Statistical Yearbook* [22]. In the carbon emissions calculation method mentioned above, the consumption of fossil fuels, electricity consumption and average low calorific value of energy also come from *China Statistical Yearbook* [22], and the carbon content of fossil fuels, carbon oxidation rate and factors that carbon emissions generated by electric power (Table 2) are obtained from *IPCC 2006 National Greenhouse Gas Inventory Guide* to study the contribution to transportation carbon emissions [23, 24].

2.3 ESDA method

2.3.1 *Hot spot analysis*

Hot spot analysis, namely Getis-Ord local statistics, compares the local sums of a certain element and its neighboring elements to determine the clustering situation of this element. Getis-Ord local statistics can be expressed as:

$$G_i^* = \frac{\sum_{j=1}^{n} w_{i,j}x_j - \bar{X}\sum_{j=1}^{n} w_{i,j}}{S\sqrt{\frac{[n\sum_{j=1}^{n} w_{i,j}^2 - (\sum_{j=1}^{n} w_{i,j})^2]}{n-1}}} \quad (6)$$

$$\bar{X} = \frac{\sum_{j=1}^{n} x_j}{n} \quad (7)$$

$$S = \sqrt{\frac{\sum_{j=1}^{n} x_j^2}{n} - \sum_{j=1}^{n} x_j^2 - (\bar{X})^2} \quad (8)$$

Among them, x_j represents the attribute value of element j, $w_{i,j}$ represents the spatial weight of elements i, j and n represents the total number of elements.

2.3.2 *Bivariate Moran's I*

Bivariate Moran's *I* index is in the global scope. The Moran's *I* index is calculated by Geoda spatial analysis tool to evaluate the degree of global aggregation, dispersion or random distribution of data. In this study, considering that there are multiple variables that may affect carbon emissions, in order to describe the spatial correlation characteristics of two variables of geographical elements, bivariate spatial autocorrelation is introduced. Formulas (9)-(11) are given as follows:

$$I_{kl}^{i} = Z_k^i \sum_{j=1}^{n} W_{i,j} Z_l^j \quad (9)$$

$$Z_k^i = \frac{(X_k^i - \overline{X_k})}{\sigma_k} \quad (10)$$

$$Z_k^i = \frac{(X_l^j - \overline{X_l})}{\sigma_l} \quad (11)$$

In the formula, X_k^i and X_l^j represent the numerical values of two attributes k, l in two spatial units i, j respectively, $\overline{X_k}$ and $\overline{X_l}$ represent the average values of the two attributes k, l respectively, and σ_k and σ_l represent the variance of the two attributes k, l respectively.

2.4 Geographically weighted regression

Geographically weighted regression (GWR) model is usually analyzed by spatial heterogeneity. Generally, a regression equation is established based on the spatial position of data to estimate the spatial distribution characteristics. The expression of GWR model is as follows:

$$y_i = \beta_0(u_i, v_i) + \sum_{k=1}^{n} \beta_k (u_i, v_i) x_{ik} + \varepsilon_i \quad (12)$$

where y_i is the dependent variable, x_{ik} is the independent variable (i.e. the influencing factor), (u_i, v_i) is the spatial position of the first observation point, $\beta_k(u_i, v_i)$ is the regression coefficient of the influencing factor k at the regression point k, and ε_i is the independent random error.

3. Results

3.1 Temporal variation trend of provincial carbon emissions from transportation

From 2000 to 2019, the average annual carbon emissions showed a monotonic increasing trend (Table 2). In the past 20 years, the average annual carbon emissions increased from 180,758,300 tons in 2000 to 670,113,100 tons in 2019, with an average annual growth of 25,755,600 tons and a year-on-year growth rate of 7.14%. From 2000 to 2019, the growth of average carbon emissions ranges from 4,131,900 tons to 53,336,800 tons, which shows that the average annual carbon emissions have increased year by year in the past 20 years, and it also means that the total amount of carbon emissions in China has increased year by year. From 2001 to 2019, the year-on-year growth rate of carbon emissions has generally shown a downward trend, and the growth rate of carbon emissions has shown a slowing trend. However, it is worth noting that before 2012, the year-on-year growth rate was high, at about 10%, but after 2012, the growth rate slowed down obviously, and the year-on-year growth rate was less than 4%, and it gradually stabilized.

Table 2 The average carbon emissions, annual average growth rate and annual average growth rate of Chinese provinces from 2000 to 2019

Year	Average carbon emissions (104 ton)	Annual growth (104 ton)	Year-on-year growth rate (%)	Year	Average carbon emissions (104 ton)	Annual growth (104 ton)	Year-on-year growth rate (%)
2000	18075.83	-	-	2010	48342.33	4752.85	10.90
2001	18949.10	873.27	4.83	2011	53676.01	5333.68	11.03
2002	20694.40	1745.30	9.21	2012	55443.52	1767.51	3.29
2003	23831.69	3137.29	15.16	2013	56361.84	918.32	1.66
2004	27647.97	3816.28	16.01	2014	57413.78	1051.94	1.87
2005	31911.64	4263.67	15.42	2015	57826.97	413.19	0.72
2006	35521.76	3610.12	11.31	2016	59172.02	1345.05	2.33
2007	39272.11	3750.35	10.56	2017	61372.06	2200.04	3.72
2008	41186.24	1914.13	4.87	2018	64451.75	3079.69	5.02
2009	43589.48	2403.24	5.84	2019	67011.31	2559.56	3.97

Comparing the inter-annual variation curves of average carbon emissions between the provinces and cities covered by the three major urban agglomerations and China's provinces (Figure 2), it is found that the carbon emissions of Hebei, Jiangsu, Guangdong and Zhejiang provinces are much higher than the national average, while the carbon emissions of Anhui Province are slightly lower than the national average. Because the area and population of Beijing, Shanghai and Tianjin are smaller than those of other provinces, it is too one-sided to compare and analyze their total carbon emissions, and it is necessary to analyze the per capita carbon emissions and carbon emissions intensity of those provinces.

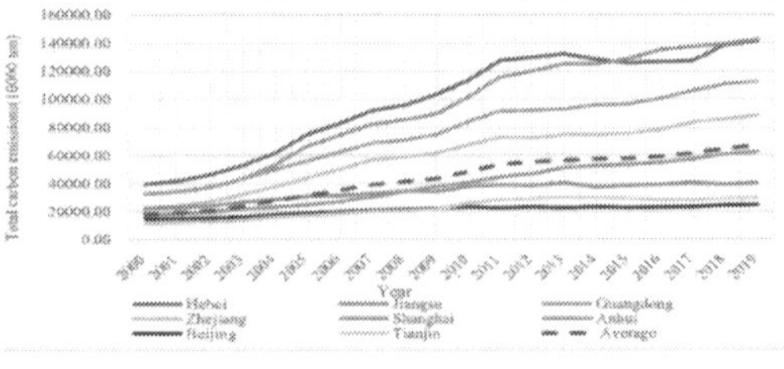

FIG 2 The change trend of carbon emissions of provinces and the national average carbon emissions in the three major urban agglomerations

3.2 Spatio-temporal heterogeneity analysis of carbon emissions from transportation

3.2.1 Spatio-temporal heterogeneity of provincial carbon emissions

In 2019, the provincial *TE*、*PCCE*、*CEI* distribution of China mainly showed the spatial distribution characteristics of high in the north and low in the south, high in the east and low in the west (Figure 3). The high total carbon emissions were mainly distributed in Inner Mongolia Autonomous Region, Shanxi Province, Hebei Province, Shandong Province, Jiangsu Province, Zhejiang Province and Guangdong Province, and the carbon emissions in the northern region increased significantly as time goes by. The provinces with higher per capita carbon emissions were roughly consistent with those that have higher total carbon emissions. The carbon emissions intensity shows the most obvious difference between the north and the south. Through the further analysis of temporal variation analysis, it is found that the carbon emissions intensity decreased obviously from 2010 to 2019, and the rate of decrease in the south was faster than that in the north, while the decrease in the southeast coastal areas was much greater than that in the inland areas in the west and north. It is worth mentioning that the total carbon emissions from transportation, per capita carbon emissions and emissions intensity in Beijing are at a low level in the northern cities.

FIG 3 Temporal and spatial variation characteristics of provincial carbon emissions in China in 2019

3.2.2 Spatio-temporal heterogeneity of the three major urban agglomerations

This study focuses on the temporal and spatial evolution of carbon emissions from transportation in Beijing-Tianjin-Hebei, Yangtze River Delta and Pearl River Delta urban agglomerations. Because the data of energy consumption in some prefecture-level cities are difficult to collect, and there is a positive correlation between regional economic strength and regional energy consumption. For some cities that are difficult to collect energy consumption, their carbon emissions are estimated according to the proportion of urban GDP in the province (Figures 4 and 5). In the past 20 years, the total carbon emissions and per capita carbon emissions of the three major urban agglomerations have generally shown a sharp upward trend, with the Beijing-Tianjin-Hebei Region and Yangtze River Delta Region regions showing a larger increase, while the Pearl River Delta Region has a relatively small increase. Among them, only Zhaoqing City has shown a decreasing trend in total carbon emissions and per capita carbon emissions in the past 20 years. The high-value centers of carbon emissions in the Yangtze River Delta Region and the Pearl River Delta Region are the cities with the most developed economy and transportation in the region, and radiate from the igh-value centers to the surrounding areas. However, what is unusual is that the high value of carbon emissions per capita in the Beijing-Tianjin-Hebei Region urban agglomeration appears in the peripheral areas of Beijing with the most developed economy and transportation.

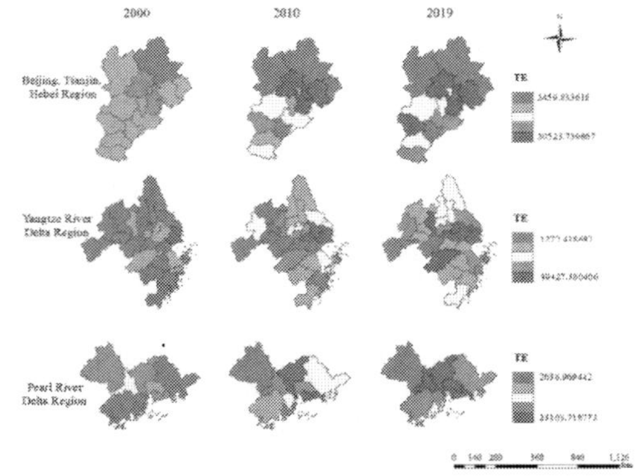

Figure 4 ***TE*** of major urban agglomeration in 2000, 2010 and 2019

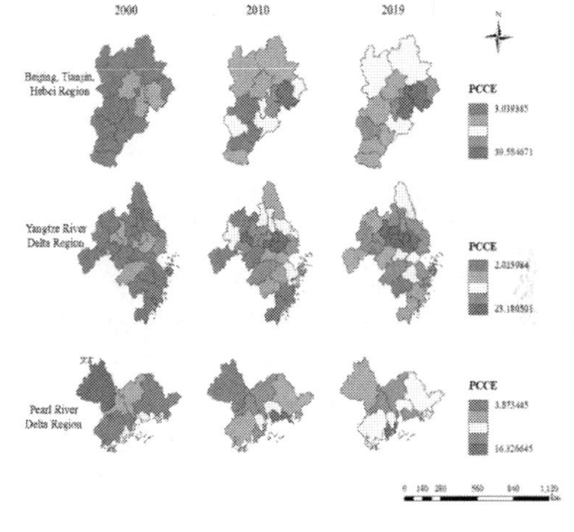

Figure 5 ***PCCE*** of major urban agglomeration in 2000, 2010 and 2019

3.3 Hot spot analysis

3.3.1 Analysis of hot spots of provincial carbon emissions

The Z-score of total carbon emissions, per capita carbon emissions and carbon emissions intensity of each province is calculated by cluster analysis, and its spatial distribution characteristics are analyzed (Figure 6). The spatial distribution of total carbon emissions (Figures 6a, d, g) has obvious clustering characteristics. The hot spots of carbon emissions are mainly distributed in the east of Hexi Corridor, while the cold spots are on the contrary, appearing in Xinjiang, Qinghai, Sichuan, Yunnan and so on. Both the per capita carbon emissions and the intensity of carbon emissions show the clustering distribution characteristics of high in the north and low in the south. The high-value concentration areas are mainly in Inner Mongolia Autonomous Region and North China, and the low-value concentration areas are mainly in the Pearl River Basin. From 2000 to 2019, the high-value concentration areas of per capita carbon emissions have not changed much, while the area of low-value concentration areas has shown a shrinking trend, gradually transitioning to lower-value concentration areas, and Xinjiang, Qinghai and other provinces have gradually transformed into higher-value concentration areas, indicating that the population growth rate in the above areas in the past 20 years is higher than its carbon. The concentration of carbon emissions intensity is gradually weakening, and the carbon emissions intensity has dropped significantly in the past decade. Northwest China is a high-value gathering area, which shows that the economic level in northwest China is relatively backward and the energy consumption per unit output value is relatively large. Guangdong Province and Zhejiang Province present the situation of low-value aggregation.

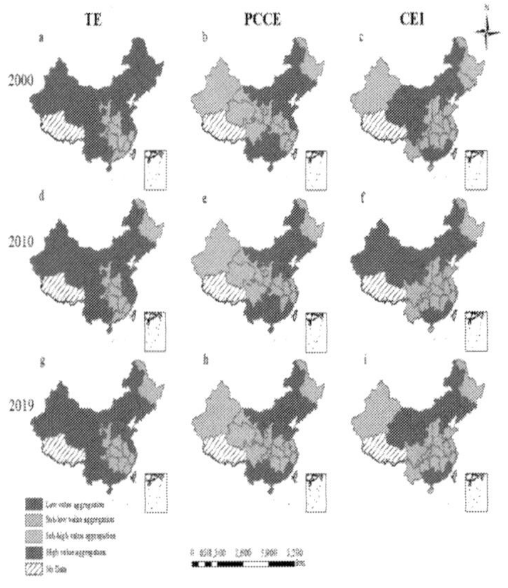

FIG 6 Analysis of hot-spots of carbon emissions in China's provinces in 2000, 2010 and 2019

3.3.2 *Hot-spot analysis of three major urban agglomerations*

Figure 7-8 reflect the aggregation of total carbon emissions and per capita carbon emissions in Beijing-Tianjin-Hebei Region, Yangtze River Delta Region and Pearl River Delta Region respectively. Generally speaking, there are similarities between Beijing-Tianjin-Hebei and Yangtze River Delta urban agglomerations, and the high-value agglomerations are mainly centered on a few cities and distributed in groups. The high-value agglomerations in Beijing-Tianjin-Hebei urban agglomeration are mainly centered on Beijing-Tianjin-Tangshan, while the high-value agglomerations in Yangtze River Delta Region urban agglomeration are mainly centered on Suzhou and Shanghai, and the clustering state of carbon emissions in coastal areas is generally stronger than that in inland areas. From the perspective of time, from 2000 to 2019, the number of high-value carbon emissions gathering areas in Beijing-Tianjin-Hebei urban agglomeration decreased, mainly changing to the second-highest value gathering areas. The carbon emissions gathering degree in this area in 2010 and 2019 was significantly lower than that in 2000. The difference is that the polarization of the Yangtze River Delta urban agglomeration gradually strengthened, showing more obvious regional differentiation. Some cities in the east of the Yangtze River Delta Region gradually turned into high-value gathering areas, and some cities in the inland areas in the west gradually turned into low-value gathering areas. In the Pearl River Delta Region, Zhongshan, Dongguan and Zhaoqing showed obvious changes, while other cities showed less obvious changes.

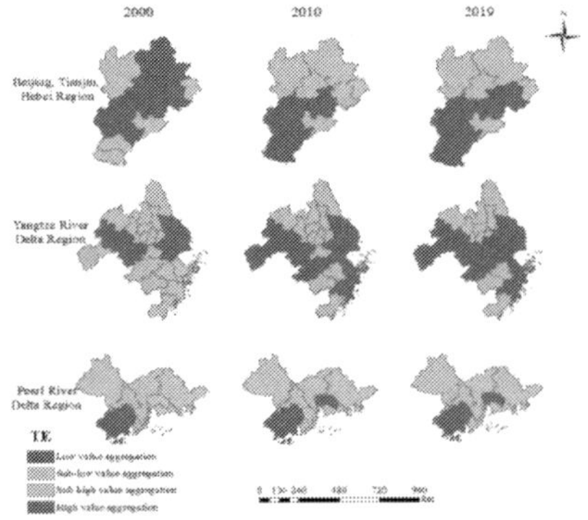

FIG 7 Analysis of hot-spots of ***TE*** of major urban agglomeration in 2000, 2010 and 2019

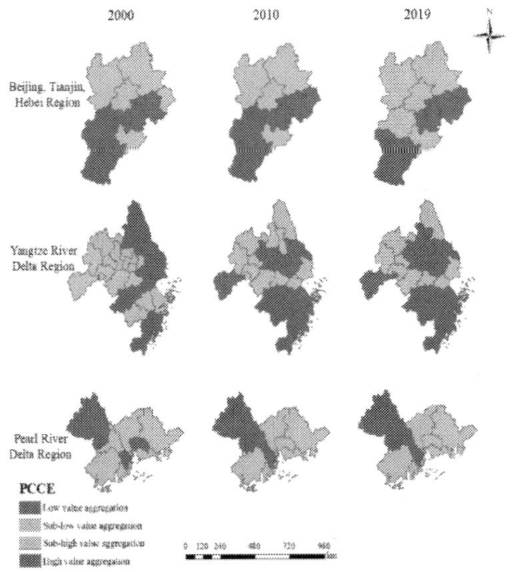

FIG 8 Analysis of hot-spots of $\boldsymbol{PCCE}$ of major urban agglomeration in 2000, 2010 and 2019

3.4 Spatial Correlation of Provincial Carbon Emissions from Transportation

3.4.1 *Bivariate spatial autocorrelation of provincial carbon emissions from transportation*

In order to explore the spatial correlation between carbon emissions from transportation and its influencing factors, this study uses Geoda for spatial analysis, taking provincial carbon emissions in 2000, 2010 and 2019 as the second variable, and substituting GDP, vehicle ownership, passenger turnover and freight turnover as the first variables. According to the calculation results (Table 3), except passenger turnover in 2000 and 2019, Moran's *I* under other influencing factors are all greater than 0, and the P value is all less than 0.05, which is statistically significant. According to Table 3, the spatial autocorrelation between provincial carbon emissions and freight turnover is the strongest, and the spatial autocorrelation between provincial carbon emissions and passenger turnover is the weakest. In addition, from 2000 to 2010, the spatial correlation between carbon emissions and these indicators tends to increase, and after 2010, the spatial correlation tends to decrease.

Table 3 Moran's I values of China's provincial carbon emissions under various influencing factors in 2000, 2010 and 2019

	Motor vehicle ownership	GDP	Freight turnover	Passenger turnover
In 2000	0.006	0.085	0.249	-0.013
In 2010	0.098	0.097	0.336	0.026
In 2019	0.063	0.017	0.141	-0.011

Figure 9 shows the local spatial autocorrelation of carbon emissions in China's provinces. In this figure, the high-high concentration areas of each index in each year are mainly concentrated in northern regions of China and the middle and lower reaches of the Yangtze River in eastern China. High-low concentration areas are mainly in Guangdong province and Sichuan province. Low-high concentration areas are mainly concentrated in Anhui and Jiangxi. Low-low concentration areas are mainly concentrated in the sparsely populated and underdeveloped provinces in northwest China, such as Xinjiang province and Gansu province. In addition, from 2000 to 2019, the number of high-high concentration areas tends to decrease, and some provinces with high-high concentration gradually turned into low-high concentration areas.

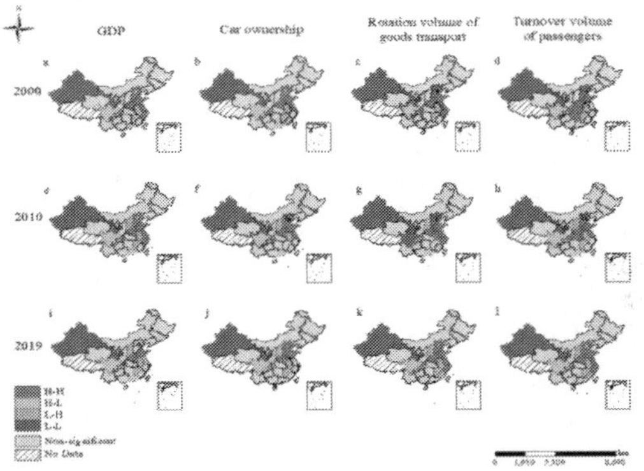

FIG 9 Distribution diagram of $\boldsymbol{TE}$ and bivariate aggregation of 4 driving factors in China in 2000, 2010 and 2019

3.4.2 *Spatio-temporal heterogeneity of driving factors of carbon emissions from in transportation in China's provinces*

According to the regression result of GWR model (Figure 10), the spatio-temporal heterogeneity of four factors of carbon emissions in China's provinces is analyzed. Generally speaking, the influence of four factors on carbon emissions is strong in the northwest and weak in the southeast. China, Xinjiang, Inner Mongolia and other northwestern provinces have a more significant impact on carbon emissions and these four factors, while southwest provinces such as Sichuan and Yunnan have a less significant impact on these factors. As for the influence of the factors, the regression coefficient of motor vehicle ownership is the largest, which shows that the number of cars has a greater impact on traffic carbon emissions. Therefore, when analyzing the spatio-temporal heterogeneity of carbon emissions in the three major urban agglomerations, we will focus on the analysis of motor vehicle ownership.

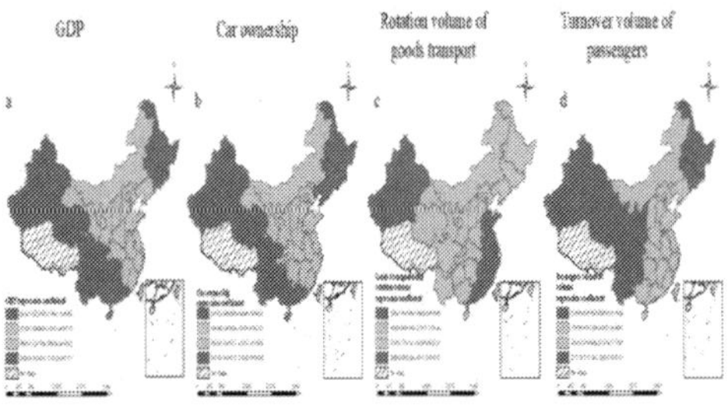

FIG 10 The spatial distribution of regression coefficients of the 4 driving factors in China in 2019

3.5 Spatial Correlation Analysis of Traffic Carbon Emissions in Three Urban Agglomerations

3.5.1 Bivariate spatial autocorrelation of traffic carbon emissions in three major urban agglomerations

According to the results of 3.4.2, the regression coefficient of motor vehicle ownership with carbon emissions in China's provinces is the largest, so this paper will focus on the analysis of motor vehicles. Figure 11 shows the local spatial autocorrelation of carbon emissions and car ownership in three major urban agglomerations in China. For the Beijing-Tianjin-Hebei Region urban agglomeration, the impact of motor vehicle ownership on traffic carbon emissions is not significant. As of 2019, only Langfang and Chengde are L-H clusters. The Pearl River Delta Region urban agglomeration is also not obvious, H-H appears in Dongguan and L-H appears in Huizhou. For the Yangtze River Delta Region urban agglomeration, the spatial differentiation between high-value agglomeration and low-value agglomeration is obvious, with high-value agglomeration in the east and low-value agglomeration in the western cities, and as time goes by, high-value agglomeration has strengthened and low-value agglomeration has weakened.

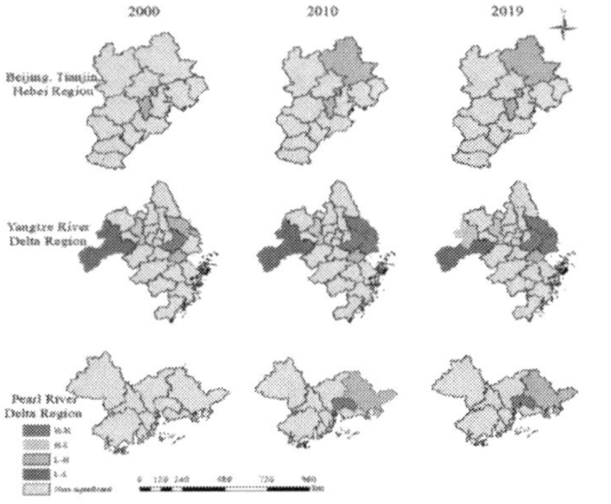

FIG 11 Spatial autocorrelation of carbon emissions and motor vehicle ownership in three major urban agglomerations in China

3.5.2 Spatio-temporal heterogeneity analysis of driving factors of three urban agglomerations

Figure 12 reflects the GWR of motor vehicle ownership in the three major urban agglomerations. The regression coefficients of vehicle ownership in the three major urban agglomerations are all positive, indicating that vehicle ownership has a positive impact on regional carbon emissions. Among them, the regression coefficient of Beijing-Tianjin-Hebei Region and Yangtze River Delta Region regions is small and has little change, while the regression coefficient of Pearl River Delta Region is the fastest in reduction. In 2019, the regression coefficient of Pearl River Delta Region is only one third of that in 2000.

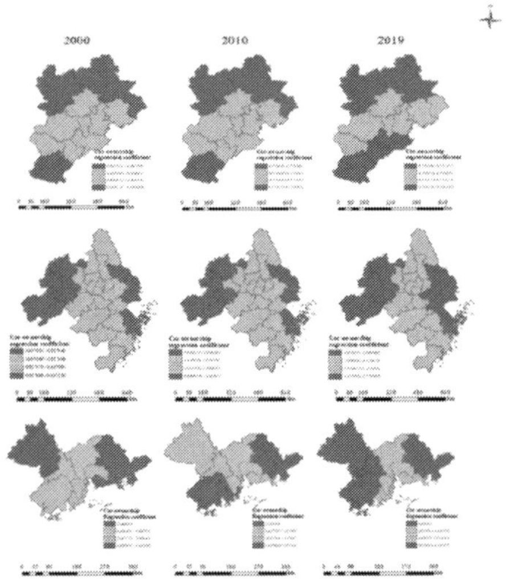

FIG 12 The spatial distribution of regression coefficients of the 4 driving factors of 3 urban agglomerations

4. Discussion

There are obvious spatio-temporal differences in carbon emissions from transportation, per capita carbon emissions and carbon emissions intensity of urban units in China's provinces and three major urban agglomerations. In terms of time, since the 21st century, China has invested heavily in social and economic construction [25], and the economy has maintained the development trend of medium or high speed. At present, the government has set up 19 large and small urban agglomerations [26], which have made great contributions to China's economy. Traffic activities are closely related to economic development [27, 28]. It is precisely because of the establishment of these urban agglomerations that regional traffic activities have been strengthened [29]. Just like the economic scale of China, the traffic industry has also developed vigorously in China [30], and the total carbon emissions from transportation of provinces and urban agglomerations have reached record highs since 2000. As an important part of global economic integration, China was naturally hit by the global economic crisis in 2008. In 2008, the growth rate of road transport freight turnover and port cargo throughput in China decreased [31]. These phenomena are also reflected in China's carbon emissions from transportation, that is, the increase of China's carbon emissions slowed down in 2008 and 2009. From the perspective of growth rate, after 2012, the growth rate of carbon emissions in China has dropped sharply and gradually stabilized. Since the "11th Five-Year Plan", China has gradually devoted itself to low-carbon pilot projects: in 2010, the National Development and Reform Commission identified Guangdong, Liaoning, Tianjin and Chongqing as the first batch of low-carbon pilot provinces and cities [32], and in 2012, it identified the second batch of low-carbon pilot cities including Beijing and Shanghai [32]. During the "13th Five-Year Plan" and "14th Five-Year Plan" period, the state devoted itself to the study of low-carbon economy, and promoted peaking carbon dioxide emissions and carbon neutrality of cities from the aspects of low-carbon production, low-carbon life and low-carbon transportation [33], for example, formulating new vehicle emissions standards, promoting new energy vehicles, guiding citizens to travel green, and planting trees to increase carbon sinks [34]. These low-carbon measures are of great significance to slow down the growth of carbon emissions and weaken the high-value aggregation of carbon emissions, and they all show results in practical work.

At the level of urban agglomeration, the overall carbon emissions level of Beijing-Tianjin-Hebei Region urban agglomeration is slightly lower than the average level in the northern region, and the per capita carbon emissions in the inner and outer areas of Beijing-Tianjin-Hebei Region urban agglomeration is higher than that in Beijing, showing the phenomenon of high outside and low inside. The carbon emissions level and spatial agglomeration within the Yangtze River Delta Region urban agglomeration are basically consistent with the economic development of its cities. However, the Pearl River Delta Region urban agglomeration, which has the second strongest economic strength, has the lowest level of per capita carbon emissions and carbon emissions intensity among the three major urban agglomerations, and the spatial aggregation and spatial correlation of carbon emissions within the urban agglomeration are not significant compared with the other two urban agglomerations. The reason may be that Beijing has formulated the strictest policy to restrain air pollution, which makes some high-carbon industries in Beijing transfer to the surrounding areas. This industrial layout will strengthen the connection between the surrounding areas and Beijing, and strengthen the traffic between the surrounding areas and Beijing. Commuting, material transfer and other aspects cannot be isolated from traffic activities. However, the industrial structure of the nine cities in the Pearl River Delta Region is relatively perfect, and the relative balance between supply and demand can be achieved within each city or within a short distance, and the material mobilization between cities is relatively infrequent, so the traffic activities among cities are relatively weak compared with Beijing-Tianjin-Hebei Region, resulting in a low level of carbon emissions in the Pearl River Delta Region. In view of the present situation of Beijing-Tianjin-Hebei Region urban agglomeration, we should devote ourselves to adjusting the industrial structure and layout. In planning, we should consider more scientific industrial layout and appropriately narrow the gap between cities, so that we can achieve a relative balance between supply and demand, weaken the traffic activities between cities to a certain extent, and thus reduce the carbon emissions from transportation to a certain extent. In addition, according to the regression results of Moran's I and GWR, it is determined that motor vehicle ownership has the most significant impact on carbon emissions, but the impact of motor vehicle ownership on the three major urban agglomerations is less than its surrounding areas, and within urban agglomerations, megacities such as Beijing, Shanghai, Guangzhou and Shenzhen have smaller impact coefficients, which may be due to the control measures adopted by the state to deal with air pollution in these key cities. It also advocates citizens to develop the concept of green travel. In addition, the urbanization level of these three major urban agglomerations is relatively high, new energy sources are popularized earlier and to a high degree, urban public transport networks are mature, and public transport vehicles are more advanced and environmentally friendly, thus weakening the impact of motor vehicle ownership on carbon emissions to a certain extent. In space, the distribution of carbon emissions from transportation in China is uneven, showing obvious characteristics of high in the north and low in the south, high in the east and low in the west, and has obvious spatial aggregation characteristics. Areas with large economic development scale have greater traffic demand [35]. Therefore, the total carbon emissions from traffic, like economic development in space, shows a trend of high in the east and low in the west. The three major urban agglomerations in China studied in this paper are all concentrated in the eastern region, and their traffic activities are very strong, which drives the traffic carbon emissions of China. However, the difference is that the level of carbon emissions from transportation of the Pearl River Delta Region urban agglomeration, which ranks second in GDP, is not obvious compared with the other two urban agglomerations, and the carbon emissions intensity of the southern region with a relatively high GDP ratio is generally lower than that of the northern region. It is speculated that the GDP of the southern region is growing faster and the energy used is cleaner than that of the northern region. While paying attention to economic development, the carbon emissions from transportation of the southern region is well controlled. Therefore, it is considered that in the future, especially in the northern region, energy structure can be actively adjusted according to local conditions. For example, some areas can make full use of prevailing wind and terrain conditions to develop wind energy, and tidal energy can be developed in coastal areas. Key cities need further follow-up policies and develop new fuel vehicles.

There are also some limitations of this study. First, for some cities that are difficult to collect energy consumption, the proportion of urban GDP in the province is estimated, so there is a certain error in the carbon emissions data at the level of urban agglomeration.

In addition, the calculation formula of total carbon emissions selected in this study is applicable to the calculation of direct carbon emissions. However, with the development and evolution of vehicles, there are more vehicles such as electric vehicles, subways and high-speed railways driven by electricity, so the indirect carbon emissions driven by electricity may not be reflected in this formula, and the completeness of calculation needs to be improved. In the follow-up study, data collection, carbon emissions calculation, influencing factors and other aspects can be further improved, so as to provide a stronger basis for carbon emissions reduction strategies.

5. Conclusion

(1) The carbon emissions from transportation in China provinces show obvious spatial distribution characteristics of high in the north and low in the south. Especially for the northern region, while developing the economy, we should devote ourselves to carbon emissions reduction, actively implement energy structure adjustment, vigorously develop clean energy, and pay attention to the prevention and control of pollutants.

(2) The carbon emissions level of the Pearl River Delta Region urban agglomeration is lower than that of the other two urban agglomerations, and the influence of each driving factor on the Pearl River Delta Region urban agglomeration is relatively smaller than that of the other two urban agglomerations. Therefore, in the carbon emissions reduction work of urban agglomerations, the Yangtze River Delta Region and Beijing-Tianjin-Hebei Region urban agglomerations can learn from the relevant policies of the Pearl River Delta Region urban agglomerations and coordinate the relationship between carbon emissions and their influencing factors.

(3) Among the four driving factors, the motor vehicle ownership has the greatest influence on carbon emissions, but its influence on the Pearl River Delta Region is gradually decreasing, indicating that the carbon emissions reduction work in the Pearl River Delta Region is relatively good in recent years, which can provide theoretical basis for carbon emissions reduction work in other regions.

References

1. Lin Jinhai, Li Haifeng. Study on the impact of human activities and climate change on hydrology and water resources [J]. Science and Technology Communication, 2013(10):2.
2. Zhu Wenqi, Lee Tan-Lo. Thinking caused by global climate issues [J]. Kehai Story Expo: Science and Technology Exploration, 2011(2):1.
3. A word daily | United Nations Framework Convention on Climate Change (UNFCCC) http://cn.chinadaily.com.cn/a/202211/22/ws637c8f90a3109bd995a5170c.html.
4. Boykoff M, Pearman O. Now or never: How media coverage of the IPCC special report on 1.5 ℃ shaped climate- action deadlines? [J]. One Earth, 2019, 1(3): 285-288.
5. Qiu Zhaoyi. Impact of international vertical specialization on CO2 emissions in China [J]. Ecological Economy, 2012(10):5.
6. Shi Dan, Ye Yunling. Study on carbon emissions trend of urban traffic and countermeasures for emissions reduction-taking Shanghai as an example [J]. Modern Management Science, 2022,4(8):3-14.
7. Lin D T, Zhang L Y, Chen C, et al. Understanding driving patterns of carbon emissions from the transport sector in China: evidence from an analysis of panel models [J]. Clean Technologies and Environmental Policy, 2019, 21(6):1307-1322.
8. Niu Yawen, Zhao Xianchao, Hu Yijue, et al. Study on Spatial Differentiation of Carbon Emissions from Land Use at County Level in Changsha, Zhuzhou and Xiangtan Based on NPP-VIIRS Night Lighting [J]. Journal of Environmental Science, 2021, 41(9):3847-3856.
9. Yan Yanmei, Wang Zheng, Wu Leying, Liu Changxin. Analysis of the influence factors of carbon emissions intensity on regional differences in China [J]. Journal of Environmental Science, 2016, 36(9):3436-3444.
10. LIU, Q., WU S., LEI Y., et al. Exploring spatial characteristics of city-level CO2 emissions in China and their influencing factors from global and local perspectives [J]. Science of the Total Environment, 2020, 754: 142206.
11. Li Yuling, Shiping Li, Qi Jingjing. Analysis of influencing factors and decoupling effect of carbon emissions from land use in Shaanxi Province [J]. Soil and Water Conservation Research, 2018, 25(1):382-390.
12. Wang M, Feng X Z, An Q, et al. Study on green and low-carbon development in Qinghai province based on decoupling index and LMDI [J]. Climate Change Research, 2021, 17 (5): 598-607.
13. LI Y H, WANG A Y, YANG D. Driving factors and carbon emissions reduction potential of China ' s transportation industry based on the combination of LMDI and international analogies[J]. Transport Research, 2022, 8(4): 36-45.
14. ZHANG P L, DU Q J. On Influencing Factors of Carbon Emissions in Beijing-Tianjin-Hebei Region: Based on the Extended STIRPAT Model [J]. Scientific and Technological Management of Land and Resources, 2022, 8(4): 36-45.
15. ZHANG W. Prediction and Scenario Analysis of Carbon Footprint in Xi'an Based on the STIRPAT Model [J]. Ecological Economy, 2021,37(04):25-29.
16. LYU W, GONG L L. Grey correlation analysis of logistics development and agricultural industrialization in China [J]. Hubei Agricultural Science, 2022, 61(2):5-9.
17. SHEN K J, SHEN Z Y Grey Relational Analysis of the Influence of Transportation Development on

Industrial Structure [J]. Special Zone Economy 2022, 398(03):110-113

18. Hu Huaimin, Zuo Wei, Xu Shiyuan. Study on the decoupling effect and driving factors of carbon emissions from transportation energy in the Yangtze River Economic Belt [J]. Resources and Environment in the Yangtze River Basin.2022, 31 (4): 862-877.
19. Yuan Changwei, Jordan, Yang Yingfang, et al. Spatial differentiation and cluster analysis of transportation carbon emissions intensity in China province [J]. Environmental Engineering, 2018, 36(7):6.
20. Zeng Xiaoying, Qiu Rongzu, Lin Ting, Hou Xiuying, Zhang Lanyi, Hu Xisheng. Spatial-temporal heterogeneity of traffic carbon emissions and its influencing factors in China. China Environmental Science [J].2020, 40(10):4304~4313.
21. Gao Wei, Wei Jianfei, Li Qiang. Spatial and temporal evolution characteristics and influencing factors of construction land efficiency in Beijing-Tianjin-Hebei Region under carbon emissions constraints. Geography and Geographic Information Science [J].2022, 38(05):88-95.
22. National Bureau of Statistics. China Statistical Yearbook [M]. Beijing: China Statistics Publishing House, 2001~2021.
23. Shan Y, Guan D, Zheng H, et al. China CO2 emissions accounts 1997~2015 [J]. Scientific Data, 2018, 5(1):1-14.
24. Shan Y, Huang Q, Guan D, et al. China CO2 emissions accounts 2016~2017 [J]. Scientific Data, 2020, 7(1):1-9.
25. Wang Mengkui. China's economic development trend and policy orientation [C]// World Productivity Congress Beijing Stage.2001.
26. Tong Yufen, and Ming Jie, Yang Huan, Liu Zhili. Study on the population pattern classification of major urban agglomerations in China [J]. Journal of Beijing Administration College, 2022(1):114-121.
27. Zhao Lina. On the road transport to promote local economic development [J]. China Storage and Transportation, 2023(1):2.
28. Tang Zhixiao. The relationship between the national economy and the development of transportation-the argument of "if you want to be rich, build roads carefully" [J]. Industrial Innovation Research, 2022(22):3.
29. Zhu Leizhou, Huang Yaping, Ding Yichen, et al. "Commuting circle" or "traffic circle": Re-cognition of the connotation and scope of metropolitan area in the new era [J]. Urban Development Research, 2022, 29(10):9.
30. Gao Darong. Study on the Impact of Transportation Development on Economic Growth [J]. China Market, 2023(3):3.
31. Official website of National Bureau of Statistics https://www.gov.cn.
32. Song Qijiao, Wang Yufei, Qiye. Carbon emissions status of low-carbon pilot cities in China [J]. China Population, Resources and Environment, 2015, 25(1):5.
33. Guo Shihong, Wang Xuechun. Research on the policy implementation guarantee mechanism towards the goal of "double carbon" in China —— Empirical experience from low-carbon pilot cities [J]. Journal of Beijing University of Technology (Social Science Edition), 2021, 021(006):57-68.
34. Yang Fengwei, Wang Bing, Zhou Mei. Measures and effects of forestry industry to cope with climate change [J]. Meteorology and Disaster Reduction Research, 2009, 32(3):7.
35. Fangyi Li, Bofeng Cai, Zhaoyang Ye, Zheng Wang, Wei Zhang, Pan Zhou, & Jian Chen (2019). Changing patterns and determinants of transportation carbon emissions in Chinese cities Energy.

Research on the application of large area 3D geological modeling splicing technology

Yinghong Li [1], Ruigang Zhang [1], Jinhua Luan [1], Deju Zhang [1], Wei Li [1], Donger Zou [1], Tingting Zhang [2], Jiahui Duan [3], Mohuan Ning [3], Guodai Wu [1]

[1] Chongqing Institute of Geology and Mineral Resources, Chongqing Key Laboratory of Exogenic Ore and Mine Environment, Chongqing, 401120, China
[2] China University of Geosciences, Beijing 100083, China
[3] Chongqing Geological Survey Institute, Chongqing 401122, China

Abstract: Three-dimensional geological modeling is widely used in the field of geological research. A successful three-dimensional geological model is a three-dimensional anatomy and integration of the study area. Large-scale three-dimensional geological modeling is limited by software performance and technical difficulties. There are disadvantages such as unequal model splicing elements, morphological distortion and attribute incompatibility. Based on the discrete storage method of DepthInsight software, this paper explores the large-area model fusion and splicing technology to deal with various geological conditions, and carries out large-area multi-block model fusion and splicing from the aspects of element type, quantity and attribute. The large-area 3D geological modeling technology scheme of discrete storage and block splicing is realized, which provides a reference for establishing a large-area 3D geological model.

Keywords: Large area 3D geological modeling; discrete storage; fusion splicing.

1. Introduction

Three-dimensional geological modeling is an organic combination of geological theory and computer three-dimensional visualization[1], the comprehensive interpretation of geological data and the establishment of geological model in three-dimensional space environment can clearly and intuitively express the superposition relationship of geological bodies on spatial scale. As early as the 1970 s, developed countries have begun to develop three-dimensional geological modeling software. After decades of continuous development and update technology, it has become increasingly mature[3], among them, there is GoCAD software developed by Nancy University in France[4], Micromine software developed by Micromine Australia[5], And Petrel software developed by Schlumberger[6]. The research and development of domestic 3D modeling software began in the 1990s, and the MapGIS series software of Zhongdi Digital Group [7], Creatar Xmodeling geological modeling software developed by Beijing Chaowei Chuangxiang Information Technology Co., Ltd[8], DepthInsight software developed by Beijing Grid Tiandi Software Technology Co., Ltd[9]. In 2012, China Geological Survey deployed three-dimensional geological survey pilot work, and deployed 13 pilot sub-projects in three typical areas : important orogenic belt, metallogenic belt and ore concentration area, economic zone and urban agglomeration. In 2017, the China Geological Survey formulated the construction of the ' underground transparent Xiong 'an ' project ; since 2019, Chongqing has successively carried out three-dimensional spatial and temporal information database pilot of natural resources, three-dimensional geological model construction of Jinyun Mountain area in Chongqing, three-dimensional geological modeling project of Chongqing demonstration mining area, three-dimensional visualization model of urban geology in Tongnan District of Chongqing, three-dimensional basic geological survey project of Wulong District of Chongqing, etc.

With the increasing demand for modeling software performance and the continuous expansion of the research area, the three-dimensional modeling of large areas is seriously limited by the performance and technical difficulties of modeling software, and it is difficult to realize the construction of three-dimensional geological model of large areas at one time. Aiming at the difficulties of large-area 3D geological modeling, this paper proposes the modeling idea of ' discrete storage + block splicing ', and discusses the multi-block 3D model fusion splicing technology.

2. Rationale

Discrete storage is a combination scheme of big data distributed storage technology and relational database. The basic idea is to store the model data discretely in the database first, and then to fuse the type, quantity and attribute of the elements. It supports the call extraction of any local model, avoids the location deformity and attribute impassability caused by the model stored in file form, and is the basis for constructing the update mechanism of three-dimensional geological model[10].

Block splicing is to integrate two adjacent models into an overall model one by one according to the type, quantity and attributes of elements. The information storage form in the large work area module is tables and fields. In the actual research, the cells are specified, that is, the block size is set to create a new model, and the partition results are saved to the database file (Figure 1a). Each cell is stored in the database, and the information set is called the slice element (Figure 1b). Generally, the information is recorded in the form of fields. When the local model is called, the slice element is judged to be locked, and the slice element data and grid within the call range and the extension area data and grid are guaranteed. The data is edited, and the model is established under the constraints of the extension area data and grid, so as to establish the local model (Figure 1c). The export of the local model is also derived by some kind of processing, which makes it into multiple slice models (cell models). The information of each slice model is recorded in the database (Figure 1d) and recorded in the corresponding cell table in the form of fields to realize storage (Figure 1e).

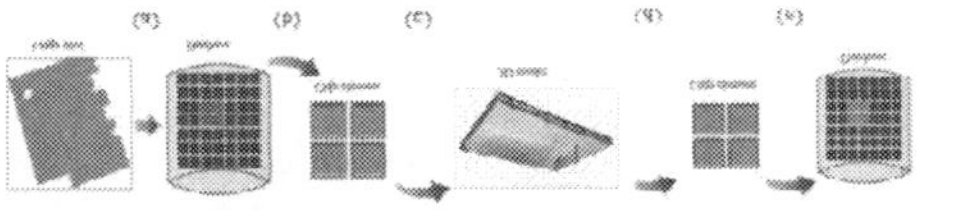

Fig.1 Flow chart of large area 3D model splicing technology

3. Model Integration

The three-dimensional geological model fusion splicing is divided into three cases : stratum splicing, structural splicing and attribute splicing. The general steps are four steps : (1) Call the working area in the large work area for editing, and the working area needs to be larger than the range of a single model ; (2) The common area needs to ensure the consistency of the number of data elements and attributes to form a determined standard profile ; (3) Trend adjustment of adjacent model dataThe boundary of two patches is connected to the standard profile. (4) Return the modified data to the large work area to complete the fusion splicing.

Of course, the following situations will occur when the model is spliced:

(1) When the occurrence of adjacent stratigraphic model elements is inconsistent, it is necessary to adjust the occurrence of both sides to be consistent with the common profile (Fig.2a).。

(2) The names of layered elements of adjacent geological models are inconsistent, such as phase change zones, which are unified according to the modeling unit standard (Fig.2b).。

(3) When the occurrence of fold structure is not coordinated in the adjacent stratigraphic model, the occurrence of both sides is adjusted according to the common section (Fig.2c).

(4) When there is inconsistency in the stratification criteria of adjacent stratigraphic models, if there is segmentation ambiguity in different models, they are unified to the group level according to the division criteria of sedimentary units (Fig.2d).

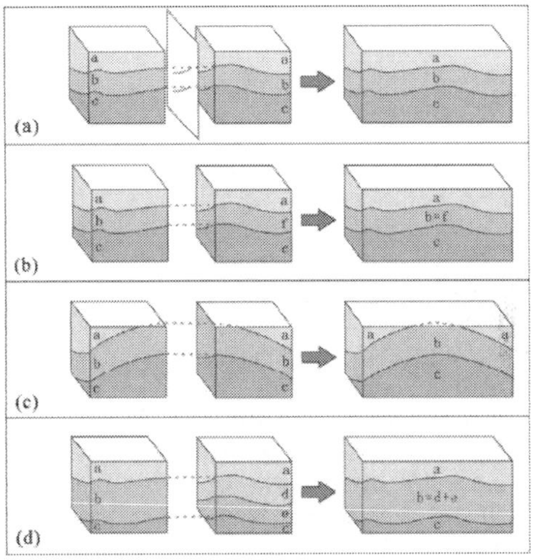

Fig.2 Different situations of model splicing (letters represent stratum code)

3.1 Strata Model Splicing

The geological data of different ranges and different fields construct a three-dimensional stratigraphic model under a unified framework. Firstly, the anisotropy of geological data between different ranges and different fields is eliminated, and the geological structure characteristics of the modeling area are uniformly recognized to ensure the consistency and correctness of the data. Secondly, it is necessary to establish a geological model database to realize the unified management and application of geological data, set up a unified grid size in the geological model database and generate an overall geological model. For applications in different ranges and different fields, models in different ranges and different fields can be mined from the model database for professional research (Fig.3).

 https://doi.org/10.1051/e3sconf/202339401028

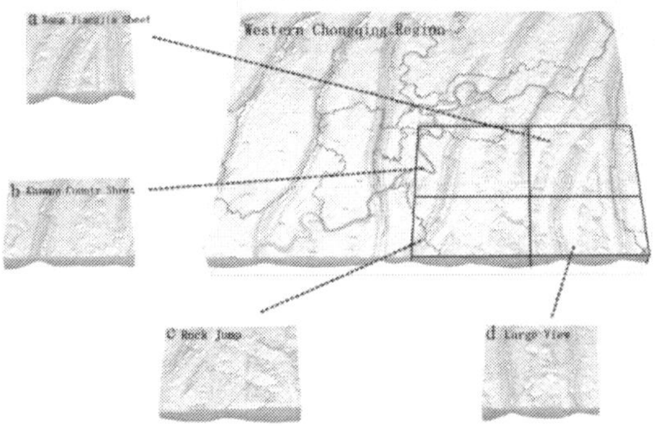

Fig. 3 Stratum model splicing results

3.2 Constructing Model Splicing

When the structural model is spliced, there are cases where the fault elements only expose a certain model or the fault elements are distributed in multiple models. Therefore, the structural model splicing is dominated by data, that is, the fault surface generation is not constrained by the boundary of the sub-region model and is only related to the data of the fault itself(Fig.4).

(1) Because the data source unit and time are different, there are different model numbers, and the modeling needs to be uniformly named.

(2) It is necessary to integrate all model data to clarify the fault occurrence, fault properties and strata of the upper and lower plates when interpreting the structure.

(3) The model boundary fault only exposes a certain map at the surface boundary, but may cross multiple models below the ground.

(4) When the fault crosses multiple models, it is necessary to call all fault and stratum data in the working area to carry out modeling.

(5) The modification of the fault model must be accompanied by the reconstruction of the stratum model, so it is necessary to modify the fault surface to reconstruct the stratum model.

(6) When modeling back, the fault can be directly spliced without modification ; fault changes have no effect on adjacent blocks, and fusion splicing can be automatically realized. Fault changes affect adjacent blocks, and blocks need to be remodeled to achieve fusion splicing.

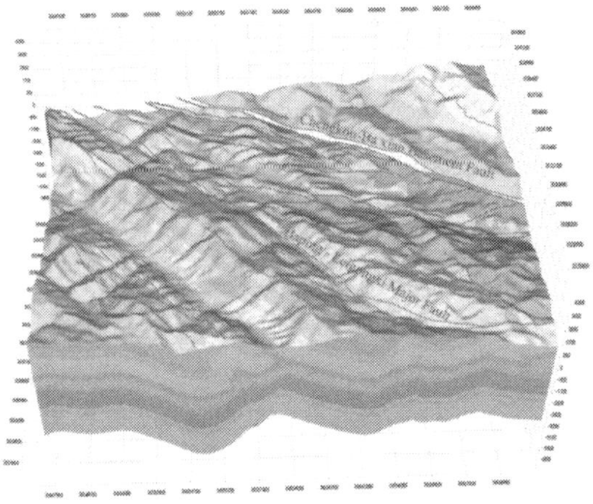
Fig.4 Structural model splicing results (Chengkou three-dimensional geological model)

3.3 Attribute Model Splicing

The attribute model is the same as the splicing method of the construction model. The whole work area is divided into many cell blocks, and the grid model, vector model and attribute model are established and stored in the model database. There is a certain splicing grid when meshing between the block attribute models. The number of grid splicing can be set by interaction. The splicing attributes can be extracted from adjacent blocks. A certain algorithm is used internally to ensure the fusion and splicing of adjacent regions of attributes. The block attributes can be updated at any time, and the model and data are synchronized in real time.

The attribute model is stored in the database in the form of three-dimensional discrete data volume. The grid structure is similar to the seismic data format. The data matrix of the attribute model information depends on the grid division in the X direction, Y direction and Z direction. When modeling, there is a certain overlap between the call modeling area and the completed area. When the local attribute modeling is performed, the reference data volume is used to extract the existing attribute data as the known data to participate in the attribute interpolation, and the boundary attributes of the control area are the same to achieve the splicing effect (Figure 5).

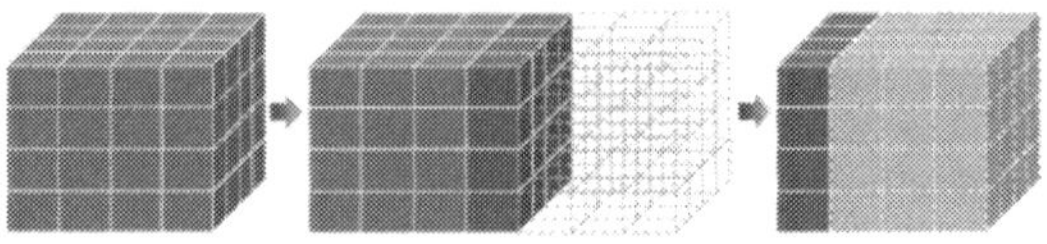
Fig.5 Attribute model splicing principle ((a) completed area, (b) overlapping area (c) overlapping area control interpolation)

4. Conclusion

In this paper, the method of large regional geological model fusion and splicing technology is discussed, and the modeling idea of ' discrete storage + block splicing ' is proposed. Aiming at the problems of stratum model splicing, structural model splicing and attribute model splicing, the corresponding solutions are given, which can effectively solve the contradiction between model accuracy and scale, and establish a large regional high-precision provincial 3D geological model. Fusion splicing technology can splice adjacent models into an overall model with unified elements and consistent attributes, which can be extended infinitely outward, and support local calls and updates inward to improve model reusability.

Acknowledgements

Fund project: 3D geological modeling of 1:50,000 regional Geological survey in Chongqing; cstc2020jcyj-msxmX0962.

References

1. Wan Jiuxiang. GIS technology and its application and development prospect [J]. Jiangxi communication technology,2003(03):34-37.
2. Li Minghui, Liang Xiaolong, Gai Yuguo. Preliminary study on the main direction of agricultural geology [J]. Deposition and Tethys geology,2001(02):108-112.
3. Pan Mao, Fang Yu, Qu Honggang. Discussion on Some Basic Problems of 3D Geological Modeling [J].Geography and Geographic Information Science, 2007(03): 1-5.
4. Zhang Yanfei, Zhu Jieyong, Zhang Wei. Construction of 3D geological model based on GOCAD [J]. Journal of Hebei University of Engineering (Natural Science Edition), 2011,28(04):69-73.
5. Application of 3D mining software in China [C] / / China Metallurgical Mining Association. Ten years review and prospect of China mining technology. Metallurgical Industry Press (Metallurgical Industry Press),2012:143-149+157.
6. Application of Petrel software in fine geological modeling [J]. Petroleum geology of Xinjiang, 2007(06):773-774.
7. Xia Yanhua, Bai Shiwei. Research on drilling data preprocessing of 3D stratum modeling [J]. Geotechnical mechanics, 2012, 33(04):1223-1226+1239.
8. Liu Zhen, Wu Gengyu, Yang Bo, Pan Mao, Liu Peigang, Jia Zhibin. Research on in-situ stress field simulation automation system based on Creatar X Modeling [J].Science technology and engineering, 2017,17(15):19-25.
9. Guo Zhaowu, Song Xiaoxia, Ren Haiqing and so on. Multi-scale 3D geological modeling of Dongzhouyao mine [J].Coal mine safety,2023,54(01):161-171.
10. Yu Danping. Design and Implementation of a Distributed Data Storage Scheme [J].Computer Knowledge and Technology,2022,18(33):68-70.

A Review on Spatio-temporal Pattern Evolution of Shrub Fire Caused by Human Disturbance in Yunnan Province

Yu Liu, Zhongliang Gao*, Chenyang WangHe, Wentian Yu, and Yufei Cao

School of Civil Engineering, Southwest Forestry University, 650224 Kunming, China

Abstract. Artificial fire is the main cause of forest fires, and human intervention can reduce the occurrence of forest fires. Yunnan province is a province with a high incidence of forest fires. A large number of flammable shrubbery are distributed in many regions. The study on the time scale and space scale of shrubbery fires caused by human disturbance is conducive to the study on the evolution of the space-time pattern of shrubbery fires and the reduction of forest fires. Based on Yunnan province from 1990 to 2018 shrubbery fire data, combined with ArcGIS geographic information system, BP neural network analysis of the impact of human interference on Yunnan shrubbery fire and shrubbery fire time and space changes. The research results can provide some theoretical support for forest fire management departments.

1 Introduction

Chinese forest fires happen frequently and cause serious casualties and economic losses every year. Among the many factors causing forest fires, man-made fire occupies the majority. Yunnan Province is a mountainous plateau region, has always been known for the complex and diverse natural geographical environment, rich and colorful plant species, wide forest coverage rate, high forest density. Shrubbery is widely distributed and of various types, because shrubbery has clumpy structure, shrub or jungle appearance, unique cause, ecological environment and specific plant species composition, once on fire, uncontrollability. Brush fires occupy a large proportion of forest fires in Yunnan province, which causes great losses and is very unfavorable to the protection of forest resources. The climatic conditions of hot and dry for as long as half a year in Yunnan forest, the overload of forest fuels, the high density of fire sources and the threat of foreign fire sources are all potential dangers of heavy and very large forest fires. The complex terrain conditions of high mountains and deep valleys make it impossible for the rescue team to arrive at the fire site in the first time and delay the best time to put out the fire.

It is of great significance to study the prevention and control of shrub fire, forecast and prediction, fire spread and fire potential behavior to reduce the fire frequency and fire loss in Yunnan forest area.

2 Research status

At present, the study on spatial and temporal distribution of forest fires in foreign countries mainly uses statistical analysis method and geographic information system spatial analysis method to study the spatial distribution characteristics of forest fires in different regions and at different levels. In 2009, A Parajuli et al. used Moderate Resolution Imaging Spectroradiometer (MODIS) active data to show forest fire activity in Nepal and to assess fire size, distribution, seasonality, and risk zones in different geographic areas. In 1971, Hanes made a study on the heavy fire disaster areas in California, the United States, and showed that the shrubs had strong adaptability to fire. After fire, the seed germination made the rapid regeneration of the bushes and completed the automatic succession of the population, and fire was the regulator of this succession. The occurrence of forest fire is a complicated system science problem, which is related to many factors, and the human factors have the most serious impact on forest fire. Domestic scholars, experts and forest management departments pay much attention to the temporal and spatial distribution characteristics of forest fires and the effects of human disturbance, and use different theoretical methods and software to support the study of the effects of different degrees of human disturbance on forest fires. In 2018, Li Shangyi et al use Bivariate method to analyze correlation between sample biomass, tree species diversity index and soil nutrient content. According to the bivariate method, different degrees of human interference affect the biomass and its spatial distribution pattern of subtropical forest. In 2018, You Yujie et al. used Nemerow index and correlation coefficient method to carry out protective disturbance and moderate human disturbance respectively with intensity human disturbance as the contrast, and concluded that moderate human disturbance had better effects on soil water storage, water retention, fertility and carbon sequestration. In 2016, Zhang Chen et al. used correlation analysis and Bootstrap significance test to study the relationship between forest fire area and landscape index in Yunnan Province, and concluded that the increase of

* Corresponding author: winalite@swfu.edu.cn

the average patch area of coniferous forest, broad-leaved forest, grassland and shrub would promote the spread of forest fire. In 2010, Wang Bin et al. used geographic information software to analyze the second-class survey data of forest resources and concluded that high-intensity external disturbances in Huoditang forest area increased the landscape diversity and richness of the region. The change of landscape pattern in Huoditang forest area fully reflected the influence of external disturbances on forest landscape pattern and the dynamic process of forest autonomous restoration. In 2009, Yang Guangbin et al. analyzed the statistical data of forest fires in Beijing from 1986 to 2006 by using statistical analysis methods and geographic information system spatial analysis methods, and concluded that the fire sources of grassland, dry land, rural residential land and other land around forest land may spread to form forest fires. In 2007, Tian Xiaorui et al. analyzed the temporal and spatial patterns of forest fires according to the statistics of forest fires in the Tibetan Autonomous Region from 1992 to 2005, and concluded that the fires mainly occurred in southeast Tibet, especially in Mangkang, Zayu, Nyingchi, Milin and other counties, and that the fires were mainly for production and domestic use. In 2007, Tang Lihua et al. analyzed and evaluated the risk degree of anthropogenic fire sources and the influence of positive anthropogenic interference through GIS spatial analysis method, and determined that the indicators of negative anthropogenic interference fire risk were agricultural land burning, agricultural mountain making and afforestation, grave visiting, tourism and basic activities, and the indicators of positive anthropogenic interference fire risk were input manpower and capital.

3 Research method

3.1 Study area profile and data sources

Yunnan is located in the southwest border of China and is called "Dian" for short. The whole province is located in a low latitude region south of 300° north latitude. According to the data of the fourth forest resources survey of Yunnan Province, the forest land area of the whole province is 26.07 million hectares, accounting for 68.0% of the total land area. Shrub land covers about 15% of the country.

In this paper, the data of brush fires in Yunnan Province from 1990 to 2018 were collected through forest fire prevention Office, yearbook and news reports.

3.2 Research procedure

This paper mainly uses BP neural network model in MATLAB and ArcGIS GIS for data collation and analysis.

The principle of BP neural network model construction is that the learning process consists of two processes: the forward propagation of signals and the reverse propagation of errors. BP network is a multilayer network composed of input layer, output layer and one or more hidden layer nodes interconnection. This structure enables the multilayer feedforward network to establish a suitable linear or nonlinear relationship between input and output, but does not cause the network output to be limited between -1 and 1. See Figure 1 for details.

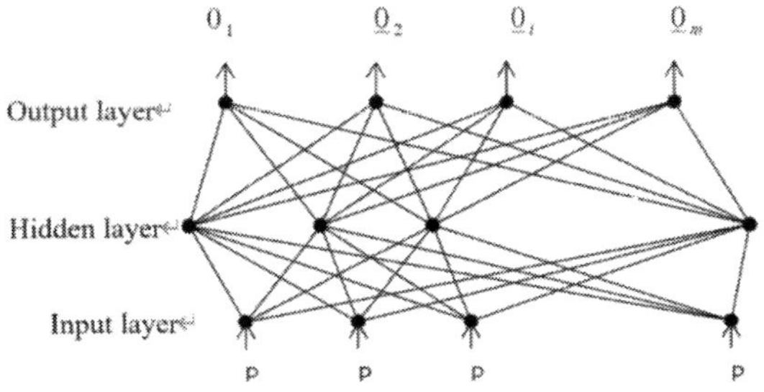

Fig. 1. BP network model.

The BP algorithm "trains" this event to get this input, and the output will get a suitable linear or nonlinear relationship.

4 Conclusion analysis

4.1 Analysis of bush fire time variation

The data of brush fires in Yunnan Province from 1990 to 2018 were collected. The number of brush fires and the number of fires in each month were different, and the fire peak was mainly concentrated in March and April.

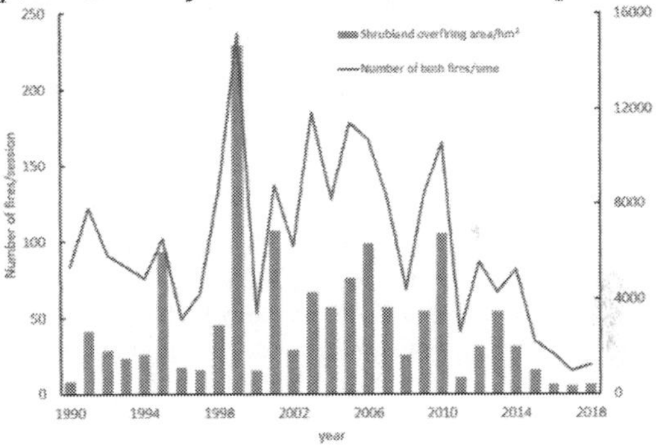

Fig. 2. Comparison of shrub fire occurrence and burning area.

Figure 2 is a comparison chart of the number of brush fires and the burned area in Yunnan Province from 1990 to 2018. It can be seen that the number of brush fires in Yunnan Province fluctuates greatly with the passage of years, but it has been on the decline since 2010. Especially, the number of brush fires has become less and less since 2016, and 2017 is the year with the least number of brush fires. The main reason may be that the government has increased the economic investment in forest fire prevention and the improvement of population quality in recent years. It can also be seen from the figure that the number of bush fires and the fire area on the whole is proportional to the number of bush fires, the fire area increases along with the number of bush fires, there are also some years of fire more times but the fire area is not large, the reason may be different areas, some years of the area where the bush fire may be more distributed.

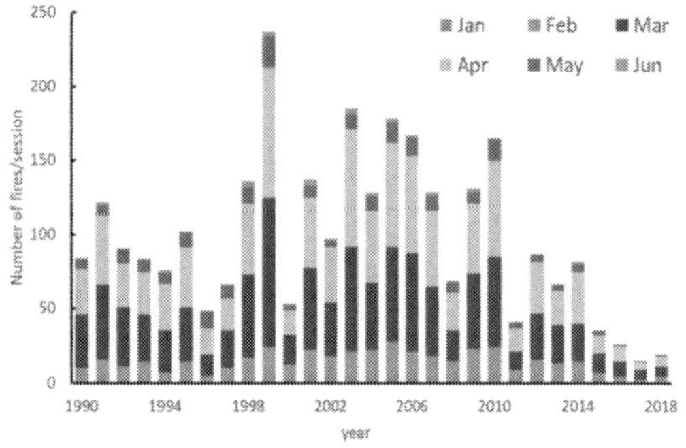

Fig. 3. Bush fires have monthly variations.

As can be seen from Figure 3, there were more brush fires in 1990 ~ 2010, but on the whole, the fires occurred in March and April, which were the most frequent. The forest fire prevention period in Yunnan Province was from December to June of each year, and the forest fire martial law period was from March to April. Brush fires in Yunnan province occur in months coinciding with fire prevention periods, with frequent occurrences in March and April.

4.2 Change of shrub fire area under human disturbance

Most forest fires in Yunnan Province are caused by man-made fire, so are brush fires. Man-made fire is a part of human disturbance, which can be divided into negative human disturbance and positive human disturbance. The index of positive human disturbance is mainly economic input. That is, when economic input is high, the number of bush fires will decrease correspondingly. Before 2005, the government's investment in shrubland was relatively small, so the number of shrubland fires was relatively high. From 2006, the trend of decline began, and the economic investment was higher than that of previous years. However, from 2008 to 2010, the shrubland fires briefly increased, mainly because Yunnan was in a period of drought at that time, and the number of fires was also relatively increased. However, the overall trend of shrub fire is decreasing, especially in recent years, the shrub fire has reached the lowest, which indicates that the economic investment of the government and the improvement of population quality in recent years play a role in the prevention and control of shrub fire.

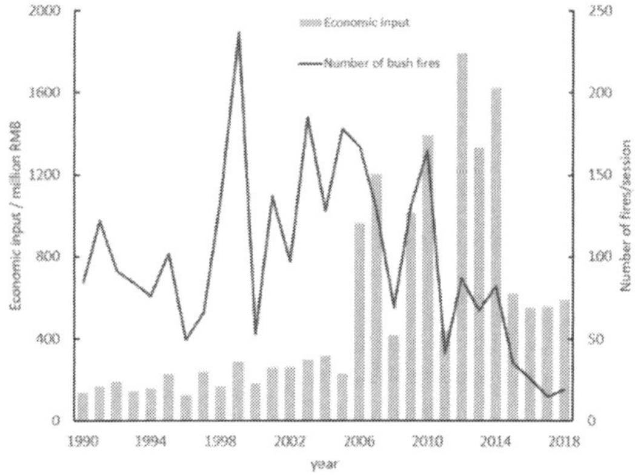

Fig. 4. Comparison of fire frequency and economic input.

Figure 5 reflects the relationship between the shrub fire area and economic input in Yunnan Province from 1990 to 2018. It can be seen from the figure that the economic input and the shrub fire area show a negative correlation. When the economic input is high, the fire area will decrease to some extent, which is mainly reflected in the increase of the government's economic input after 2005, especially in 2012. Economic input reached the highest, the fire area is relatively reduced.

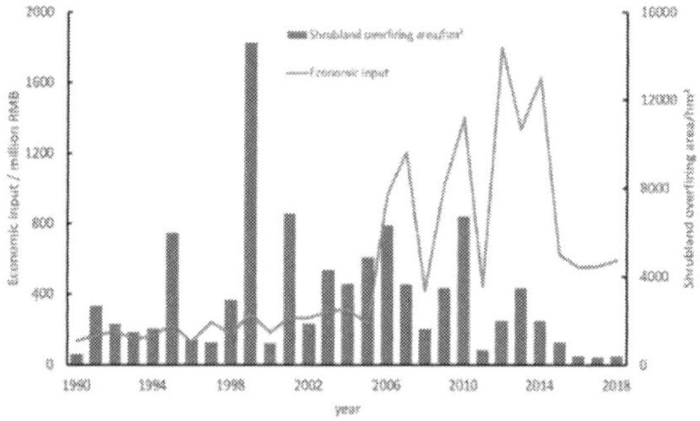

Fig. 5. Comparison of fire area and economic input

4.3 Spatial variation of shrub fire under human disturbance in Yunnan Province

The shrub is widely distributed in Yunnan Province. Among the 16 cities and towns, Dali, Lijiang and Kunming occupy a large area of shrub. From the brush fire data in the past 30 years, it can be seen that the number of fires in these areas is relatively high. Human ignition source is the leading factor of shrub fire. Man-made fire sources can be divided into two categories: productive fire source and non-productive fire source. Productive fire source mainly refers to burning land and charcoal, making mountains and afforestation and burning pasture. Non-productive fire sources mainly refer to the field smoking, heating and cooking, burning paper, burning mountains to drive animals, children playing with fire, dementia and fire and wire caused by these factors lead to the frequent occurrence of bush fires.

4.3.1 Analysis of productive fire source

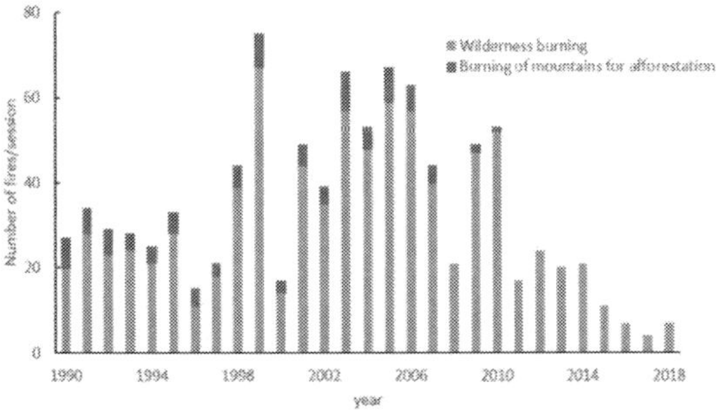

Fig. 6. Analysis of productive fire source of bush fire

As can be seen from Figure 6, the main productive fire sources of shrubland fires in Yunnan Province are wilderness burning and hill-making and afforestation, of which wilderness burning accounts for most, reaching 90%. Wildland burning is the main means of farmers' cultivation. The frequent occurrence of forest fires indicates that farmers' awareness of forest fire prevention has not been improved. However, the number of wildland fires caused by wildland burning has decreased significantly in recent years, so the government may strengthen management.

4.3.2 Analysis of non-productive fire sources

It can be seen from Figure 7 that the largest proportion of non-productive fires in the shrub fires in Yunnan province is outdoor smoking, which is the non-productive fire source and plays a leading role. Children playing with fire, heating and cooking, burning paper in the grave, dementia fire, electric wire caused by a certain proportion, but is less dominant than outdoor smoking, it can be seen that non-productive fire sources of outdoor smoking has the greatest impact on the brush fire in Yunnan province, but since 2010, all kinds of fire sources have been gradually reduced, the main reason is that the quality of the population is improved, the government to strengthen fire source management, In addition, with the progress of science and technology, people change the way of farming and the control of wildland burning, which has brought the fire source under control in recent years.

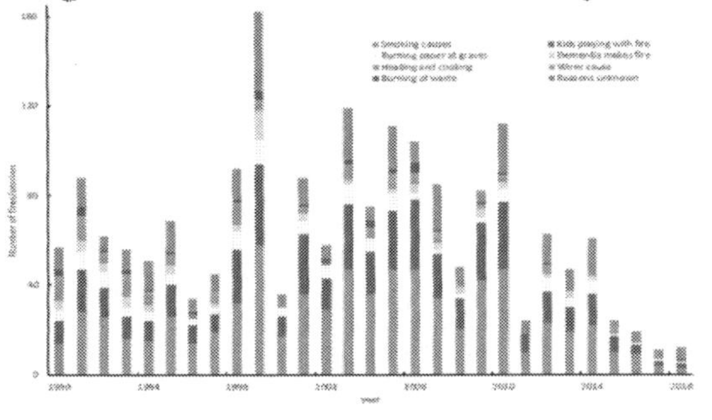

Fig. 7. Analysis of unproductive fire source of bush fire.

4.4 Spatial distribution characteristics of shrub fires in Yunnan Province

The distribution of shrubland in 16 prefectures and cities of Yunnan Province is different, and the number of shrubland fires in each city is also different. FIG. 8 is the comparison of the number of brush fires and the burned area in each city of Yunnan Province from 1990 to 2018. It can be seen that the number of brush fires in Kunming, Dali and Lijiang is the highest, Qujing, Dehong, Lincang and Wenshan is the least, Kunming has the largest burning area, Dali has more fires than Kunming, but the burned area is smaller than Kunming. The hazard degree of shrub fire in Kunming is higher than that in Dali.

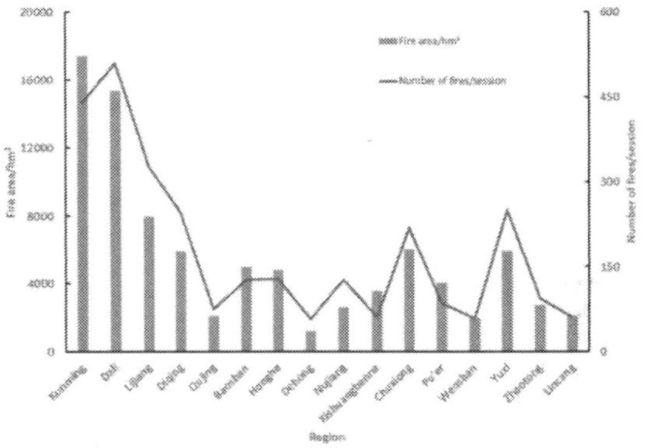

Fig. 8. Comparison of fire frequency and fire area in each city of Yunnan Province

4.5 Brush fire modeling in Yunnan province based on BP neural network

The number of brush fires in Yunnan Province tends to be stable in recent years. The simulation and model construction based on BP neural network are as follows: Data are selected as the training data, test data and prediction data of BP neural network. The error learning target of BP neural network is set as 1e-2, the maximum number of iterations is 50,000, the display step size is 50, the learning speed is 0.01, and the hidden layer is set as 100 neurons. BP neural network model is constructed by using neural network function provided by MATLAB.

The model diagram is as follows:

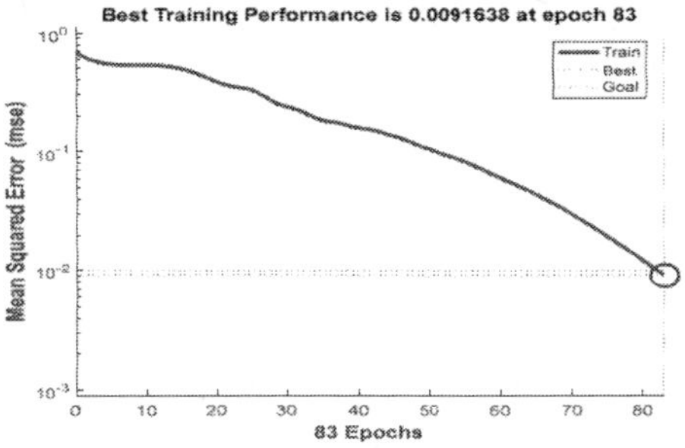

Fig. 9. Neural network error adjustment diagram

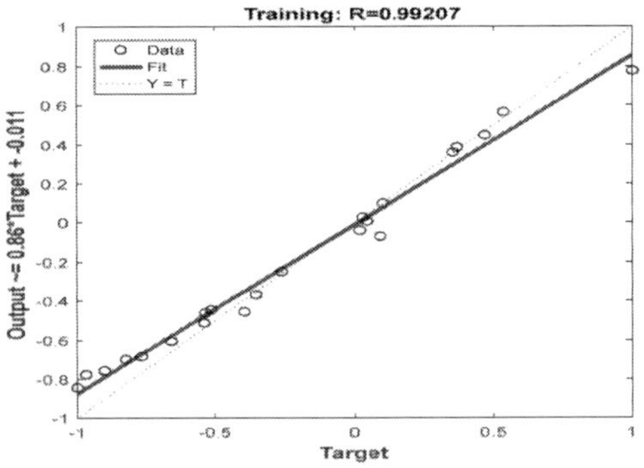

Fig. 10. Neural network goodness of fit graph

Figure 9 shows that the error of setting the target is 0.01, and the error of each iteration will be reduced. After 83 iterations of training, the target error is reached, so the iteration is stopped. Figure 10 shows the goodness of fit graph, in which each data point fluctuates little above and below the matching line, indicating that the model has well learned the data and has a high goodness of fit.

5 Conclusion

Based on the brush fire data of Yunnan Province from 1990 to 2018, the following conclusions can be drawn by ArcGIS and BP neural network prediction:

The brush fires in Yunnan Province showed a decreasing trend on the whole, with a substantial decrease in the past three years. The main reason is the increase in government investment.

The analysis of fire source indicates that the main factors causing shrub fires in Yunnan Province are burning land, burning charcoal and smoking in the field.

Shrub fires in Yunnan Province showed a decreasing trend, mainly in March and April. In terms of spatial scale, it is mainly concentrated in Kunming and Dali, which have the largest area of fire.

Acknowledgements

This work is supported by Joint Agricultural Project of Yunnan Province (202101BD070001-094), National Nature Science Foundation of China (31860214), Science and Technology Innovation Project for University Students of Yunnan Education Department (202110677052).

References

1. Li Shangyi, Fang Li, Chen Jinlei et al. Effects of anthropogenic disturbance on forest biomass and its spatial distribution pattern in the central subtropics[J]. Journal of Ecology,2018,38(17):6111-6124.
2. You Yu J,Wang Yixiang,Zhang Huafeng et al. Effects of different anthropogenic disturbance measures on soil fertility and water storage capacity of natural secondary scrub[J]. Journal of Ecology,2018,38(3):1097-1105.
3. Zhang C, Niu Shukui, Chen F et al. Impact of GIS-based landscape patterns on forest fires in Yunnan Province[J]. Journal of Beijing Forestry University,2016,52(7):48-56.
4. Wang B,Zhang SHX,Yang CS. Effects of human disturbance on landscape patterns in the Tierra del Fuego forest area of the Qinling Mountains[J]. Forestry Science,2010,23(4):544-553.
5. Yang Guangbin, Tang Xiaoming, Ning Jinjie et al. Spatial and temporal distribution patterns of forest fires in Beijing from 1986 to 2006[J]. Forestry Science,2009,45(7):90-95.
6. Chai Zhaopo, Tian Changlan, Li Fengzhi et al. Distribution patterns of forest fire in the Heihe region[J]. Forestry Science,2009,34(4):38-41.
7. Tian Xiaorui, Shu Lifu, Wang Mingyu et al. Spatial and temporal distribution patterns of forest fires in Tibet[J]. Fire Science,2007,16(1):10-14.
8. Tang LH, Fang LM, Xia K et al. Research on fire risk evaluation based on GIS spatial analysis technology of human interference intensity[J]. Journal of Zhejiang Forestry Science and Technology,2007,27(4):43-45.
9. Zhang Qiang, Li Jiaxiang, Xu Wenting et al. Biomass allocation of Rhododendron scrub and its carbon density estimation in a subtropical mountainous area of China[J]. Journal of Plant Ecology,2017,41(1):43-52.
10. Yang Y, Wang Yixiang, Bai Shangbin et al. Response of plant diversity in Lin'an secondary scrub to spatial heterogeneity in forest fire intensity[J]. Journal of Ecology,2016,36(14):4438-4446.
11. Wang Qiuhua,Li Wei,Li Shiyou et al. Fire environment study of forest fires in Kunming area, central Yunnan[J]. Journal of Jiangxi Agricultural University,2015,37(1):108-113.
12. Wang JH, Long XianHua, Niu ShuKui et al. Spatial distribution characteristics of forest fires in Sichuan Province based on GIS[J]. Journal of Northeast Forestry University,2012,40(9):51-56.
13. Song K, Mi X.C., Mark P., et al. Effects of different levels of anthropogenic disturbance on the genealogical structure of forest communities in the Gutian Mountains[J]. Journal of Ecology,2011,19(2):190-196.
14. Sun Yurong, Zhang Gui. Spatial and temporal distribution characteristics of anthropogenic fire sources in Hunan Province[J]. China Agricultural Bulletin,2011,27(28):103-108.
15. Du Chunying, Li Shuai, Liu Dan et al. Spatial and temporal distribution of forest lightning fires in the Daxinganling region[J]. Journal of Natural Hazards,2010,19(3):72-76.

Dynamic Response of Shrubbery Fire to Meteorological Changes in Yunnan Province

Qingqing Wang, Zhongliang Gao*, Chenyang WangHe, Yufei Cao, and Wentian Yu

School of Civil Engineering, Southwest Forestry University, 650224, Kunming, China

Abstract. This paper takes shrubland fires as the research object and selects typical shrubland forest fire cases that occurred throughout Yunnan Province from 1954 to 2018 as the research samples, and the shrubland fire loss elements and meteorological data are studied by SPSS correlation analysis and MATLAB gray prediction model to analyze shrubland fires and meteorological elements. The results showed that there were significant correlations between shrubland fires and meteorological data; the shrubland fire over fire area, forest damage area, and economic loss composite data index showed positive correlations with annual maximum temperature, annual average minimum temperature, annual average temperature, and annual sunshine hours; the annual average relative humidity, precipitation days, and the composite index of the same day meteorological data showed significant negative correlations, etc. In exploring the main meteorological factors affecting the distribution of forest fires, we establish a meteorological fire model to provide a real and credible scientific basis and theoretical foundation for future forest fire prevention and prediction.

1 Introduction

Forest fire is a natural disaster that occurs widely, suddenly and destructively in the world today, and is more difficult to deal with and fight. Control of forest fires is a global problem and is receiving more and more attention from the world. The high temperature and drought conditions in Yunnan forest area for half a year, the overload of combustible materials on forest land, the high density of fire sources and the threat of foreign fire sources are all potential hazards for serious and large forest fires, the complex terrain conditions of high mountains and deep valleys make it impossible for firefighters to reach the fire site at the first time and miss the best time to put out the fire, the complex fire conditions of surface fire, canopy fire, underground fire burning in three dimensions and the influence of fire microclimate are very easy to The complex fire conditions of surface fires, canopy fires, underground fires and fire microclimates can easily lead to casualties. In recent years, the number of forest fires in Yunnan, the area of affected forests and the number of casualties due to forest fires have been decreasing significantly, but it is still the province with the highest number of fires in China.

Depending on the data of the fourth forest resources survey in Yunnan Province, the forest land in the province covers 26.07 million hectares, accounting for 68.0% of the national land area; shrubland accounts for about 15% of the national land area. Several major forest fire caused by shrublands occurs in Yunnan every year, bringing huge economic losses and serious damage to forest resources. Forest fire in Yunnan is concentrated in areas such as Kunming, Yuxi, Dali, Chuxiong and Qujing, where large areas of flammable shrubs are distributed. The dense irrigation area is mostly composed of ground pine and dong type shrubs, with a high density of forest trees and concentrated contiguous, once a fire occurs, the burning intensity is high and direct fighting is difficult, which can easily lead to major fires and cause casualties. In addition, the region belongs to the major urban centers in Yunnan Province, with high population density and complex human activities, increasing the possibility of forest fires. It is important in order to conduct research on shrubland fire prevention and control, forecasting and prediction, and fire spread and potential fire behavior to reduce the number of fires and fire losses in Yunnan forest areas.

2 Research Status

2.1 Status of foreign research

The combination of meteorological data with forest fire warning was initially explored by experts as early as the 1940s in Japan, and after the 1950s, further research on forest fire forecasting began in an increasing number of countries.M. D. Flannigan[1]and others also used a Canadian forest fire risk rating system in the future to simulate changes in forest fire, and the results matched paleontological information, showing that the system can predict future fire. Showed that the system can predict future fire conditions.Timothy J. Brown[2]used a parallel climate model to forecast changes in forest fire risk due to climate change in the western United States in the 21st century.Jesper Rydén et al[3]used the Poission model and the zero-inflated Poisson model to simulate the

* Corresponding author: winalite@swfu.edu.cn

relationship between forest fire occurrence and meteorological factors. Long-term and ultra-long-term trend projections of forest fire risk states based on the prediction results of the General Circulation Model (GCM) and Regional Climate Model (RCM), combined with the forest fire risk hierarchy through mathematical analysis and inference, are also part of climate change and forest fire research.

In shrub research, overseas research has also been conducted earlier and certain research results have been achieved. Burner J M et al[4]shrub forests are characterized by high regeneration ability, rapid reproduction and certain fire resistance after fire. In addition, Clarke P J[5]et al. studied the fire resistance of shrubs and showed that the fire resistance of shrubs is not only related to the structure of the tree species, but also closely related to the shrub density and shrub age. Finances et al[6]found that understory shrubs connect to surface combustibles and can increase the intensity of the fire head and the rate of spread of forest fires, thus increasing the risk of surface fire evolving into canopy fire. Shive et al[7]showed that the growth status and spatial distribution of shrubs have a momentous influence on forest fire spread.

2.2 Status of domestic research

Meteorological information has been used recently in China to develop the fire risk law method, the wind speed correction composite index method and the standard method of double finger method, etc. Chen Feng et al[8]selected annual average wind speed and annual average daily temperature difference as meteorological factors indicators to study the relationship between meteorological factors and forest fires in various ecological zones in Yunnan Province from 1982 to 2008. In addition, it was shown that precipitation, temperature and wind speed were the three main meteorological factors that had significant effects on forest fire occurrence. Drought, high temperature, and high wind is important conditions for major forest fire [9-11].

2.3 Development trends

The advantages of satellite monitoring system, such as comprehensive monitoring range, high temporal frequency and timeliness, make it widely used in forest fire monitoring [12]. Kezhen Liu et al[13]established a fire trend regression model for hotspot concentration areas by extracting spatio-temporal factors of files and combining them with meteorological data. The results showed that the forest fire data from satellite monitoring statistics were in good agreement with data from local forestry bureaus.

3 Shrubbery status in Yunnan Province

Shrubbery is a vegetation type was controlled by shrubs. It has a definite canopy/story height of about 5 m, no main trunk but clusters, and a cover of more than 30-40%. The ecological range is wider than that of tree forests, and the distribution range is often larger than that of three forests. Shrubbery are usually found in dry or cold climates where trees are not suitable for growth.

Shrubbery is dwarf forests with well-developed root systems, strong sprouting ability, tolerance to barrenness, cold, drought, and adaptability. Shrubs are woody perennials, and their growth and distribution also affect durability. Usually clumps of shrubs spread more quickly than scattered shrub forests, and shrub forests combined with weeds are more flammable.

Depending on statistics, the earlier shrubbery area in the province increased from 1.814 million hectares in 1964 to 4.985 million hectares in 1975, and to 5.530 million hectares in 1980, showing an increasing trend year by year. 2017 saw an increase in the shrubbery area in Yunnan Province. Such a trend of change is caused by the weak management of forest resources, for forest fire prevention, lack of awareness of resource protection, relevant protection measures are not in place, etc. After the destruction of the forest, the secondary species dominate, the residual forest phase increases and the storage volume per unit area decreases. The shrubbery area has increased considerably and the forest quality has decreased.

4 Research method

4.1 Correlation analysis method

4.1.1 Data sources

Shrubbery fire data in Yunnan Province from 1954-2018: books such as "Typical Cases of Forest Fires in China" and "Review of Typical Battle series of Forest Fire Fighting", Yunnan Forestry Bureau website, Fire Prevention Office website and other news media.

Meteorological data of some areas (Kunming, Yuxi, Dali) in Yunnan Province from 1954-2018: annual value dataset of Chinese toponymy climate data from China Meteorological Data Network, including monitoring statistics of annual maximum temperature, annual average minimum temperature, annual average temperature, annual average relative humidity, sunshine days and other relevant regional ground stations.

4.1.2 Data processing

This paper studies the correlation between brush fire and meteorological data as well as the prediction of meteorological trends, and then a comprehensive prediction of future trends of a brush fire. The first step is to start with the brush fire cases occurring in Yunnan Province, select typical brush fire cases as the analysis samples, and collect and organize their relevant data. The meteorological data were divided into two major groups: current daily meteorological data and current year meteorological data. The correlation analysis is done with the forest fire data respectively.

The following is a detailed description of each of these elements:

(1) brush fire data: fire area (hm2), forest damage area (hm2), and economic loss (million yuan).

(2) Meteorological data of the year: annual maximum temperature (°C), annual average temperature (°C), annual average relative humidity (percentage), annual average minimum temperature (°C), daily precipitation ≥ 0.1 mm days (days), annual sunshine hours (hours).

(3) Meteorological data of the day: the highest temperature of the day, the highest wind of the day, the fire danger level of the day.

4.1.3 Analysis methods

The first group: the current day meteorological data and forest fire data are analyzed, and the current day meteorological data, the highest wind of the day, and the current day fire danger level are regarded as three vectors and measured by length. A comprehensive descriptive data x1 was calculated and named as meteorological data. Then, large comprehensive indicator Y of the occurrence loss and probability of occurrence of the brush fire under study is named as fire occurrence loss. The correlation between meteorological data x1 and fire occurrence loss y was studied by spies correlation analysis method, and icons were made.

The second group: correlation of each small index such as meteorological data of the year: Annual maximum temperature x2, annual average temperature x3, annual average relative humidity x4 with fire occurrence loss y was analyzed and icons were made.

Data processing approach: two kinds, fire occurrence loss indicators, choose information that integrates three small indicators (burnt area, forest damage area, economic loss); meteorological data indicators, choose information that integrates three small indicators (highest temperature of the day, highest wind of the day, fire danger level of the day); specific collated data are shown in the appendix.

Using the data in the summary table of brush fire losses and climatological data can be analyzed using SPSS for correlation analysis to derive descriptive statistics, person correlation coefficient table for analysis to derive results.

4.2 Trend Forecast Analysis

4.2.1 Analysis objects

The forecast indicators are forest damage area, annual average temperature, annual average relative humidity, and annual sunshine hours. The predicted areas are Kunming, Yuxi and Dali in Yunnan Province.

4.2.2 Analysis method

Gray forecasting refers to the use of GM models to estimate and predict the development pattern of system behavioral characteristics, as well as to estimate and calculate the time when anomalies in behavioral characteristics occur, and to study the future time distribution of events occurring within a specific time zone. In essence, these works treat "stochastic processes" as "gray processes" and "random variables" as "gray variables These works essentially treat "stochastic processes" as "gray processes" and "random variables" as "gray variables" and deal with them mainly with the GM(1,1) model in gray system theory.

According to the GM (1,1) gray prediction model for software programming, the programming data used in the program are as follows, where the data for the next three years were predicted using matlab programming, and the original data and the predicted data were plotted to see the fit effect, and then analyzed and illustrated in detail.

5 Result analysis

5.1 Results of correlation analysis

5.1.1 Descriptive Statistics

Using the attached table, the data in the summary table of using the attached table, the data in the summary table of brush fire losses and meteorological data can be analyzed by SPSS. The results of correlation analysis are as follows: losses and meteorological data can be analyzed by SPSS. To keep the table simple, the following abbreviations are used instead ： Fire loss(FL), Annual maximum temperature(AMMT), Annual mean temperature (AMT), Annual mean relative humidity(AMRH), Annual mean minimum temperature (AMNT), Precipitation Days(PD), Annual sunshine duration(ASD), Meteorological data(MD).The results of correlation analysis are as following.

Table 1. Table of person correlation coefficients

		FL	AMMT	AMT	AMRT	AMNT	PD	ASD	MD
FL	C	1	0.53	0.82	-0.878	0.415	-0.824	0.54	-0.922
	S		0.034	0.024	0.011	0.045	0.044	0.043	0.087
AMMT	C	0.53	1	0.088	0.233	0.102	-0.093	-0.223	-0.092
	S	0.034		0.675	0.263	0.629	0.659	0.284	0.663
AMT	C	0.82	0.088	1	0.039	0.15	0.072	0.086	0.395
	S	0.024	0.675		0.853	0.475	0.732	0.682	0.05
AMRT	C	-0.878	0.233	0.039	1	-0.012	.451*	-0.04	-0.063
	S	0.011	0.263	0.853		0.954	0.024	0.851	0.765
AMNT	C	0.415	0.102	0.15	-0.012	1	0.05	-0.042	0.363
	S	0.045	0.629	0.475	0.954		0.813	0.842	0.075
PD	C	-0.824	-0.093	0.072	.451*	0.05	1	-0.123	-.399*
	S	0.044	0.659	0.732	0.024	0.813		0.558	0.048
ASD	C	0.54	-0.223	0.086	-0.04	-0.042	-0.123	1	0.12
	S	0.043	0.284	0.682	0.851	0.842	0.558		0.567
MD	C	-0.922	-0.092	0.395	-0.063	0.363	-.399*	0.12	1
	S	0.087	0.663	0.05	0.765	0.075	0.048	0.567	

Significance test results, namely sin (bilateral). Can be used to show whether the correlation analysis results you get are statistically significant. Usually, sign<0.05 is considered significant and has statistical significance. If it is not meaningful, even if the correlation coefficient is large, it cannot show that the correlation is significant, and the correlation may be caused by sampling error. The asterisk after the correlation value also reflects the significance, with a * indicating significant at the 0.05 level and ** indicating significant at the 0.01 level.

The correlation coefficient indicates that all the diagonal elements are 1, indicating the correlation between the variable and the variable itself. The elements in the lower left corner and the upper right corner are symmetric, so only one side is needed in order to be be looked at. Since the research is on the correlation between the fire loss of dependent variable and each independent variable, only the first column of data needs to be looked at, Pearson correlation is the correlation coefficient, which is used to measure whether two data sets are on the same line. It determines the linear relationship between the distance variables. The larger the absolute value of the correlation coefficient is, the stronger the correlation is: the closer the correlation coefficient is to 1 or -1, the stronger the correlation degree is, and the closer the correlation coefficient is to 0, the weaker the correlation degree is. In general, the correlation strength of variables is assessed by the value range in the following table.

Table 2. Correlation intensity judgment

Correlation coefficient	Correlation intensity
0.0-0.2	Extremely weak correlation or no correlation
0.2-0.4	Weak correlation
0.4-0.6	Moderate correlation
0.6-0.8	Strong correlation
0.8-1.0	Extremely strong correlation

Through correlation data analysis of 100 brush fire samples in Yunnan Province from 1954 to 2018, the annual maximum temperature, the annual mean temperature, annual mean relative humidity, number of precipitation days, and sign significance test of daily meteorological data have statistical significance. The meteorological data of that day is less than 0.05, indicating that the meteorological index is significantly higher than that of brush fire loss data of 0.05. The average annual relative humidity was close to 0.01 level.

Through data processing, among various elements of meteorological data, the correlation index of annual maximum temperature, annual mean minimum temperature and fire loss is 0.53 and 0.415, indicating that there is a certain positive correlation with brush fire loss. The correlation indexes between annual mean temperature and annual mean relative humidity and fire loss are 0.82 and -0.878, indicating that the two indexes have a strong linear correlation with the bush fire loss, indicating that the annual mean temperature and annual mean relative humidity have a great impact on the bush fire in the fire area. There was a significant negative correlation between the annual mean relative humidity and the bush fire loss, and a significant positive correlation between the annual mean temperature and the fire loss. The correlation index of precipitation days and meteorological data on that day is -0.824 and -0.922, which are significantly negative correlated with the brush fire loss. The results showed that the fire loss in the sample area was negatively interfered by the number of annual precipitation days and the comprehensive meteorological data of that day.

5.2 Dali City Trend Forecast Results

Grey prediction results of forest affected area in Dali in the next three years are 114.7302, 99.61871 and 86.49757. The affected area of the forest in Dali was significantly different from that in previous years, and the affected area of the forest decreased dramatically from 1954 to 2018. Grey forecast in the next three years predicted value has reduced trend, but the base is still large, the situation is not optimistic.

Grey prediction results of the annual mean temperature in Dali in the next three years are as follows: 15.0544, 15.101 and 15.1476. From the grey prediction results, the average annual temperature in Dali from 1954 to 2018 showed a discrete division between 13 and 16 degrees Celsius. The grey prediction curve of the average annual temperature showed a slow rising trend, while the grey predicted value of the average annual temperature in the next three years showed a slight rising change rule.

Grey prediction results of the average annual relative humidity in Dali in the next three years are 75.0958, 75.777 and 76.4643. From 1958 to 1980, the average annual relative humidity in Dali was mostly between 68% and 72%. From then on, the average relative humidity showed a slow rising trend, and the three-year Gray predicted value was close to 76%, showing little change.

Grey prediction results of sunshine duration in Dali in the next three years are: 2046.8112, 2033.5312, 2020.3374. The annual sunshine duration in Dali fluctuated greatly from 1962 to 2005, with a maximum of 2650 hours and a minimum of about 1600 hours. In the next three years, the annual sunshine duration showed a downward trend, but basically stabilized at about 2000 hours, with little fluctuation.

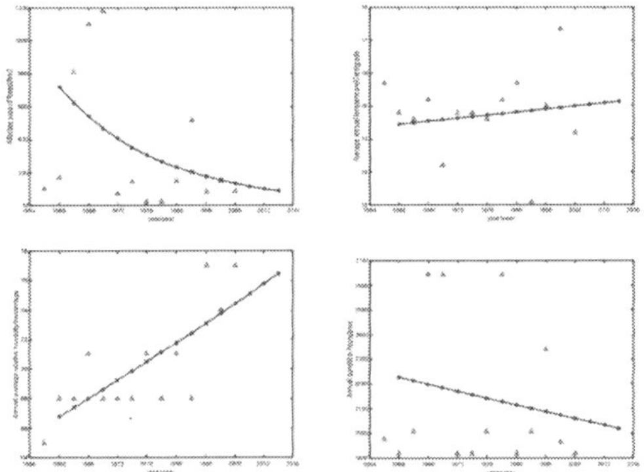

Fig. 1. Trend and forecast of indicators in Dali City

Grey prediction results of the bush damage area in Yuxi in the next three years are : 153.22679, 404.40529, 1067.3306. Yuxi shrubbery affected area from 1990 to 2018 showed multiple growth. Grey prediction value changes greatly, and the data analysis results show that the prediction data of the victim area will increase greatly in the next few years, which needs to focus on prevention and protection management.

Grey prediction results of the average temperature in Yuxi City in the next three years are as follows:14.5286, 13.8891, 13.2777. Yuxi annual average temperature between 13℃-16. Results of grey prediction show that the annual average temperature of Yuxi City will decrease in the future.

Grey prediction results of annual average relative humidity in Yuxi City in the next three years are as follows: 64.2072, 60.6411, 57.2731. In general, the average annual relative humidity in Yuxi City shows a certain downward trend, and the forecast results in the next three years are still consistent with this trend.

Grey prediction results of annual sunshine duration in Yuxi City in the next three years are: 2060.6617, 2115.5688, 2171.9388. The annual sunshine duration of Yuxi City has shown a certain upward trend since 1950, which is exactly opposite to that of Dali City, and the predicted results in the next three years are consistent with this trend.

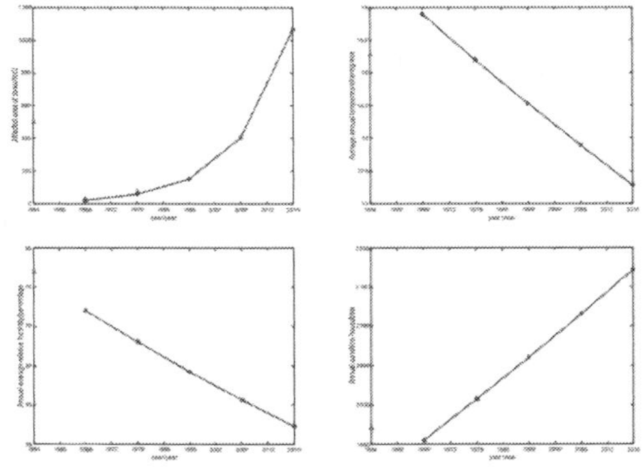

Fig. 2. Trend and forecast of indicators in Yuxi City

Grey prediction results of shrub damage area in Kunming City in the next three years are: 297.0346, 328.2397, 362.723. The affected area of shrubbery in Kunming fluctuates greatly over the years. The grey forecast curve shows a slow rising trend, and the grey forecast value in the next three years is about 300 hectares, showing an increasing trend year by year.

Grey prediction results of the annual mean temperature in Kunming City in the next three years are: 16.2663, 16.252, 16.2376. In general, the annual mean temperature in Kunming is relatively stable except for some extreme weather years.

Grey prediction results of annual sunshine duration in Kunming in the next three years are: 2490.857, 2528.0767, 2565.8527. It is similar to the sunshine duration in Dali, and has certain fluctuation. But in general, there is a certain upward trend.

Grey prediction results of the average annual relative humidity in Kunming in the next three years are: 74.1295, 75.5726, 77.0438. The average annual relative humidity in Kunming has a great difference over the years, but it can be analyzed from the forecast results that it has a certain rising trend.

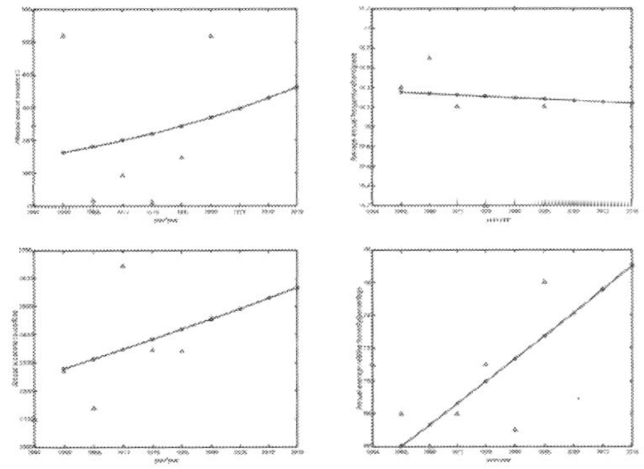

Fig. 3. Trend and forecast of indicators in Kunming City

5.3 Verification of Prediction Results

In order to verify the accuracy of the prediction results, another group of brush fire forest affected areas in Kunming, Yuxi and Dali from 1954 to 2017 and three indexes of meteorological data (average annual temperature, average annual relative humidity and annual sunshine duration) were analyzed with the same software data processing method to obtain the predicted values in 2018. After comparing the predicted value with the actual value in 2018, it was found that the prediction error of the gray prediction GM (1.1) model was large. The prediction error of annual sunshine number and annual mean temperature is small. The analysis error of annual mean relative humidity is minimum. The results show that the prediction results of yearly sunshine number, annual mean temperature and annual mean relative humidity have a certain accuracy, and the prediction trend of future forest affected area has a certain reference value for forest fire prediction.

6 Conclusion

This paper analyzed the change pattern and correlation between brush fire data and meteorological data in Yunnan Province from 1954-2018. The temperature in Yunnan Province has increased significantly and rainfall has decreased in the last 50 years. Against the background of the overall warm and dry trend of the province's climate, shrubbery vegetation types are used to study the dynamic relationship between changes in meteorological data and forest fire. Due to the complex topography and local microclimate differences in Yunnan, in order to reduce the combined effects of regional climate, altitude, latitude and longitude differences as well as data errors, the brush fire case in Yunnan province was used as a data sample to extract each relevant index with the climate data of the case area for analysis and research.

Through the correlation analysis, the brush fire area, forest damage area, economic loss composite data indicators and annual maximum temperature, annual average minimum temperature, annual average temperature, annual sunshine hours show positive correlation, with the annual average temperature correlation is the most significant; annual average relative humidity, precipitation days, the day meteorological data composite index shows a significant negative correlation.

under the climate change scenario of Yunnan Province, trend prediction results of mathematical nature were obtained by the software analysis of gray prediction mathematical model for three regions, Kunming, Dali and Yuxi. The annual average temperature, annual average relative humidity and sunshine hours of the three regions were predicted mathematically in nature. The gloomy prediction model indicated that the shrubbery forest damage area in Kunming and Yuxi showed a multiplicative increase in the next three years. The forest damage area in Dali shows a decreasing tendency, but the base is still large.

Acknowledgments

This work is supported by Joint Agricultural Project of Yunnan Province (202101BD070001-094), National Nature Science Foundation of China (31860214), Science and Technology Innovation Project for University Students of Yunnan Education Department (202110677080), National Key Research and Development Program (2017YFD0600106).

References

1. M. D. Flannigan, K. A. Logan, B. D. Amiro, W. R. Skinner, B. J. Stocks. Future area burned in Canada [J]. Climatic Change,2005,72(1-2).
2. Timothy J. Brown, Beth L. Hall, Anthony L. Westerling. Twenty-first century climate change impacts on wildland fire risk in the western United States: an applied perspective[J]. Climatic Change, 2004, 62(1-3).
3. Jesper Rydén,Igor Rychlik. A note on estimating fire intensity with incomplete data[J]. Journal of Fire Safety,2006,41(5).
4. Varner J M , Kane J M , Hiers J K , et al. Suites of Fire-Adapted Traits of Oaks in the Southeastern USA: Multiple Strategies for Persistence[J]. Fire Ecology, 2016, 12(2):48-64.
5. Clarke P J , Lawes M J , Midgley J J , et al. Resprouting as a key functional trait: How buds, protection and resources drive persistence after fire[J]. New Phytologist, 2012, 197(1).
6. The fire ecology and management of maritime pine (Pinus pinaster Ait.) [J]. Forest ecol manage, 2007, 2007,241(1-3)(-):1-13.
7. Shive K L , Sieg C H , PZ Fulé. Pre-wildfire management treatments interact with fire severity to have lasting effects on vegetation response after wildfire [J]. Forest Ecology & Management, 2013, 297:75-83.
8. Chen F, Lin XD, Niu Shukui, Wang S, Li DE. Impact of climate change on forest fires in Yunnan Province[J]. Journal of Beijing Forestry University,2012,34(06):7-15.DOI:10.13332/j.1000-1522.2012.06.010.
9. Wang Shuhua,Sun Pengfei,Cheng Chunxiang. Correlation study between meteorological factors and forest fires in Yichun City[J]. Forestry Science and Technology,2008(02):24-26.
10. Du Yulong,Lei Chongdian. Analysis of spring forest fire meteorological conditions in northern Shaanxi[J]. Meteorological Science,2006(03):3341-3345.
11. Liu Yuanchun. Impact of climate change on the spatial and temporal distribution pattern of forest fires in China[D]. Northeast Forestry University, 2007.
12. Liu Yungui, Nie Yong, Tian Jingrong, Zheng Shan, Ao Weiwu. A method for determining detailed information of satellite forest fire monitoring hotspots using forest rights management software[J]. Forestry Survey and Planning,2009,34(04):65-67.
13. Liu Kezhen, Shu Lifu, Zhao Fengjun, Zhang Yunsheng, Li Yanyun. Forest fire distribution and occurrence forecasting model based on satellite monitoring hotspots[J]. Journal of Forestry Engineering,2017,2(04):128-133.DOI:10.13360/j.issn.2096-1359.2017.04.021.

https://doi.org/10.1051/e3sconf/202339401031

Implementation Principles of Optimal Control Technology for the Reduction of Greenhouse Gases in Semiconductor Industry

Yuling Liang [1], KonTee Tan [2], Yuanzhe Li [3,*]

[1] EUNCET Business School, Universitat Politecnica de Catalunya, Barecelona, 08034, Spain
[2] Facility Engineering and Environment Department, AFPD Pte Ltd, Singapore 528798, Singapore
[3] Sustainability Design, China Academy of Art, Hangzhou, 310002, China

Abstract. Climate change is occurring at a much faster rate than in the past and it cannot be explained solely by natural causes. Humans, specifically the greenhouse gases (GHG) emissions produced by human activity, are the primary cause of the earth's rapidly changing climate today. As one of the most GHG emission industries, semiconductor, at present, there is no clear guideline for the implementation of significant reduction method for many fluorinated compounds (FCs), N2O and NF3 greenhouse gases in the domestic and overseas semiconductor industry. Therefore, this report, implementation principles of optimal control technology for the GHG emissions have been proposed based on the Intergovernmental Panel on Climate Change (IPCC) and old reduction method. This reduction method is suitable for the installation of efficient exhaust gas destruction equipment for the removal of FCs and N2O from Etching, Thin Film, including chemical/physical vapour deposition, and Diffusion processes in the semiconductor industry.

Keywords: Greenhouse Gas (GHG); GHG Emission Reduction; Methodology; Semiconductor Industry; Gas Abatement System

1. Background

In response to the World Semiconductor Council's (WSC) post-2010 voluntary fluorinated greenhouse gas (PFC) emission reduction plan and the global goal of achieving net zero emissions, the Taiwan Semiconductor Industry Association (TSIA) Environmental Safety Committee member companies have been actively engaged in implementing effective strategies. To support these efforts, the WSC has released crucial guidelines and methodologies aimed at reducing PFC emissions in the semiconductor industry.

The WSC has introduced the Methodology for Reduction of Fluorinated and N_2O Greenhouse Gas Emissions from the Semiconductor Industry (TM002) and the Best Practice Guidance for Emission Reduction of Semiconductor Fluorinated Greenhouse Gas (BPGER). These guidelines serve as invaluable references for TSIA Environmental Safety Committee member companies in their F-gas reduction and testing initiatives. Furthermore, the latest IPCC 2019 Emission Guidelines have been incorporated to provide comprehensive guidance for the reduction of greenhouse gas emissions.

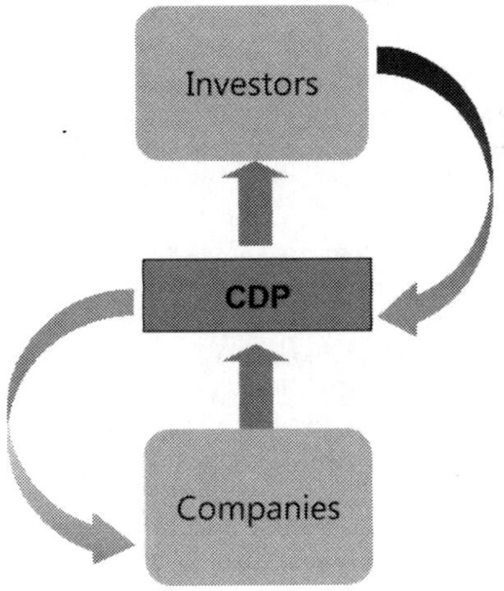

Figure 1. Carbon Disclosure Project (CDP) implemented by international investment institutions in 2003.

The development of these guidelines involved collaborative efforts between the Semiconductor Association member companies and the esteemed Industrial Technology Research Institute (ITRI). Drawing

* Corresponding author: yuanzhe001@e.ntu.edu.sg

upon their collective expertise and knowledge, these organizations have formulated practical standards and recommendations that enable TSIA member companies to effectively mitigate and monitor fluorinated greenhouse gas emissions. Moreover, the Carbon Disclosure Project was also initiated by international investment institutions in 2003 (Figure 1). The CDP survey report mainly provides investors information to understand the status of carbon risk management of an enterprise, and the scale is getting bigger year by year. In the report published recently, a total of 9,331 enterprises disclosed their current situations to CDP, and 219 of which were Taiwanese enterprises. CDP provides five major projects and integrates the information disclosure platform. Carbon disclosure information is integrated from the perspective of Corporate Social Responsibility (CSR).
By adhering to these guidelines and methodologies, TSIA Environmental Safety Committee member companies demonstrate their unwavering commitment to environmental sustainability and their proactive approach to meeting emission reduction targets. Through their collaborative efforts and adherence to international standards, these companies contribute significantly to the semiconductor industry's global endeavors in addressing climate change and working towards a future of net zero emissions.

2. Future GHG Emission Targets

2.1 Standard Emission Rate and 2030 Reduction Target

In line with its dedication to sustainability, the semiconductor industry, under the leadership of the World Semiconductor Council (WSC), is actively pursuing significant reductions in fluorine-gas emissions. The Taiwan Semiconductor Industry Association (TSIA), as a prominent member of the WSC, is committed to implementing the best control technology to achieve a standard emission rate (NER) of 0.22 kg CO_2e/cm^2 by 2020. This ambitious target represents a remarkable 30% reduction in NER compared to the baseline established in 2010. By setting such a challenging objective, TSIA aims to demonstrate its industry leadership in mitigating greenhouse gas emissions. The semiconductor sector recognizes the urgent need for sustainability practices and acknowledges its responsibility in contributing to global emission reduction goals. Through the implementation of cutting-edge control technologies and the adoption of best practices, TSIA is committed to driving significant improvements in emission rates.
Furthermore, TSIA recognizes the importance of long-term sustainability planning. In alignment with the WSC's vision, TSIA will establish its 2030 reduction target to ensure continuous progress towards reducing fluorine-gas emissions. By aligning its goals with the WSC's long-term objectives, TSIA reaffirms its commitment to sustainable practices and acknowledges the global nature of the environmental challenges at hand. The standard emission rate and the 2030 reduction target serve as critical milestones for the semiconductor industry's efforts to combat climate change. By striving to achieve these targets, TSIA aims to make a substantial contribution to global emission reduction efforts and demonstrate its commitment to environmental stewardship. Through collaboration, innovation, and a shared vision, TSIA and the entire semiconductor industry are working towards a more sustainable and environmentally responsible future [4].

2.2 Expansion of Control Techniques

In its pursuit of comprehensive emission reduction, the semiconductor industry recognizes the need to expand the implementation of best control techniques. This expansion encompasses not only emission reporting and new plants but also includes the inclusion of "Rest of the World" plants. While these plants are situated outside the World Semiconductor Council (WSC) region, they are operated by WSC Association member companies, making them an integral part of the industry's efforts to minimize environmental impact. By extending the application of control techniques to "Rest of the World" plants, the semiconductor industry demonstrates its commitment to global environmental sustainability. These plants, although located in different regions, play a crucial role in the overall environmental footprint of semiconductor manufacturing. Recognizing this, the industry seeks to ensure that rigorous control measures are applied consistently across all facilities, regardless of their geographical location.
Through the implementation of best control techniques, the industry aims to minimize the release of fluorine-gas emissions and other related pollutants. This expansion serves as a testament to the industry's collective responsibility in addressing environmental challenges on a global scale. By promoting sustainability practices beyond regional boundaries, the semiconductor industry aims to foster a more cohesive and environmentally responsible approach to semiconductor manufacturing. The inclusion of "Rest of the World" plants within the purview of control techniques highlights the industry's commitment to comprehensive emission reduction and environmental stewardship. It signifies a collective effort to minimize the environmental impact of semiconductor manufacturing operations worldwide. Through collaboration, knowledge sharing, and the adoption of best practices, the industry aspires to create a sustainable future for semiconductor production, one that balances technological advancements with responsible environmental practices [1,4]. Define kilogram carbon equivalent (kg CO_2e/cm^2) per wafer area processed, based on the Global Semiconductor Association's NER as the unit of measure, which will be a WSC target for the global semiconductor industry. This target should not be applied to a single individual region, and companies or facilities that break ground on semiconductor fabs after May 2011 are considered new fabs and must adopt the WSC best execution procedures [5]. This TSIA Semiconductor Fluorine Gas Emission Reduction BAT Implementation Principles will also be used in conjunction with the WSC review and update of the Fluorine Gas Emission Reduction Target.

2.3 Standardized Measurement and Best Execution Procedures

The global semiconductor industry places significant emphasis on standardized measurement as a means to evaluate and mitigate greenhouse gas emissions. Recognizing the need for consistency and comparability, the industry has embraced the kilogram carbon equivalent (kgCO2e/cm2) per wafer area processed as the unit of measure for assessing emissions. This standardized measurement, based on the NER (Normalized Emission Rate) developed by the Global Semiconductor Association, serves as a common target for the entire industry.

It is crucial to highlight that this target is not intended for implementation within a specific region but rather on a global scale. By adopting a unified measurement approach, the industry aims to facilitate meaningful comparisons and benchmarking of greenhouse gas emissions across semiconductor manufacturing facilities worldwide. This global perspective underscores the industry's commitment to addressing environmental challenges holistically and collaborating on sustainable practices. Furthermore, the industry recognizes the significance of best execution procedures in establishing and operating semiconductor fabs. To ensure the highest environmental standards, the World Semiconductor Council (WSC) has mandated that any semiconductor fabs initiated after May 2011 are classified as new fabs. These new fabs must adhere to the WSC's best execution procedures, which encompass a comprehensive set of guidelines and practices for minimizing environmental impact throughout the lifecycle of a facility.

The WSC's best execution procedures encompass various aspects, including site selection, facility design, construction, and ongoing operation. These procedures promote the integration of sustainable technologies and practices into all aspects of semiconductor manufacturing, ranging from energy efficiency and waste management to water conservation and emissions control. By adhering to these procedures, semiconductor fabs are positioned to achieve optimal environmental performance and contribute to the industry's broader sustainability goals. The establishment of standardized measurement and the adoption of best execution procedures reflect the semiconductor industry's proactive approach towards environmental stewardship. By setting global targets and implementing comprehensive guidelines, the industry strives to harmonize its environmental efforts and ensure the highest level of sustainability across semiconductor manufacturing operations. Through continuous improvement and collaboration, the industry aims to advance responsible practices, reduce its environmental footprint, and drive positive change within the semiconductor ecosystem [5].

2.4 TSIA Semiconductor Fluorine Gas Emission Reduction BAT Implementation Principles

The TSIA Semiconductor Fluorine Gas Emission Reduction BAT (Best Available Techniques) Implementation Principles serve as a vital roadmap for the semiconductor industry in achieving the ambitious fluorine gas emission reduction targets set by the World Semiconductor Council (WSC). These principles provide a comprehensive framework that guides industry stakeholders in adopting best practices, implementing advanced control technologies, and fostering collaboration to effectively reduce fluorine gas emissions. By adhering to these implementation principles, semiconductor companies actively embrace cutting-edge techniques and strategies to minimize their environmental impact. The principles outline a range of practices, such as the optimization of process technologies, the utilization of innovative gas abatement systems, and the integration of sustainable design approaches. These measures ensure that semiconductor manufacturing processes are conducted in an environmentally responsible manner. Moreover, the TSIA Semiconductor Fluorine Gas Emission Reduction BAT Implementation Principles work in tandem with the WSC's ongoing review and updates. This collaborative approach ensures that the semiconductor industry remains at the forefront of sustainable practices and continually improves its emission reduction strategies. The principles serve as a dynamic guide, allowing industry stakeholders to adapt to emerging technologies and evolving regulatory requirements while maintaining a strong commitment to environmental stewardship.

The implementation principles also foster collaboration among industry players, academia, and research institutions. By sharing knowledge, expertise, and research findings, the semiconductor industry can collectively address challenges and identify innovative solutions to further reduce fluorine gas emissions. This collaborative approach promotes the exchange of best practices and encourages the adoption of state-of-the-art technologies, ultimately driving continuous improvement within the industry. The TSIA Semiconductor Fluorine Gas Emission Reduction BAT Implementation Principles play a pivotal role in ensuring that the semiconductor industry achieves its emission reduction targets and contributes to a more sustainable future. By aligning with these principles, semiconductor companies demonstrate their commitment to environmental responsibility and their dedication to mitigating the environmental impact of their operations. Through the adoption of best practices and the continuous improvement of emission reduction strategies, the industry strives to minimize its environmental footprint and contribute positively to global sustainability efforts [4].

3. GHG Emission Reduction Program

The GHG Emission Reduction Program encompasses various strategies and methodologies aimed at minimizing greenhouse gas emissions within the semiconductor industry (Figure 2). This section highlights key approaches, including reduction methodology, process optimization, greenhouse gas substitution, advanced abatement methodology, and the implementation of remote plasma cleaning systems.

 https://doi.org/10.1051/e3sconf/202339401031

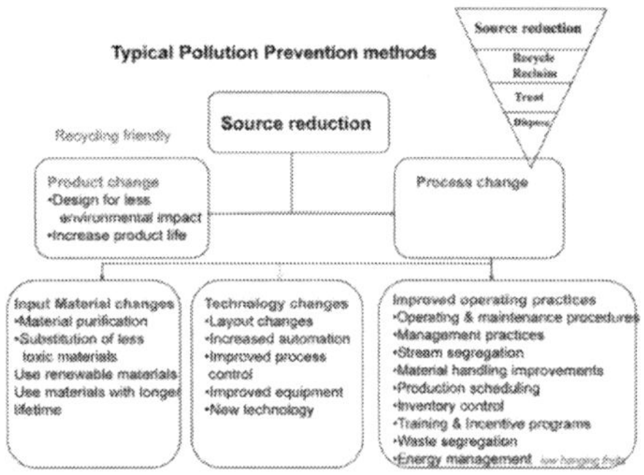

Figure 2. Guideline for sustainable process and prevention of pollution.

3.1 Process Optimization

Process optimization is a key strategy employed in the semiconductor industry to reduce greenhouse gas consumption and minimize fluorinated greenhouse gas emissions. By fine-tuning various process variables, such as chamber pressure, temperature, plasma power, cleaning gas flow rate, gas flow time, and gas mixture ratios, semiconductor companies can achieve significant reductions in carbon emissions [6]. The optimization efforts extend to different processes within semiconductor manufacturing, including chemical vapor deposition (CVD) chamber cleaning and etching procedures. These processes are carefully analyzed and modified to enhance their efficiency and environmental performance. By making adjustments to the process variables, companies can achieve a more sustainable operation while maintaining high-quality production standards.

To support process optimization, advanced inspection systems are employed to monitor changes and gather crucial data. Techniques such as mass spectrometry (MS), infrared (IR) spectroscopy, optical emission spectroscopy (OES), and radio frequency (RF) impedance monitoring are utilized to accurately assess process conditions and identify areas for improvement [7]. These inspection systems provide real-time feedback on gas composition, concentrations, and other relevant parameters, enabling semiconductor companies to make informed decisions in optimizing their processes.

Endpoint inspection, a widely adopted technique for cleaning CVD chambers, can also be applied to other fluorinated greenhouse gas plasma operations, including etching processes. By monitoring the endpoint of these operations, semiconductor companies can determine the optimal timing for terminating the process, minimizing waste and reducing greenhouse gas emissions. Process optimization efforts are driven by the industry's commitment to sustainability and the reduction of environmental impact. By continuously improving process variables and parameters, semiconductor companies can achieve substantial reductions in carbon emissions, contributing to global efforts in mitigating climate change and achieving a more sustainable future.

3.2 Greenhouse Gas Substitution

In the pursuit of reducing net fluorine-gas emissions, the semiconductor industry recognizes the importance of exploring alternative solutions. One approach involves substituting high global warming potential (GWP) gases with lower GWP or non-GWP gases, as well as optimizing the efficiency of gas usage in the plasma process. By adopting these alternative chemical methods, semiconductor companies can significantly decrease greenhouse gas emissions [2]. The substitution of high GWP gases with lower GWP or non-GWP alternatives is a proactive measure to mitigate the environmental impact of semiconductor manufacturing. This strategy involves identifying and implementing gases that have a reduced potential to contribute to global warming. By replacing high GWP gases with more environmentally friendly options, semiconductor companies can make substantial strides in reducing their greenhouse gas emissions.

When evaluating alternative chemicals for substitution, it is essential to consider various factors beyond their environmental impact. Safety and health implications on fab operations, employee protection, and the external environment must be thoroughly assessed. This comprehensive evaluation ensures that the chosen alternative chemicals not only minimize greenhouse gas emissions but also meet stringent safety standards and contribute to the overall well-being of the semiconductor manufacturing ecosystem [8]. By prioritizing greenhouse gas substitution and conducting thorough assessments, the semiconductor industry strives to make sustainable choices that promote environmental stewardship without compromising safety and operational efficiency. These efforts pave the way for a greener and more sustainable future in semiconductor manufacturing.

3.3 Advanced Abatement Methodology

The semiconductor industry has demonstrated its commitment to reducing fluorine gas emissions through the development and implementation of advanced abatement technologies. These innovative solutions prioritize localized emission reduction, recognizing that it is more effective to tackle emissions at the source before they disperse and contaminate the environment further. Through the utilization of advanced abatement methodologies, the industry aims to achieve precise and accurate calculations of Fluorinated Compounds (FCs) emissions by connecting each emission stream to a dedicated local scrubber [9-10].

https://doi.org/10.1051/e3sconf/202339401031

Table 1. Important features of the advanced abatement methods.

General practice of the Advanced Abatement system	Installation of effective abatement system for the removal of Fluorinated Compound (FCs) and N_2O, such as combustion, electric, or plasma type, etc., for the removal of fluorinated and N_2O GHG from semiconductor manufacturing processes such as Etching process, ThinFilm process, include chemical/physical vapour deposition, and Diffusion processes.
Type of greenhouse gas reduction	Reduce emissions of fluorinated compounds (FCs) and N_2O greenhouse gases. These FCs include CF_4, C_2F_6, C_3F_8, c-C_4F_8, CHF_3, CH_2F_2, CH_3F, NF_3 and SF_6.

Table 1 outlines the important features of these advanced abatement methods, shedding light on their role in the semiconductor manufacturing process. These methods encompass a range of practices aimed at removing fluorinated compounds (FCs) and N2O greenhouse gases from the production system. The installations of effective abatement systems, such as combustion, electric, or plasma-based systems, are instrumental in achieving significant reductions in emissions (Figure 3). These abatement systems target specific fluorinated compounds, CF_4, C_2F_6, C_3F_8, c-C_4F_8, CHF_3, CH_2F_2, CH_3F, NF_3 and SF_6, as well as N_2O greenhouse gases [11-12].

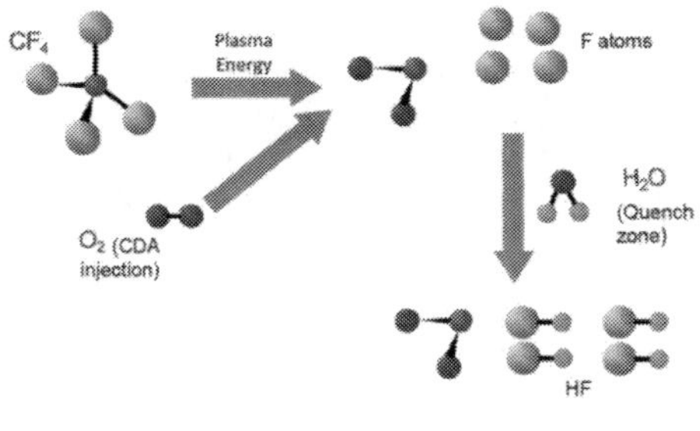

Figure 3. Chemistry of PFCs abatement with proteus.

The effective operation of the abatement systems relies on various factors, including venting equipment and process conditions. Temperature control, fluorinated greenhouse gas inlet concentration, flow rate, and the composition of the total inlet flow all play a crucial role in determining the performance and efficiency of the venting system. By optimizing these parameters and implementing advanced abatement technologies, the semiconductor industry is making significant progress in minimizing its environmental impact and contributing to a greener future [11-12].

The ongoing exploration and evaluation of advanced abatement methodologies highlight the industry's commitment to continuous improvement. As new technologies emerge and are assessed within the framework of the World Semiconductor Council (WSC), the best practices document will be updated to reflect the latest advancements. Furthermore, reliable measurement protocols are essential to accurately assess emissions and evaluate the effectiveness of new technologies, ensuring that the industry maintains a robust and transparent approach to emission reduction [16]. Through collaborative efforts and knowledge-sharing within the WSC, the semiconductor industry is driving advancements in abatement technologies and solidifying its position as a responsible global citizen committed to sustainable manufacturing practices.

3.4 Remote Plasma Cleaning System

The remote plasma cleaning system has emerged as an effective alternative to in-situ CVD chamber cleaning technology, offering a sustainable solution for removing residues left in the chamber after deposition [6]. This innovative technology involves the installation of a plasma generation unit at the front of the CVD chamber. The remote cleaning process typically begins with the introduction of NF3 into the plasma, triggering the generation of fluorine radicals and ions within the remote plasma unit. These reactive species are then directed into the processing chamber, where they chemically react with the deposited material [13]. The by-products of this reaction, including substances like SiF4, are carried away in gaseous form. Remote plasma cleaning has become a widely adopted practice for cleaning CVD chambers, showcasing its effectiveness in reducing fluorine gas emissions [12].

To facilitate the implementation of remote plasma cleaning technology, equipment suppliers have developed and integrated remote plasma systems that can be retrofitted into existing processing tools. These systems replace the original chemistry used for fluorine gas cleaning, enabling semiconductor companies to transition to more environmentally friendly cleaning practices [14]. The semiconductor industry, through collaboration within the World Semiconductor Council (WSC), continually evaluates and shares new technologies and advancements related to fluorine gases [15]. As knowledge and experience grow, the Best Practices document is regularly updated to reflect the latest developments and provide industry stakeholders with comprehensive guidelines for emission reduction. To accurately measure emissions or evaluate the effectiveness of new technologies, companies must adhere to reliable measurement protocols, ensuring consistent and reliable data collection [16].

The adoption of remote plasma cleaning systems represents a significant step forward in the semiconductor industry's commitment to reducing greenhouse gas emissions. By embracing sustainable cleaning technologies and updating best practices, the industry continues to drive innovation and foster a more sustainable manufacturing ecosystem. Through the collective efforts of equipment suppliers, semiconductor manufacturers, and industry organizations like the WSC, the industry is well-positioned to make substantial progress in reducing its environmental impact and promoting a greener future.

4. Basis of Preparation & Monitoring Plan

The Taiwan Environmental Protection Agency (EPA) has established specific objectives and requirements for estimating greenhouse gas (GHG) emissions based on internationally recognized guidelines and values (Figure 4). These objectives and needs align with the "2019 IPCC Guidelines for National Greenhouse Gas Inventories, Volume 3, Chapter 6" (IPCC GL) and the Fourth Assessment Report (AR5) Global Warming Potential (GWP) values [7, 17].

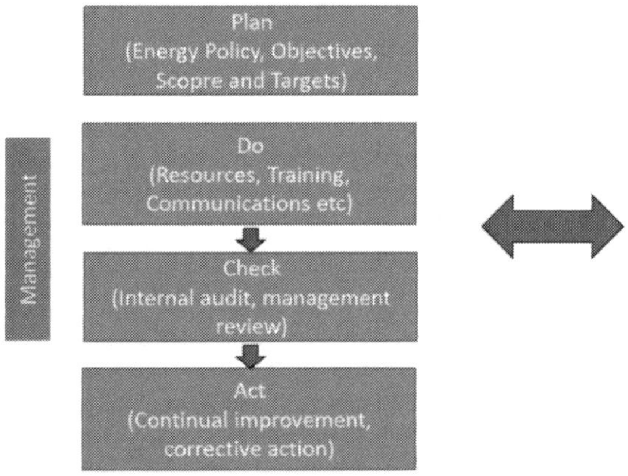

Figure 4. Requirement to achieve the purpose of adopting the standards for basis of preparation.

The EPA's objectives and needs for estimating GHG emissions are as follows:

1. Accuracy and Consistency: The EPA aims to ensure that GHG emissions estimates are accurate and consistent across different sectors and sources. By following the IPCC GL, which provides internationally recognized methodologies and guidance, the EPA can promote consistency in GHG inventories and enhance the accuracy of emissions estimates.
2. Comprehensive Coverage: The EPA seeks to include all relevant GHGs and sectors in its estimation methods. The IPCC GL offers comprehensive guidance on the measurement and estimation of various GHGs, enabling the EPA to account for emissions from different gases such as carbon dioxide (CO_2), methane (CH_4), nitrous oxide (N_2O), and fluorinated gases.
3. Transparency and Documentation: The EPA places importance on transparency and documentation of estimation methods to facilitate peer review, enhance credibility, and ensure the reproducibility of results. By relying on the IPCC GL, which provides detailed guidelines and methodologies, the EPA can document its estimation procedures effectively and provide transparent information to stakeholders.
4. Scientific Rigor: The EPA aims to base its estimation methods on scientifically rigorous approaches and the latest scientific findings. The IPCC GL, along with the AR5 GWP values, reflects the most up-to-date scientific knowledge on GHG emissions, their impacts, and the calculation of GWPs. By incorporating these scientific advancements, the EPA ensures that its estimation methods are robust and reliable.
5. Comparability and International Cooperation: The EPA recognizes the importance of comparability and international cooperation in addressing global climate change. By aligning with the IPCC GL and the AR5 GWP values, the EPA's estimation methods can be easily compared and harmonized with those of other countries, promoting international cooperation and facilitating the assessment of global emission trends.

By adhering to the guidelines and values outlined in the IPCC GL and the AR5, the EPA can fulfill its objectives and meet the needs for accurate, comprehensive, transparent, scientifically rigorous, and internationally comparable estimation of GHG emissions. These efforts contribute to the development of effective climate change mitigation strategies and informed decision-making for a more sustainable future.

5. Future Trends and Development of Semiconductor Industry

The semiconductor industry is experiencing significant growth and is poised to play a crucial role in shaping the future. The increasing demand for consumer electronics products and the emergence of transformative technologies like artificial intelligence, machine learning, and the Internet of Things are driving the expansion of the semiconductor market [18]. Factors such as rising household incomes, population growth, digitization, and urbanization are fueling the demand for consumer electronics devices, leading to a rapid market expansion. However, it is important to address the sustainability challenges associated with semiconductor manufacturing. The industry's current capacity may struggle to meet the surging demand in a sustainable manner [19]. To mitigate the environmental impact of semiconductor production and move towards a greener future, companies need to explore and adopt greener alternatives and practices.

One key area of focus is the adoption of greener materials and processes. Semiconductor manufacturers can invest in research and development to find alternatives to hazardous substances and chemicals with high global warming potential. By incorporating eco-friendly materials and implementing cleaner production techniques, the industry can minimize its environmental footprint and reduce greenhouse gas emissions. Additionally, the semiconductor industry can strive to enhance its energy efficiency and transition to renewable energy sources. Investing in energy-efficient technologies and optimizing manufacturing processes can help reduce energy consumption. Moreover, embracing renewable energy sources such as solar and wind power can further reduce the industry's reliance on fossil fuels and contribute to a lower carbon footprint.

Collaboration and knowledge sharing among industry stakeholders are vital for driving sustainable development in the semiconductor sector. Companies can work together to establish best practices, share innovative solutions, and collaborate on research and development projects focused on sustainability. Industry associations and organizations can play a crucial role in facilitating such collaborations and promoting sustainable practices across the semiconductor value chain. Furthermore, policymakers and regulatory bodies can incentivize and support the adoption of sustainable practices in the semiconductor industry through regulations, tax incentives, and funding programs. Encouraging investments in research and development of greener technologies and providing support for renewable energy adoption can accelerate the industry's transition towards sustainability.

Additionally, Taiwan Semiconductor Industry Association (TSIA) and the semiconductor industry in Taiwan have a crucial role to play in driving sustainable development. Taiwan is a major hub for semiconductor manufacturing, hosting some of the world's leading semiconductor companies. As the industry continues to grow, it presents a unique opportunity for TSIA and Taiwanese semiconductor companies to lead by example in implementing sustainable practices. TSIA can take the lead in promoting and coordinating sustainability initiatives within the Taiwanese semiconductor industry. By establishing guidelines, sharing best practices, and facilitating collaboration among member companies, TSIA can drive the adoption of greener alternatives and practices throughout the industry.

Taiwan, with its advanced technology and manufacturing capabilities, can leverage its strengths to develop and deploy innovative solutions that contribute to sustainability. The government and industry stakeholders can work together to invest in research and development of eco-friendly materials, energy-efficient processes, and renewable energy technologies. This collaboration can foster the growth of a sustainable semiconductor ecosystem in Taiwan.

6. Conclusion

In conclusion, the semiconductor industry in Taiwan, supported by TSIA and its member companies, has an opportunity to lead the way in driving sustainability. By embracing greener alternatives, optimizing energy efficiency, and collaborating with stakeholders, Taiwan can contribute to the global goal of achieving net-zero emissions by 2050. Through these collective efforts, the semiconductor industry in Taiwan can continue to thrive while making significant strides towards a more sustainable future.

Competing Interests

All authors declare no Competing Financial or Non-Financial Interests.

Data Availability

The data that support the findings of this study are available from the corresponding author upon reasonable request.

Acknowledgements

This research was funded by Enerstay Sustainability Pte Ltd (Singapore) Grant Call (Call 1/2022) _GHG (Project ID VS1-001), Singapore.

References

1. Hilaire, J. et al. Negative emissions and international climate goals—learning from and about mitigation scenarios. Clim. Change 157, 1–31 (2019).
2. Rose, S. & Scott M. Review of 1.5°C and Other Newer Global Emissions Scenarios: Insights for Company and Financial Climate Low-Carbon Transition Risk Assessment and Greenhouse Gas Goal Setting (EPRI, 2020).
3. Luderer, G. E. A. Residual fossil CO2 emissions in 1.5–2 °C pathways. Nat. Clim. Change 8, 626–633 (2018).
4. Smith, P. et al. Biophysical and economic limits to negative CO2 emissions. Nat. Clim. Change 6, 42–50 (2016).
5. IPCC. Special Report on Global Warming of 1.5 °C (Intergovernmental Panel on Climate Change, 2018).
6. van Vuuren, D. et al. Alternative pathways to the 1.5 °C target reduce the need for negative emission technologies. Nat. Clim. Change 8, 391–397 (2018).
7. Van Vuuren, D. et al. The role of negative CO2 emissions for reaching 2 °C— insights from integrated assessment modelling. Clim. Change 118, 15–27 (2013).
8. Czerniak, M.R., Tang, K. and Li, S-N. Has the Challenge of HFCs/PFCs _really Been Solved? Semiconductor International. Oct 2007.
9. Li, S-N, G-H Leu, S-Y Yen, S-F Chiou, S-J Yu and C-F Hsu.Controlling Contaminants with Enhanced Gas Leak Detection.Solid-State Technology, July 2007.
10. Hsu J-N, Tsai C-J, Chiang C. and Li, S-N. Silane Removal at Ambient Temperature by Using Alumina-Supported Metal Oxide Adsorbents. J. Air & Waste Management Association. 57: 204-210, 2007.
11. C-J Tsai, C-H Lin, Y-M Wang, C-H Hunag, Li, S-N, Z-X Wu and F-C Wang. An efficient venture system to remove submicron particles in exhaust gas. Air & Waste Management Association, 55: 319-325, 2007.

12. Li, S-N, H-Y Shih, Wang K-S, Hsieh K, Chen Y-Y, Chou J.Preventive Maintenance Measures for Contamination Control. Solid-State Technology, Dec. 2005.
13. Li, S-N, Lin, C-N, Shih, H-Y, Cheng, J-H, Hsu, J-N and Wang, K-S. Default values appear to be overestimating F-GHG emissions from fabs. Solid-State Technology, Sep. 2004.
14. Li, S-N, Yin-Yung Chen, Hui-Ya Shih and J. L. Hong. Using an Extractive Fourier Transform Infrared (FTIR) Spectrometer for Improving Cleanroom Air Quality in a Semiconductor Manufacturing Plant. American Industrial Hygiene Association Journal, June 2003.
15. Li, S-N, Hsu, J-N, Lin, S-J, Hong, J-L and Shih, H-Y. FTIR spectrometers measure scrubber abatement Efficiencies. Solid-State Technology, July 2002.
16. Li, S-N, Hsu, J-N, Leu, G-H, Tang, K. and Chiu, C-L . Improved Technique for Evaluating Point-of-use Abatement Systems. Semiconductor Fabtech, 14th Edition, June 2001.
17. Ju-Hsiu Cheng, Study the uncertainty of FCs process utilization rates (Tier 2b) in the semiconductor industry, IHTESH, 2012.
18. Myhre, G., D. Shindell, F.-M. Bréon, W. Collins, J. Fuglestvedt, J. Huang, D. Koch, J.-F. Lamarque, D. Lee, B. Mendoza, T. Nakajima, A. Robock, G. Stephens, T. Takemura and H. Zhang, 2013: Anthropogenic and Natural Radiative Forcing. In: Climate Change 2013 (IPCC AR5): The Physical Science Basis.Contribution of Working Group I to the Fifth Assessment Report of the Intergovernmental Panel on Climate Change [Stocker, T.F., D. Qin, G.-K.Plattner, M. Tignor, S.K. Allen, J. Boschung, A. Nauels, Y. Xia, V. Bex and P.M. Midgley (eds.)]. Cambridge University Press, Cambridge, United Kingdom and New York, NY, USA.
19. US EPA Greenhouse Gas Mandatory Reporting Rule, Subpart I , Electronic Code of Federal Regulations ,Table I-16 to Subpart I of Part 98—Default Emission Destruction or Removal Efficiency (DRE) Factors for Electronics Manufacturing, http://www.ecfr.gov/cgi-bin/text-idx?SID=a8058176db91462ef40c742dd2e236d4&mc=true&node=pt40.23.98&rgn=div5#ap40.23.98_198.16.

Comparative Study of Sub-23nm Particle Number Emission Characteristics for Non-road Diesel Engine on RMC and NRSC Test

Yongzhan Huo, Xiaowei Wang *, Tao Gao, Kang Liu and Zhenyu Wang

CATARC Automotive Test Center (Tianjin) Co., Ltd., Tianjin 300300, China

Abstract. A non-road diesel engine meeting China IV emission regulation was selected, the RMC and NRSC test procedures of the EU non-road Stage V emission regulation were run on the engine test bench. The emissions of particle number (PN) with particle sizes above 23 nm and 10 nm were tested simultaneously using direct exhaust sampling and full-flow dilution channel CVS dilution sampling equipment. The results showed that the emission characteristics of particulate matter above 23 nm and above 10 nm in the RMC test had the same trend, and the cumulative PN emissions were 1.9×1012 and 2.3×1012 for the whole test cycle, respectively. The PN23 values of dilution sampling were higher than those of direct sampling. The main reason for the difference in PN emission values at the same working modes in RMC and NRSC is the change in DPF capture efficiency caused by the different order of working modes and conditions.

Keywords: Non-road diesel engine, Sub-23nm, Particle number (PN), RMC, NRSC.

1. Introduction

Sub-23 nm particles are classified as exhaust particulate matter with a geometric mean diameter (GMD) smaller than 23 nm[1]. A growing number of studies have found that the prevalence of sub-23 nm particles in the exhaust and their volatile, as well as solid nature [2-3], pose even stronger harms to human health than the legislated due to larger surface/volume ratio and higher reactivity [4-5].

Focusing on the internal combustion engine sector, along with the commercialization of advanced combustion concepts, emission reduction technologies, and renewable fuels, sub-23 nm particles in the total particle number (TPN) could further climb [6-7].

According to the "China Mobile Source Environmental Management Annual Report (2022)" released by the Ministry of Ecology and Environment [8], China's PM emissions from non-road mobile sources will be 234,000 tons in 2021, much larger than the 64,000 tons of PM emissions from automobiles, so particulate matter emissions from non-road mobile machinery need to be controlled urgently.

EU stage V regulation introduced a new limit for particle number emissions. The PN limit is designed to ensure that a highly efficient particle control technology—such as wall-flow particulate filters—be used on all affected engine categories. The Stage V regulation also tightened the mass-based PM limit for several engine categories, from 0.025g/kWh to 0.015g/kWh. Adopting particle number (PN) emission limits for several categories of CI engines between 19 and 560 kW.The regulation of "Limits and measurement methods for exhaust pollutants from diesel engines of non-road mobile machinery(China's Fourth Phase)" [9] was revised and updated in 2020, which not only sets limitation on particulate matter quality, but also requires to count the particle number with a particle size above 23nm(PN emission) for 37-560kW diesel engine. The limitation of PN emission is 5×10^{12}#/kWh at the non-road steady cycle(NRSC) and the non-road transient cycle (NRTC), respectively.

However, laboratory tests reveal that particles from vehicle engine in sub-23nm region should not be neglected[10].Particle filter in current exhaust aftertreatment units may have to improve the efficiency of trapping fine particles smaller than 23 nm. However, there are few studies on PN10 for non-road diesel engines, and none of them consider the adaptability of non-road diesel engines meeting China emission regulations to EU regulation cycles and the differences in test cycles.

In this study, the PN10 and PN23 emission characteristics of non-road diesel engine running the RMC test meeting the China IV regulation was comparatively studied, and the PN emissions at each operating point of the RMC and NRSC test cycles were compared and analyzed.

* Corresponding author: wangxiaowei@catarc.ac.cn

 https://doi.org/10.1051/e3sconf/202339401032

2. Test bench and method

2.1 Test bench

The following is a schematic diagram of the engine full-flow emission test bench, which contains a non-road diesel engine, an electric dynamometer, an intake air conditioner, an intake flow meter, a fuel flow meter, a full-flow dilution channel, two particulate counters, and a test control and data logging system.Our testing system includes two particle counters, the first (PN#1) is sampled directly through the exhaust line and the second (PN#2) is sampled through the CVS dilution.

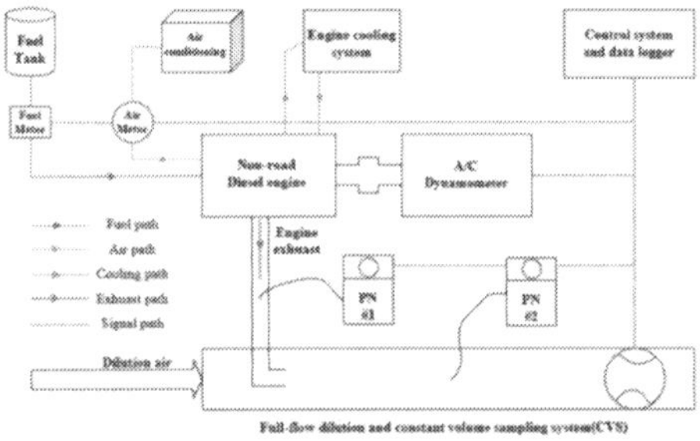

Fig. 1. The schematic diagram of test bench

2.1.1 Test engine

The tested engine was a non-road diesel engine with a displacement of 8.71 liter with a rated power of 305 kW and a maximum torque of 1850 Nm. The specific parameters of the engine can be seen in Table 1.

Table 1. Main parameters of tested engine

Parameter	Value
Engine capacity	8.71L
Cylinder number-Bore×Stroke	6-117mm×135mm
Compression ratio	15.9
Rated power/speed	305kW/2000rpm
Maximum torque/speed	1850Nm/1300-1400rpm
Idle speed	900±100rpm
Emission Control Technology Route	DOC+DPF+SCR+ASC
Emission Standard	China Non-road IV

2.1.2 Test equipment

The main test equipment is summarized in Table 2, including the equipment type, manufacturer and accuracy.

Table 2. Test equipment

Equipment name	Equipment Type and Manufacturer	Accuracy
AC Dynamometer	Horiba HD600	Torque: ± 0.3%F.S. Speed: ± 1rpm
Intake air conditioning	CAC4000-C	Temperature: ±1K Humidity:± 3(%RH)
Air flow meter	Sensyflow FMT700-P	±1%
Fuel meter	Horiba FQ-3100DP	±0.1%
Full-flow dilution sampling system(CVS)	CVS DLT-24150W	±1%
Particle counter #1 (for CVS dilution sampling)	MEXA-2000SPCS	±10%
Particle counter #2 (for exhaust pipe direct sampling)	AVL APC^{Plus} ADVANCED with AVL CPC option for sub-23nm	±8%

2.2 Test procedure and analysis method

2.2.1 Test cycle

According to Non-road EU V emission regulation, at the choice of the manufacturer, a steady-state test cycle may be run as a discrete-mode cycle or a ramped-modal cycle. Based on the engine category, we select the C1 steady-state test cycle, which consists of 8 speed and load modes. Ramped Modal non-road steady-state test Cycle ("RMC") test sequence, which is hot running cycle where emissions shall be started to be measured after the engine is started. The engine shall be continuously controlled by the test bed control unit during the RMC test cycle. The gaseous and particulate emissions shall be measured and sampled continuously during the RMC test cycle in the same way as in a transient cycle. An RMC is intended to provide a method for performing a steady-state test in a pseudo-transient manner.

Table 3. RMC-C1 test modes

RMC mode number	Mode time[s]	Engine speed	Torque[%]
1a Steady-state	126	Idle	0
1b Transition	20	Linear transition	Linear transition
2a Steady-state	159	Intermediate	100
2b Transition	20	Intermediate	Linear transition
3a Steady-state	160	Intermediate	50
3b Transition	20	Intermediate	Linear transition
4a Steady-state	162	Intermediate	75
4b Transition	20	Linear transition	Linear transition
5a Steady-state	246	100 %	100
5b Transition	20	100 %	Linear transition
6a Steady-state	164	100 %	10
6b Transition	20	100 %	Linear transition
7a Steady-state	248	100 %	75
7b Transition	20	100 %	Linear transition
8a Steady-state	247	100 %	50
8b Transition	20	Linear transition	Linear transition
9 Steady-state	128	Idle	0

Each RMC consists of a series of steady state modes with a linear transition between them. The relative total time at each mode and its preceding transition match the weighting of the discrete mode steady state cycle("NRSC"). The change in engine speed and load from one mode to the next one has to be linearly controlled in a time of 20±1 seconds. The mode change time is part of the new mode (including the first mode). In some cases modes are not run in the same order as the discrete model steady state cycles or are split to prevent extreme changes in temperature.

2.2.2 Analysis method

This section describes data processing and calculation methods for exhaust direct sampling and CVS dilution sampling.
For direct sampling through the exhaust pipe:

$$T_{PN} = G_{\text{exh}} \cdot c_{PN} / \rho_{exh} / 0.0036$$

$$G_{\text{exh}} = G_{fuel} + G_{air}$$

Where: TPN is the instantaneous particle number(#/s), Gexh, Gfuel, Gair are the exhaust, fuel and intake air mass flow rates respectively (kg/h), cPN is the PN concentration(#/cm^3) measured by the direct sampling equipment, PN#1 , ρexh is the exhaust density (as 1.2943 kg/m^3)

$$N_{PN} = \int_0^t T_{PN} dt$$

Where: NPN is the cumulative particle number(#).
For dilution sampling through CVS :

$$T_{PN-CVS} = 10^6 \times Q_{CVS} \cdot c_{PN-CVS}$$

Where: TPN-CVS is the instantaneous particle number(#/s), QCVS is the dilution channel CVS volume flow rate (m^3/s), cPN-CVS is the PN concentration (#/cm^3) measured by the dilution sampling equipment, PN#2.

$$N_{PN-CVS} = \int_0^t T_{PN-CVS} dt$$

Where: NPN-CVS is the cumulative particle number(#).
Particle number per unit cycle work, PN(#/kWh):

$$PN = \frac{N_{PN}}{W} = \frac{N_{PN}}{\int n \cdot T / 9550}$$

Where: W is the cycle work(kWh), n is the engine speed(r/min), T is the engine torque(Nm).
In the calculation, it should be noted that the RMC test is a full process including the integration of steady-state and transient transition conditions, while the NRSC test is only the calculation of steady-state conditions and does not include the part of transient transition conditions.

3. Results and discussions

3.1 PN10 and PN23 results under RMC test

Comparative analysis of PN10/PN23 recorded by the direct particle counter and PN23 recorded by the CVS dilute particle counter during the RMC test, results showed that the concentration and cumulative emission changes of PN10 and PN23 in the RMC test had the same trend, with cumulative PN emissions of 2.3×10^{12} and 1.9×10^{12} for the whole cycle, respectively. At the transition of switching between each working mode of RMC, such as the 5th to 6th working mode for the sudden drop of 100% load to 10% load under rated speed condition, the PN emission sudden increase changes obviously, and the PN emission rate of the 6th working mode is the largest. In addition, the PN emission is generally higher in the transition condition of the working mode switching.

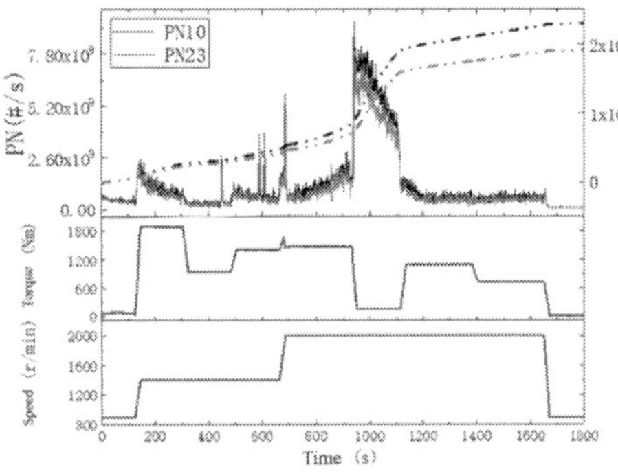

Fig. 2. Instantaneous and cumulative PN10 and PN23 under RMC test

The ratio of direct sympling PN10 to PN23 for each operating mode of the RMC is between 1.05 and 1.29, and the overall cycle PN10 emission is about 1.2 times of PN23, among which the 5/6/7/8 operating modes are rated speed, and their multiples are above 1.2, indicating that the percentage of 10-23 nm particulate matter is relatively high under high speed operating conditions. In the idle and intermediate speed low load conditions, the ratio of PN10 to PN23 is smaller, which indicates that the percentage of 10-23nm particles generated under these conditions is relatively low.

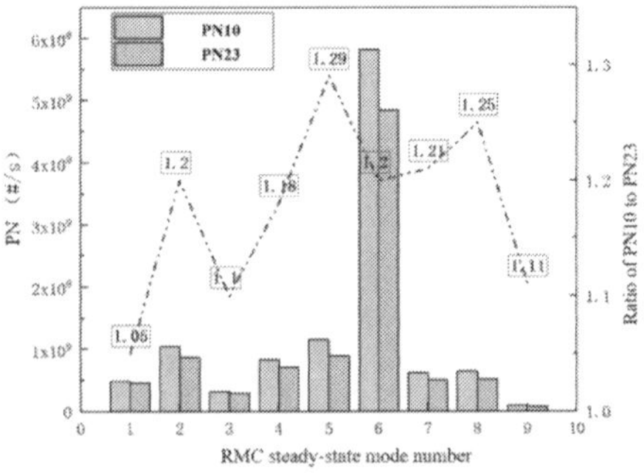

Fig. 3. Ratio of PN10 to PN23 under RMC modes

From the test results, the exhaust pipe direct sampling PN10/PN23 and CVS dilute PN23 have the same trend, and the specific emissions of the whole cycle are 2.721×10^{10}, 2.265×10^{10} and 2.741×10^{10}#/kWh, respectively. Compared with the direct PN23, the overall value of dilute PN23 is high, which may be related to the mixing process of dilute air and exhaust causing large particulate matter breaking down into smaller particles.

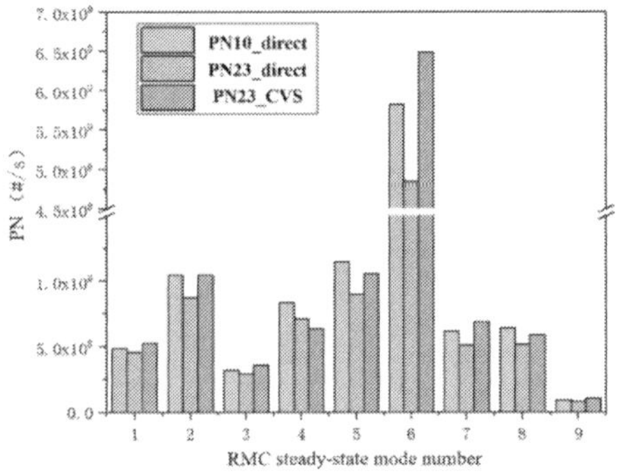

Fig. 4. Direct exhaust PN10/PN23 and dilute CVS PN23 under RMC modes

3.2 PN comparison of RMC and NRSC

The speed and torque in the steady-state operating modes of RMC and NRSC are the same, both contain idle speed, 100%-75%-50% load at intermediate speed, and 100%-75%-50%-10% load at rated speed, but the operating sequence and duration of the modes are different. The RMC cycle is shown in the figure, following the 1-2-3-4-5-6-7-8-9 mode sequence, and the NRSC cycle carries out the test according to the 5-7-8-9-2-4-3-1 (a-b-c-d-e-f-g-h) mode sequence.

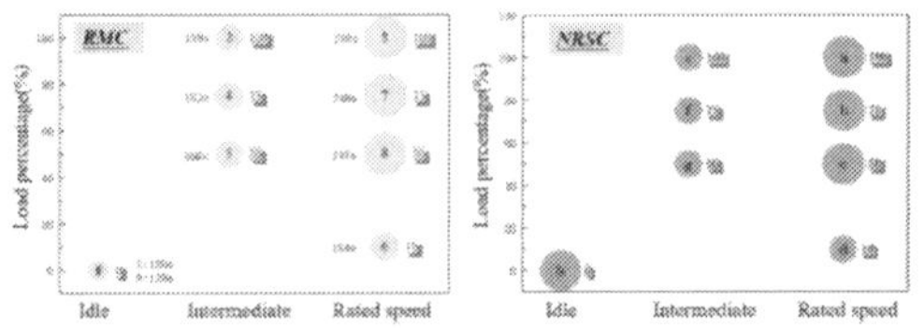

Fig. 5. Test cycle mode sequence of RMC and NRSC

The higher PN emissions in the first four operating modes and lower PN emissions in the last four operating modes in the NRSC test indicates that the PN emissions are higher in the high speed operating modes than in the low speed operating modes in the NRSC operating sequence, and the PN emissions increase with increasing load in the high speed condition.

Compared with NRSC, RMC has higher PN emissions at idle and intermediate speed operating modes, and lower emissions at high load operating modes at rated speed.The PN emissions were higher at 10% torque and 100% torque at rated speed, 100% torque at intermediate speed operating modes in the RMC test. The highest PN emissions at 10% torque rated speed may be due to the reduced capture efficiency with DPF carbon load reduction after the transition from rated condition modes with the high speed and high torque, where the high PN emissions at the first four condition points of NRSC are also influenced by this factor.

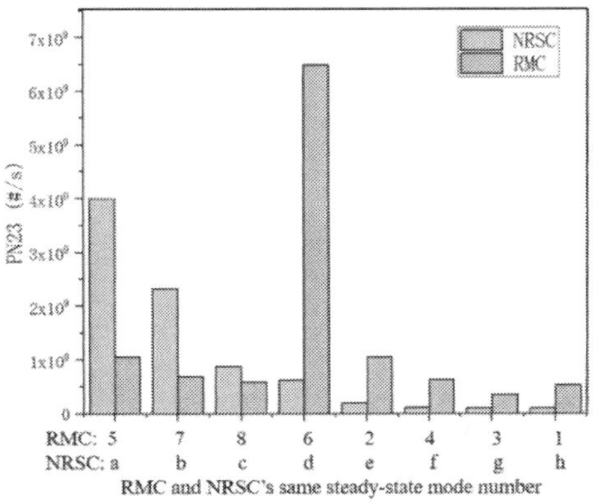

Fig. 6. PN23 comparation of RMC and NRSC

4. Conclusion

In this paper, PN tests based on RMC and NRSC were carried out on off-road diesel engines meeting China IV non-road emission regulations. The transient and each steady-state operating modes emission characteristics of PN10 and PN23 in the RMC cycle were investigated, and the PN23 emissions at each same operating mode in the RMC and NRSC cycles were compared and analyzed.

(1) The ratio of PN10 to PN23 for each working mode of RMC is between 1.05 and 1.29, and the overall cycle PN10 emission is about 1.2 times that of PN23. Particle with 10-23nm diameter accounts for a higher percentage under high speed working conditions, while the percentage of 10-23nm particle generated under idling and intermediate speed, low load working modes is relatively low.

(2) The exhaust direct PN10/PN23 and CVS dilute PN23 vary in the same trend, and the specific emissions of the whole cycle are 2.721×10^{10}, 2.265×10^{10} and 2.741×10^{10}#/kWh, respectively. Compared with PN23 in the direct exhaust extraction, the overall value of PN23 in the dilute extraction is higher, which may be related to the fragmentation of large particles caused by the mixing process of dilute air and exhaust decomposition into smaller particulate matter.

(3) The PN emissions at the same speed and torque operating modes are more affected by the sequence of the operating modes. Compared with NRSC, RMC has higher PN emissions at idle and intermediate speed operating modes, and lower emissions at high load operating moedes at rated speed.

(4) The high PN emissions at the 6th operating mode of RMC and the first 4 operating modes of NRSC may be caused by the reduced carbon loading of DPF after the transition from the high speed and high torque rated operating modes, which makes the capture efficiency lower.

References

1. Dongdong Guo, Yunshan Ge, Xin Wang. Sub-23 nm Particle Emissions from China-6 GDI Vehicle: Impacts of Drive Cycle and Ambient Temperature. Atmosphere, 2022, 13(8):1216
2. Samaras, Z.C., Andersson, J., Bergmann, A., Hausberger, S.Measuring Automotive Exhaust Particles down to 10 nm. SAE Tech. Pap, 2020, 2017:1–12.
3. Giechaskiel, B., Vanhanen, J., Väkevä, M., Martini, G.Investigation of Vehicle Exhaust Sub-23 nm Particle Emissions. Aerosol Sci. Technol, 2017, 51:626–641.
4. Donaldson, K.; Li, X.Y.; MacNee, W. Ultrafine (Nanometre) Particle Mediated Lung Injury. J. Aerosol Sci. 1998, 29, 553–560.
5. Gidney, J.T.; Twigg, M.V.; Kittelson, D.B. Effect of Organometallic Fuel Additives on Nanoparticle Emissions from a Gasoline Passenger Car. Environ. Sci. Technol. 2010, 44, 2562–2569.
6. Singh, R.; Han, T.; Fatouraie, M.; Mansfield, A.; Wooldridge, M.; Boehman, A. Influence of Fuel Injection Strategies on Efficiency and Particulate Emissions of Gasoline and Ethanol Blends in a Turbocharged Multi-Cylinder Direct Injection Engine. Int. J. Engine Res. 2021, 22, 152–164.
7. Toumasatos, Z.; Kontses, A.; Doulgeris, S.; Samaras, Z.; Ntziachristos, L. Particle Emissions Measurements on CNG Vehicles Focusing on Sub-23nm. Aerosol Sci. Technol. 2020, 55, 182–193.
8. China mobile source environmental management annual report in 2022[J]. Ministry of Ecology and Environment of the People's Republic of China, 2022.
9. Limits and measurement methods for exhaust pollutants from diesel engines of non-road mobile machinery (CHINA III, IV) GB 20891-2014 Modification order.
10. Xiaowei Wang, Lin Zhang, Mingda Wang, Xiaojun Jing and Xuejing Gu. Sub-23 nm Solid Particle Number Emission Characteristics for a Heavy-duty Engine Fuelled with Compression Natural Gas. E3S Web of Conferences 329, 01012(2021).

AUTHOR INDEX